Decentralized Systems and Distributed Computing

Scrivener Publishing
100 Cummings Center, Suite 541J
Beverly, MA 01915-6106

Decentralized Systems and Next-Generation Internet

Series Editors: Mufti Mahmud and Suman Lata Tripathi

Scope: Next Generation Internet (NGI) is the focus of countries like the US and UK towards the improvement and revolution in the present and future internet with its backend networks and infrastructure to develop faster, reliable, secure internet platforms. The objective of NGI is to develop an advanced version of the internet. The target deliverables of NGI include building network communication architecture with enhanced levels of data access, human communication and productivity and achieving substantially and faster internet bandwidth and speed. An evolution of the internet from a low-level focus to higher level focus on interconnectivity, increased user interactions, video chat, and financial and social interactions in the virtual world are the major objectives towards development of NGI. A virtual world which is not owned or controlled by a single entity or metaverse in that a computer-generated virtual environment is created for reliable user interactions. Web 3.0 is an advancement that will control tomorrow's internet and metaverse centers for better user experiences. In the metaverse, users interacting using software from different vendors will experience monetization by each vendor with seamless interactions in spite of different technologies.

This series covers the information from the ground level of requirements for internet-based user interactions, platforms and applications leading to the development of next-generation internet. Future requirements and dependencies on more online activities will need to work more on developing decentralized systems to improve user experience with speedy, reliable and secured interactions in a virtual environment. This series will provide the opportunity to the academician and industry professionals to share their knowledge and experiences with learners and practitioners relevant to diverse areas of improvements for the development of next-generation internet and decentralized systems or metaverse.

Publishers at Scrivener
Martin Scrivener (martin@scrivenerpublishing.com)
Phillip Carmical (pcarmical@scrivenerpublishing.com)

Decentralized Systems and Distributed Computing

Edited by
Sandhya Avasthi
Suman Lata Tripathi
Namrata Dhanda
and
Satya Bhushan Verma

Scrivener Publishing

WILEY

This edition first published 2024 by John Wiley & Sons, Inc., 111 River Street, Hoboken, NJ 07030, USA and Scrivener Publishing LLC, 100 Cummings Center, Suite 541J, Beverly, MA 01915, USA

For more information about Scrivener publications please visit www.scrivenerpublishing.com.

Wiley Global Headquarters
111 River Street, Hoboken, NJ 07030, USA

For details of our global editorial offices, customer services, and more information about Wiley products visit us at www.wiley.com.

Library of Congress Cataloging-in-Publication Data

ISBN 9781394204366

Front cover images supplied by Adobe Firefly
Cover design by Russell Richardson

Set in size of 11pt and Minion Pro by Manila Typesetting Company, Makati, Philippines

Printed in the USA

10 9 8 7 6 5 4 3 2 1

Contents

Preface

The use of distributed systems is a big step forward in IT and computer science. As the number of tasks that depend on each other grows, a single machine can no longer handle all of them. The computations in such a distributed systems are called distributed computing where processing and data storage is distributed across several computers or systems. Distributed computing is better than traditional computer settings in several ways. Distributed systems reduce the risks of a single point of failure, making them more reliable and able to handle mistakes. Most modern distributed systems are made to be scalable, which means that processing power can be added on the fly to improve performance. The Internet of the future is meant to give us freedom, give us choices, encourage diversity and decentralization, and make it easier for people to be creative and do research. This book will look at the different aspects of the next-generation internet, distributed systems, distributed computing, and their effects on society as a whole. Some of the emerging technologies in the field of distributed computing such as Edge Computing, Mobile Edge computing, and Fog computing is discussed in various chapters in this book. The primary motivation behind the idea of the edited book is to explore the field of distributed systems and computing due the recent development in internet technologies. The recent development in internet and internet-based applications is due to pandemic and how it changed the workplace environment and our education system.

People assert that the Internet will eventually evolve into the metaverse. In his 1992 science fiction novel Snow Crash, Neal Stephenson introduced the term "metaverse" for the first time. Facebook changed its name to Meta in October 2021 and announced it will invest at least $10 billion in the concept that year. Google, Microsoft, Nvidia, and Qualcomm, along with other tech giants, are investing billions of dollars in the concept. McKinsey & Company, a management consulting firm, estimates that the metaverse economy might be worth $5 trillion by 2030. E-commerce is anticipated to be the most important engine in the metaverse, where games, entertainment, education, and marketing will all likely grow in importance. The future belongs to internet and internet-based technologies. By making the internet more three-dimensional and immersive, the metaverse could introduce more ways to use it. Some people have expressed negative things about the metaverse, and there is much uncertainty regarding its future. Analysts in the field have pondered if the metaverse will differ much from our current digital experiences, and if so, whether people will be willing to spend hours per day exploring virtual space while wearing a headset.

This book is aimed at the Researchers, Professors, IT Experts working in the domain of Distributed systems, Internet of Things, Metaverse, Artificial Intelligence and Internet Technologies. The book addresses issues and challenges in the area of distributed systems and computing. In addition, the book gives an insight into the various application areas in this internet era.

1

Introduction to Next-Generation Internet and Distributed Systems

Swapnil Gupta[1], Rajat Verma[1*] and Namrata Dhanda[2]

[1]Department of Computer Science and Engineering, Pranveer Singh Institute of Technology, Kanpur, Uttar Pradesh, India
[2]Department of Computer Science and Engineering, Amity University Uttar Pradesh, Lucknow, India

Abstract
As digital vulnerabilities and concerns over data breaches and surveillance continue to rise, the need for an internet that provides better protection for users' sensitive information and communications becomes imperative. This document serves as a comprehensive introduction to the Next-Generation Internet (NGI) and distributed systems, highlighting their pivotal role in addressing these growing concerns. The NGI represents a paradigm shift that unlocks new frontiers for applications and services currently unattainable with the existing internet infrastructure, including virtual and augmented reality, distributed systems, advanced artificial intelligence, and machine learning systems. The development of the NGI is expected to have a profound impact on future data management practices, necessitating a significant departure from the platform-centric approach of the present. Instead, the NGI emphasizes a human-centric model, placing users at the center of the data ecosystem and empowering them with greater control over their data. By adopting this approach, the NGI offers potential solutions to mitigate the digital divide, enhance inclusivity, and improve accessibility to the internet, particularly for individuals lacking reliable internet infrastructure. The document also explores the intricate challenges associated with studying distributed systems, characterized by their complex nature and the difficulties involved in analyzing their various components. It provides an overview of the current network infrastructure, including protocols, proposed architectures, and deployment challenges. Additionally, the chapter introduces distributed systems, examining their general design, goals, and the disruptive technology of blockchain.

**Corresponding author*: rajatverma310795@gmail.com

Sandhya Avasthi, Suman Lata Tripathi, Namrata Dhanda and Satya Bhushan Verma (eds.) Decentralized Systems and Distributed Computing, (1–34) © 2024 Scrivener Publishing LLC

Throughout the document, key findings suggest that the NGI's human-centric model and distributed systems can revolutionize data management practices, fostering security, inclusivity, and user empowerment. The NGI has the potential to reshape the internet landscape and bridge the digital divide by offering innovative solutions for secure and accessible data sharing. However, realizing these advancements requires addressing technical challenges, such as scalability, performance, and security, through ongoing research and development. The overview of the NGI and distributed systems is catered extensively, offering insights into their potential implications for data management. By embracing a human-centric approach and leveraging distributed systems, the NGI represents a promising future for a more secure, inclusive, and user-centric internet. The document emphasizes the need for continued exploration and innovation to address challenges and unlock the full potential of these transformative technologies.

Keywords: CDN, CCN, consensus, DCS, NGI, P2P, scale, SDN

1.1 Introduction

Due to the extraordinary spread of the omnipresent internet as well as IoT device dispersion, the Internet is growing at an exponential rate. From the perspective of the devices, rather than at the Internet's core infrastructure, this development is mostly taking place on its outskirts. Next-Generation Internet is an umbrella term that encompasses a range of technological advancements and innovations. NGI incorporates the use of artificial intelligence and machine learning technologies to enable more intelligent, personalized, and efficient internet services. One of the key technical components of NGI is the development of advanced networking technologies, such as software-defined networking (SDN) [1] and network function virtualization (NFV) [1]. These technologies are designed to improve network flexibility, efficiency, and scalability, making it easier to manage and optimize complex network infrastructures. Although the current Internet has been phenomenally successful, its foundational network is weakened by the limitations of the IPv4 protocol, such as address space exhaustion, lack of quality of service, and security vulnerabilities. To address these issues, a transition to the Next-Generation Internet is necessary, with the Internet Engineering Task Force (IETF) [2] proposing IPv6 as a solution since 1991. IPv6 became standard in 1995, and the IETF established a free test bed called 6Bone [3] in 1996 to evaluate IPv6 before phasing it out by 2005. IPv6 provides numerous technical benefits, such as a larger address space, enhanced efficiency, improved security, and better quality of service [4].

The conventional network framework also entails significant cost implications. The expense associated with procuring network equipment, including routers, switches, and other networking devices, is often exorbitant. This prohibitive cost poses a challenge for organizations to justify the expense of upgrading or replacing their existing infrastructure. Moreover, managing a large network's complexity can be expensive in terms of personnel and training costs. Furthermore, traditional networks typically necessitate dedicated hardware and software for specific tasks like load balancing or security, which can further escalate costs. In contrast, emerging software-defined networking (SDN) and network function virtualization (NFV) technologies provide more flexible and cost-effective alternatives to traditional network architectures. In a traditional network, the control and data planes are both embedded within a single device. One of the major problems that traditional architecture faces is that it is a closed network architecture [5]. Manufacturers of network equipment have their own hardware, operating system, application of standards, and extended range of functions. This implies that timely testing and verification of the new techniques are not possible. This is because the system is only accessible to the manufacturer. In a conventional network, switches serve as both the controller software and the forwarding element. Software instruction simply imposes rules that dictate the forwarding component. These networks are intricate and challenging to control [6]. When the data plane (forwarding devices) and the control plane (one that regulates network traffic) are combined into a single network device, it becomes more sophisticated. This architecture decreases flexibility and eliminates the opportunity for innovation, preventing the advancement of network infrastructure.

A distributed system is a collection of independent computing pieces that seem to users as a single cohesive system. Distributed systems are characterized by geographically dispersed computers, which are interconnected using wired, wireless, or hybrid networks [7]. The size of a distributed system can range from a few devices to a vast network of millions of computers. These systems are highly dynamic, with computers frequently joining and leaving, causing continuous changes to the network topology and performance [8]. Building and maintaining distributed systems can be challenging, as there are several limitations and trade-offs to consider. Users may be located in various parts of the world, and that can introduce issues with network latency and communication. It is difficult to completely prevent failures of networks and nodes, and distinguishing between slow and failing computers can be tricky. It is also hard to be certain that a server performed an operation before a crash and achieving full transparency in a system can come at the cost of performance. Keeping web caches up to date with the head can be difficult, as can immediately flushing write operations to disk for fault tolerance. Despite these challenges, it is important to consider all the factors while designing and building distributed systems to ensure that they are dependable and efficient. The current internet is made up of a patchwork of different technologies, standards, and protocols, which can make it difficult to integrate new applications and services. To achieve the next generation of the internet, there will be a need for greater interoperability between different systems and devices, as well as more open and standardized protocols for data exchange and communication. Achieving the next generation of the internet will require a collaborative effort between organizations and individuals around the world, as well as significant investment in research, development, and infrastructure. To effectively construct the next iteration of the Internet, a comprehensive outlook is required, encompassing a range of topics such as advancements in the physical and transmission layers of both conventional and wireless media, in addition to the implementation of new switching and routing paradigms at the device and sub-system layer. Additionally, it is necessary to consider alternative approaches to TCP and UDP for data transport in emerging environments, as well as novel models and theoretical frameworks for comprehending network complexity [9].

1.2 Traditional Network

It is important to have an overview and outline of current-generation internet or traditional network and their architecture to better understand the next-generation internet network. In a traditional network architecture, a single device, such as a switch or router, is responsible for integrating both the control plane and data plane functions [10]. The command plane is in charge of managing network protocols, policies, and rules, while the data plane must forward network traffic to its destination based on the instructions received from the control plane. The aforementioned approach may result in complications and inefficiencies, as new network policies must be configured individually on each device making use of low-level, vendor-specific instructions [11]. Moreover, at the time both the control plane and data plane functions are integrated into a single device, it limits the network's flexibility and restricts opportunities for innovation. Figure 1.1 illustrates the structure of the traditional network.

One major limitation of traditional networks is bandwidth. With the increasing number of devices and data traffic on the network, traditional networks can quickly become overwhelmed, leading to slow data

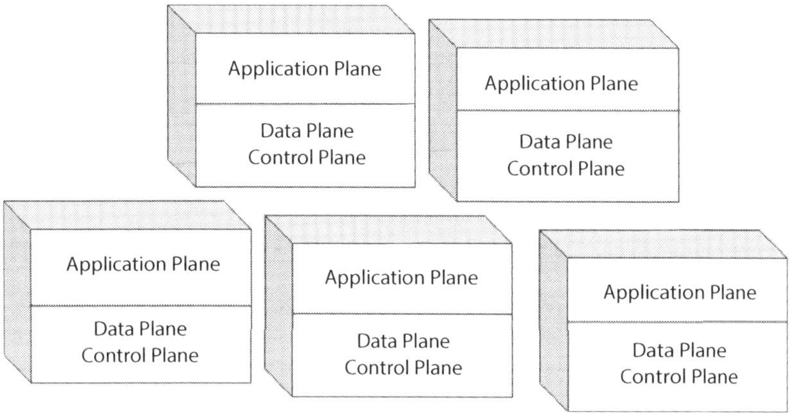

Figure 1.1 Traditional network [12, 13].

transmission, network congestion, and performance issues. Upgrading the network infrastructure to increase bandwidth can help support more devices and data traffic and improve overall network performance. Another limitation of traditional networks is security vulnerabilities. These networks are often exposed to various security threats that can exploit vulnerabilities and compromise the integrity of the network. Upgrading to newer technologies with enhanced security features can help organizations mitigate these threats and ensure the overall security of the network. Traditional networks can also struggle with scalability, making it difficult to adapt to changing business requirements. Upgrading to more flexible and scalable technologies can help organizations quickly adapt to these changes and improve overall network performance. Latency is another limitation of traditional networks, which can negatively impact the performance of applications and services. Upgrading to faster and more efficient technologies can help reduce latency and improve overall network performance. Software-defined networking (SDN) has emerged as a solution to address the challenges posed by traditional network architectures. It separates the data plane, which is in charge of forwarding network traffic, from the control plane, which is in charge of managing and making decisions about the network [14]. This allows for greater flexibility and simplifies network management by centralizing control and automating network configuration.

1.3 Next-Generation Internet

NGI stands for the Next-Generation Internet and refers to an upcoming version of the Internet that will have significantly better performance levels, with typical improvements ranging from ten to hundred times faster than the current Internet. This will be made possible by the widespread implementation of high-speed broadband technologies; this will allow average users to access this faster connection. Additionally, the user community is expected to grow steadily over time.

It is incorrect to think of the NGI as one technological undertaking with a single objective, owned by a development team, and scheduled for delivery and deployment at a later time. The term "NGI" alludes to the continued development of the current Internet. The entire size of the network continues to grow as new generations of faster routers and switches are created and extensively used, as well as gigabit fiber links becoming more common [15]. It is anticipated that the forthcoming Next-Generation Internet (NGI) would see the implementation of new generations of infrastructure elements and connection options, many of which are now accessible but have not yet been fully implemented, such as IPv6, QoS [16], and more. The deployment of complicated applications on networks, which was previously constrained by the Internet's deficient performance, can now be made possible by employing regular commercial networking technology, according to their results [17]. They also point out that while private intranet environments use the same technologies as public networks, their advantages and disadvantages are similar. Several sectors, including air traffic control, banking, power distribution, and the medical field, are under intense cost pressure, which is pushing them to embrace networked software architectures for all kinds of applications, even crucial ones. They raise concern nevertheless, that the current network infrastructure is still prone to dependability, security, and performance problems, making it underprepared to handle such delicate and important uses. The creation of mission-critical applications via this infrastructure is not supported by widely used commercial technologies.

1.3.1 Next-Generation Internet Protocol

The IP address functions as a topological marker in the network layer and object identification in the transport layer. This creates interdependence between an end-host identity and its location, leading to the absence of a consistent identifier, the use of IP addresses as both object identifiers and topological locators allows for efficient and effective routing of network traffic, contributing to the Internet's scalability and widespread connectivity. The Next-Generation Internet relies on the Internet Protocol (IP) as a fundamental component for efficient routing and object identification. The use of IP addresses as both topological locators and object identifiers enables scalable and widespread connectivity. As the NGI continues to evolve, advancements in IP protocols, such as IPv6 and innovative addressing approaches, will play a pivotal role in addressing the evolving needs of connectivity, security, and efficiency in network communication.

1.3.1.1 IPv6

IPv6 is recognized as the successor to IPv4 and is expected to offer numerous benefits, such as better address design, greater security, and more address space. While IPv4 only provides a 32-bit address space, IPv6 offers a much larger 128-bit address space, which can accommodate a greater number of users and devices and enhance address allocation and traffic routing efficiency. Nonetheless, since IPv6 and IPv4 are incompatible, they cannot communicate directly with each other. To facilitate their coexistence, various transition mechanisms have been developed. Long-standing IPv4 address space constraints led to the development of several methods, including Network Address Translation (NAT) [18], Dynamic Host Configuration Protocol (DHCP) [19], and Classless Inter-Domain Routing, to prolong the life of the current infrastructure (CIDR) [20]. Network Address Translation, for example, enables various devices within an enterprise to share one or more global IPv4 addresses for external communications while using local private addresses. Although NAT has momentarily postponed the expiration of IPv4 address space, its use can hinder bi-directional typical application communication. Also, it does away with the fundamental purpose of network address translation (NAT), which was first set up to help with IPv4 address exhaustion. The core, periphery, and access networks are the three subcategories of networks in networking [21]. The core network serves as the network's foundation, and its routers may flip and forward transmissions at high rates via a variety of telecommunication interfaces. In the meantime, an edge router that is connected to one or many core routers is placed close to the end clients. Till then, both protocols need to be inter-operated together.

1.3.1.2 Tunneling

The process of tunneling involves transporting one protocol inside another protocol. In the context of networking, tunneling enables the transmission of data packets from one network protocol to another. In the context of IPv6-over-IPv4 tunneling, IPv6 datagrams are encapsulated within IPv4 packets, allowing for their efficient transport across IPv4 networks. This approach serves as a bridging mechanism, enabling communication between IPv6 hosts and routers over existing IPv4 infrastructure. It effectively extends the reach of IPv6 connectivity by leveraging the widespread deployment of IPv4 networks. Conversely, in the case of IPv4-over-IPv6 tunneling, IPv4 packets are encapsulated within IPv6 packets, facilitating communication between IPv4 hosts or networks over an IPv6 infrastructure. This technique ensures backward compatibility and enables IPv4-only devices to connect to and communicate with IPv6 networks [22].

Tunneling plays a crucial role in the smooth migration from IPv4 to IPv6, allowing organizations and service providers to adopt IPv6 gradually without disrupting existing IPv4 networks. It provides a transitional mechanism that enables the coexistence of both protocols during the migration process, ensuring uninterrupted connectivity and service availability.

1.3.2 Quality of Service (QoS)

We are trying to establish systems that permit a range of service levels at various prices. At first, the Internet was intended to be a shared resource that was supported centrally. Broadly, more resources were made available to meet the growing demand. In such a setting, only one service class was implemented, and everyone had equal rights to the resources that were available. Yet, the demands of time-sensitive applications and the exponential increase in the number of users make it difficult to merely increase bandwidth. Consequently, strategies must now be developed to address the problem of service allocation in crowded networks in an elegant manner. The commercialization of the Internet has made it simpler to find a solution to this problem because everyone and everything is no longer required to be treated equally. It is now feasible to acquire a "want more/pay more/get more" mentality, regardless of whether it is imposed by a formal payment system, peer pressure, or financing organizations. It is crucial to remember that many current and future applications will demand a quality of service above "best effort." These applications range from emerging high-availability commercial and industrial applications to very high-performance scientific applications.

1.4 Network Middleware

Both private intranets and the public internet rely on network middleware to successfully run networked applications. It includes a set of reusable and extendible services and features that facilitate resource management, interoperability, security, and efficient and dependable communication [23].

The rapid expansion of inter-networking in recent years can be attributed to both the numerous platforms on which these protocols are available and the technical brilliance of internet protocols. The widespread acceptance of internet protocols is largely due to their usability and dependability as well as the accessibility of the Berkeley implementation [24]. Furthermore, the development of client-server computing, automated workflows, and inexpensive Ethernet technology has made it possible to connect corporate campuses. These developments have made it easier to develop message-oriented middleware, which is networked object computing environments, and databases as application frameworks for middleware. These middleware frameworks have enabled users from a variety of backgrounds to create useful and worthwhile networking apps by utilizing both private intranets and the general internet. By offering pre-built components and services that make it easier to create networked applications, they hasten the development process.

These frameworks, which speed up the creation of networking applications, are going to be the defining feature of NGI. Middleware is a key component of many applications' ability to perform well in a networked environment. It can be thought of as a collection of reusable and extensible services and functionalities. These management functions and services provide end-to-end interoperability, security, integrity, and resource management. They are essential to building a reliable system that can handle the competing needs of a complex and expanding set of end-user applications. Even though the NGI might not be able to address every problem associated with software application development, it is crucial to think about how to organize the architectural structure to accomplish reliable and efficient end-to-end operations that are comparable to the existing communication protocols that serve as the foundation for middleware and applications.

1.5 Software-Defined Network (SDN)

SDN or Software-Defined Networking is an emerging technology that represents a new paradigm for the next generation of the Internet. SDN is characterized by the division of the control plane and data plane. This involves removing the control functions from the forwarding devices, such as switches, and placing them in a centralized location known as the controller, which acts as a server. To achieve this separation, a programming interface is established between the controller and the forwarding devices. In SDN, the controller serves as the primary location for the control plane, which is segregated from the forwarding devices like switches. The controller serves as a server, managing the network's overall behavior and making strategic decisions. The network infrastructure may be controlled centrally and is programmable thanks to this division. A programming interface is built to allow communication between the controller and the forwarding devices. The controller can directly regulate the behavior of the data plane elements thanks to this interface's facilitation of information and command interchange. OpenFlow, which offers a standardized and widely used API for communication between the controller and the switches, is the most widely used programming interface in SDN.

Network administrators have never-before-seen freedom and control over their networks because of SDN. Through software-based control, they may dynamically configure and manage network resources, optimize traffic flows, and implement cutting-edge network policies and services. This programmability enables quick innovation, quick implementation of new services, and effective resource management. The Next-Generation Internet gains a lot from SDN. It makes network virtualization possible, enabling the coexistence of numerous virtual networks on a single physical infrastructure. This makes it easier to isolate resources, improve security, and streamline network administration. Network slicing, which allows for the creation of specialized network slices suited to certain applications or user requirements, is another feature of SDN.

1.5.1 Necessity of Software-Defined Network

Initially, the need for SDN arose within data centers that were struggling to manage their networks using traditional network management methods. The rise of cloud technology and the increasing data access from anywhere in the globe have become easier because of the rise in popularity of mobile devices. The proliferation of network data traffic has significantly increased as a result of the unstructured data produced by public platforms, e-commerce websites, and other sources. To manage these growing data, a new networking architecture was needed, one that could dynamically manage networks. SDN fulfills this need by offering programmability and virtualization capabilities, making it possible to manage large networks effectively. In contrast to traditional network devices, which have embedded control and data planes, SDN architecture separates the control plane and moves it to an external controller device. The controller is responsible for managing routing decisions and routing tables. Figure 1.2 illustrates the architecture of a typical SDN network [25].

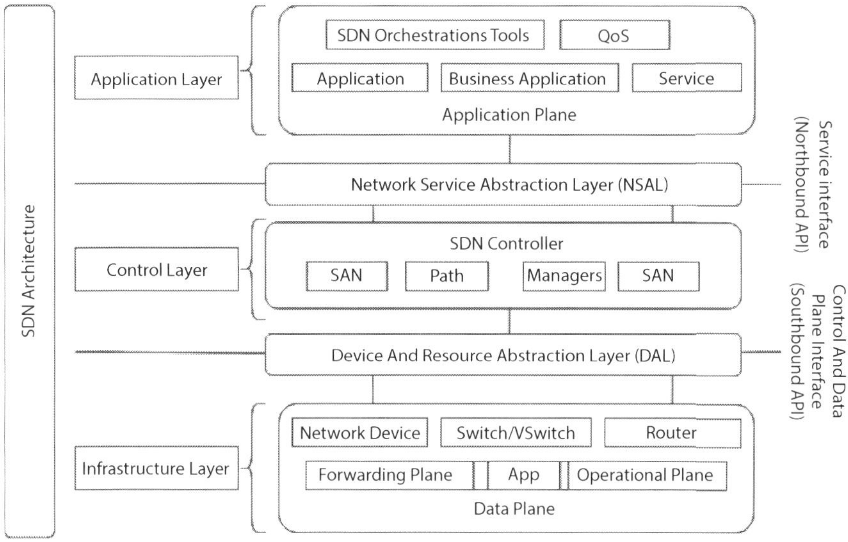

Figure 1.2 SDN network architecture [25].

A router examines its flow entries at the time it receives inbound streams from a source. The router transmits the initial packet of the stream to the controller for routing decisions if the flow cannot be located. Upon receiving the packet, the controller examines the preamble and updates the flow tables of the pertinent device. The device subsequently forwards the flow's remaining packets following the controller's instructions.

1.6 Edge Cloud-Based Next-Generation Internet

To reduce the processing burden and bring intelligence to the edge, the suggested next-generation Internet model incorporates CDN and cloud computing ideas. This Edge Cloud-based architecture intends to address significant issues with Cloud computing, such as latency, security, privacy, and high bandwidth needs for data-intensive applications. By transferring intelligence to the edge network, CDNs have already overcome several restrictions. The Internet's central processing unit, however, is built to forward IP packets simply; hence, any extra IP packet examination or caching may result in longer end-to-end delivery times [26].

1.6.1 Edge Cloud

To combine user-side virtual servers with Content-Centric Networking (CCN) capability at the edge of the network, the Edge Cloud is a valuable solution. One of the key advantages of the Edge Cloud is its ability to enable users to create "mash-ups" or custom combinations of content from various sources without overburdening their own devices. This means that instead of relying solely on centralized cloud servers for processing and data retrieval, the Edge Cloud empowers users to perform content-related operations closer to their location. This reduces the round-trip time for data requests and enables more interactive and dynamic applications.

By distributing computational resources closer to the network edge, the Edge Cloud minimizes the latency that occurs when data have to travel long distances to centralized data centers. This is particularly beneficial for real-time applications, such as video streaming, gaming, augmented reality, and IoT applications, where responsiveness and low latency are critical. With the Edge Cloud, processing can be performed in close proximity to the end-user, resulting in faster response times and improved quality of service.

1.6.2 Content Distribution Networks (CDNs)

Content distribution networks (CDNs) were initially created to relieve the load on central servers by distributing the delivery of high-bandwidth content to strategically placed servers with high-bandwidth links. By moving the content closer to the end-user, CDNs have reduced delay. A new approach to highly distributed CDNs has been proposed, where servers are located at the edge of networks, eliminating the middle mile. However, these CDNs can only serve cacheable content, and their highly distributed architecture introduces challenges with security, management, and synchronization.

Traditional centralized server-based content distribution has given way to more modern distributed methods like Content Distribution Networks (CDNs) and Peer-to-Peer (P2P) systems [27] in the Internet's content delivery mechanisms. By delivering bits of the content closer to the user at the network edge, much like cache memory does on a computer, CDNs are meant to reduce the load on content servers. Content is proactively sent to the edge networks through effective replication and caching technologies, allowing quick delivery to users. Highly requested servers can speed up response times by caching content close to the user's network edge. This reduces pressure on the primary server and lowers latency for the user while simultaneously improving response times. By moving (portions of) the material to the network's edge, much like a computer's cache memory, CDNs reflect an extension of the client-server paradigm that was first developed at the beginning of the internet. The fundamental job of a CDN is to push material as quickly as possible to edge networks from an edge node that is nearest to the user. This is done by using the right replication and caching technologies.

1.7 Network Architecture

A network's technical design, including its protocols and algorithms, is governed by a set of broad principles referred to as network architecture. This framework usually specifies the name of network entities, as well as how the state is added, withdrawn, and managed. It also controls the interrelationships and operational details of the naming, addressing, and routing services. A protocol stack is created by layering the communication functions. The network architecture also outlines how network resources are distributed across flows and how end systems, such as fairness and congestion management, react to this distribution. A network design also specifies and enforces security boundaries. Furthermore, management restrictions are set up and only crossed if necessary. Network architecture also describes how different quality of service can be demanded and obtained. Network architecture provides an essential framework for creating dependable, secure, and effective networks. The architecture ensures that all components of a network operator act in unison to achieve optimal performance by providing guidelines for the individual components.

Architecture is more comprehensive and long-lasting than a particular technical design that complies with it. The architecture itself is anticipated to remain and be adaptable to numerous generations of technology, even though a technical design generated from a specific architecture may alter in response to changing requirements. The Internet architecture, which includes both IPv4 and IPv6 as separate technical designs that adhere to the same fundamental architecture, serves as an obvious illustration of this. The architecture itself is sturdy and long-lasting despite the evolution of the technology designs contained inside it. A connection-less packet-forwarding infrastructure, fixed-size numerical addresses with a straightforward hierarchy, and a least-common-denominator data delivery service at the inter-network layer are some of the key characteristics of Internet architecture. Its architecture has changed significantly throughout time, particularly during the 1974–1980 research phase as well as in the early 1980s and beyond. Version 4 of the protocols split TCP into a transport layer (TCP) and an inter-network (IP) layer during the 1974–1980 research era. Subnetting, autonomous systems, and the domain name system (DNS) were all incorporated into the architecture from the beginning of the 1980s, bringing about hierarchical design to meet scaling concerns.

1.7.1 Some Proposed Architectures

1.7.1.1 Security Architecture for Networked Enterprises (SANE)

A security-focused framework called SANE architecture is intended to safeguard against malicious network attacks. This is accomplished by mandating that the origin and objective of all network traffic be explicitly declared at the beginning. To do this, SANE uses a private network security architecture that is specifically designed for such networks and makes use of centrally controlling network-wide policies with a domain controller. To allow end-host APIs to communicate their intentions to the wider internet in public contexts, SANE mandates adjustments to those APIs. By default, SANE requires communication between hosts and the domain controller, and client authentication is necessary to provide a secure channel for ongoing communication. The central control technique of SANE, which creates a difficulty with scalability and single points of failure in the architecture, may be a problem. The data plane may be slowed down by SANE's need that switches to carry out cryptographic operations per packet and end hosts must be upgraded to defend against malicious attacks. There are significant research issues associated with integrating middleboxes and proxies into the SANE architecture. Before they can be used in the real world, more elaborate procedures and designs to handle these problems need to be presented and confirmed [28].

1.7.1.2 Swarming Architecture

Similar to the P2P system, the envisaged swarm architecture is a swarm-based distributed file delivery paradigm for the next-generation internet. Peers are connected in this architecture in a loosely coupled, decentralized way to offer robustness and adaptability. The architecture suggests a unified swarm paradigm in which peers participate in the system and aid one another in achieving their goals. A general name and resolution method is one of the essential adjustments needed to build a swarming-based architecture. A group of peers that can give the data are identified by the Metadata Resolution Service (MRS), which resolves the metadata. To translate intent supplied in such an application-specific form, use the Intent Resolution Service (IRS), to a standardized metadata. Highly replicated tracking, in-network tracking, and peer-to-peer tracking technologies are all combined in the design of the MRS service. Some of the issues that hinder P2P systems from being the leading content delivery technology of the future are resolved by the suggested architecture. The design does, however, have several shortcomings, such as poor support for post-popular downloads if there may only be a few seeds and low demand for the content. Moreover, the architecture increases the peers' load and latency and repeated trackers lead to disorganized swarms [28].

1.7.1.3 Flexible Internet Architecture (FlexNGIA)

These restrictions might be overcome by FlexNGIA, an adaptable tactile internet's next-generation internet architecture. To enable in-network computing, FlexNGIA defines a unique internet architecture that takes advantage of the computing resources that will be made available throughout future network infrastructures. This makes it possible for the network to host cutting-edge network features that might support network applications. Moreover, FlexNGIA establishes an economic model where network administrators could provide service function chains with ambitious standards for availability, performance, and reliability in addition to data transmission. By facilitating in-network complex capabilities to be aware of the flows associated with a given proposal and the imperatives of each of them, the support for the conjunction of the transport and the internal layers enables the system to provide the best facilities for collision avoidance and reliability. FlexNGIA views packet headers as being completely flexible and adaptable to the needs of the application. While FlexNGIA affords a promising starting point for the next-generation Internet architecture, it is important to recognize that many issues with signaling, service function chain design, allocating resources, fault tolerance, effectiveness, monitoring, cost structure, and security still need to be resolved. However, significant effort needs to be made to fully utilize FlexNGIA's potential and create a reliable and effective network infrastructure for the future [29, 30].

1.7.1.4 Architecture for Mobility Support

The current design of inter-networking has some drawbacks if it comes to naturally enabling multi-homing, mobility, and security. Semantic overload in IP is one of the main problems. A new layer of identification has been suggested to be added between the transport and network layers to solve this problem. This layer will enable the deployment of additional services like mobility, multi-homing, and security in a smooth manner and give a stable cryptographic identity for end-hosts. The proposed architecture is split into external and internal modules, with the latter offering end-host features such as service discovery, an identifier to locator resolution, and legacy name resolution. The internal modules offer security, mobility, and legacy application services. A legacy application's query for name resolution is intercepted by the DNS-Handler module, which then inserts the IDH and utilizes the IDM module to locate the target IDGW. The Security Association Database, which provides end-to-end integrity and secrecy, was constructed using the Diffie-Hellman method, and the Security module searches there for the appropriate security association. The packet is sent to the Routing module for processing after encryption, where the Routing module resolves the ID to location mapping in the Peer Cache before sending the packet to the network via IP and UDP encapsulation [31].

1.7.1.5 The Data-Oriented Network Architecture (DONA)

The Data-Oriented Network Architecture (DONA) proposes a new data-centric architecture for the internet, in contrast to the current host-centric architecture. DONA places a strong emphasis on a cutting-edge mechanism for identifying content and identity resolution to build an architecture surrounding service and data access. Three essential conditions for data and service access—persistency, availability, and authenticity—that are not inherently supported by the host-centric internet design as it currently exists are provided by this method. The "principal," that is the owner or originator of the information or service, must be explicitly centered around the information or service during the use of the central DONA technique. The names are formatted as follows: P:L is the digital hash of the principal's public key and L is the identifier for the information or service that its principal has selected. The next step is to associate an information or service name with a particular location using the routing-on-name method. The elements of the routing processes responsible for directing information names (P:L) to specific data servers are referred to as "routing handlers," or RHs. By relying on data names for routing, DONA delivers reliability and self-healing characteristics for data access, which are similar to those for host access on the current internet. Principals' flat cryptographic names aid in establishing the legitimacy and provenance of the data. Data late binding through the DONA technique to the main server also accomplishes data persistence, freeing it from dependence on the persistence of the host [32].

1.7.1.6 MILSA (Mobility and Multihoming Supporting Identifier-Locator Split Architecture)

The IRTF RRG established several design goals, including routing scalability, mobility, multihoming, and traffic engineering. The MILSA architecture is an end-host-based ID locator split technique that seeks to satisfy these goals. Without relying on Provider Independent (PI) addresses for global routing, MILSA accomplishes these objectives and leverages signaling and data separation to boost performance and effectiveness. The decoupled ID space introduced by MILSA supports location privacy while enabling trust interactions and policy enforcement across various businesses. Earlier work on MILSA has proposed a secure structured ID system, numerous ID determination and mapping, multicast, and integration capabilities. Concentrate on MILSA's deployment approach in this study because it has not been covered in earlier publications. With few modifications to the current Internet infrastructure, the suggested hybrid transition method combines the two MILSA directions and progressively shrinks the size of the global routing table. Also, our architecture enables evolution in any direction the market chooses. MILSA hosts can communicate with legacy hosts for backward compatibility throughout the transition time [33].

1.7.1.7 Delay/Disruption Tolerant Networks (DTN) and Related Architectures

The Web Research Task Force (IRTF)'s working group is in charge of directing research into Data Tolerant Networks (DTN) and associated designs. DTN was initially created as part of a project to create the Interplanetary Internet for use in deep-space communication, but it quickly became apparent that the concepts created there could also be used to solve problems with other challenging terrestrial networking systems. The end-to-end message-oriented overlay known as the bundle protocol is built on top of the channel's transport layer (or another layer) and offers store-forward services by controlling persistent storage at middle nodes to help with sporadic connectivity.

The end-to-end concept is re-defined in DTN networks in such a way that it is still true in the context of a DTN environment, which is an intriguing property of DTN networks. The "custody transfer" mechanism is described in the bundle protocol. If an intermediate node accepts custody of the bundle, it can then assume responsibility for the bundle's trustworthy delivery. The bundle can now be removed from the buffer of the node that transfers custody to node N. As the source node may not be connected for long enough to guarantee end-to-end reliability, this idea of reliability is relevant in the DTN context as opposed to the end-to-end principle. To help with effective scheduling and exploitation of contacts' communication possibilities, it stores and forwards "bundles," which are variable-sized, typically lengthy messages transformed from arbitrarily long application data. Another significant problem with DTN is routing. To solve routing in diverse operating contexts, multiple routing protocols have been developed for DTNs. While it is simpler to route in deterministic environments, an "epidemic routing" approach has been developed for routing in extremely erratic circumstances. With the relay-based strategy, the node does a "controlled broadcast" of the message to its close neighbors if the path to the target is not open. Extending the principles created to a wide range of networking situations with either a high bit error rate or high delay is something that the next generation of DTN research entails. These settings use DTN techniques to facilitate or improve their data transmission scenario, and they have a lot of potential to be key components of the next-generation networking paradigm [34, 35].

1.7.2 Challenges for Architecture

Network architects face numerous challenges in designing and managing modern networks. Among these challenges are four key areas of research: network services, network management, network performance, and diversity and change [36].

1.7.2.1 Network Services

Network services are the need to design and deploy complex service architectures that are both dependable and scalable. With the proliferation of cloud computing, mobile applications, and IoT devices, the demand for network services is growing rapidly. Network architects must design systems that can handle the increasing traffic demands and provide reliable service to users. A wider definition of Digital services, including multicast and QoS capabilities, has resulted from the Internet's development. While certain potential capabilities can be anticipated, others will not emerge until much later, presenting challenges for researching the Next-Generation Internet (NGI). A key challenge in NGI research is to develop mechanisms that can accommodate emerging needs and facilitate the introduction of new services as they arise. These services may be loosely coupled with core network components such as switches and routers, or tightly integrated to maximize their benefits.

1.7.2.2 Network Management

Network management is the need to monitor and control network resources dynamically and efficiently. As networks become more complex and diverse, network administrators must be able to quickly identify and resolve issues to maintain optimal performance. This requires advanced tools and techniques for network monitoring, analysis, and optimization. The capacity of network administrators to configure (and reconfigure) network hardware, locate problems, take corrective action, and prepare for network expansion has to be improved. The rapid expansion of the Internet cannot be managed using current techniques. Network service is frequently interrupted by mistakes in manually setting routing tables or in the execution of routing protocols. It is important to comprehend how routing policies impact network scalability and how they might be used such that expansion is enabled rather than constrained.

1.7.2.3 Network Performance

Because these devices usually serve as the performance supply chain for end-user applications, the productivity of devices that are connected to the network is an issue of concern. As demands rise, frequently in unforeseen ways, so does the use of networked information servers. To fully utilize the capabilities of the underlying technology, new methods for managing data transit between network, application software, and hardware components will be required. New architectural methods for end-system design at all levels, from hardware to application programming, may be necessary as a result.

1.7.2.4 Diversity and Change

Using the Internet Protocol (IP) service models, the Internet has shown to be incredibly efficient in managing a wide range of applications and underlying network technology. Nonetheless, it is a substantial technical challenge to deliver dependable and high-quality support for applications over a network that simultaneously transmits extremely unpredictable data traffic.

1.8 Security and Safety

Isolation in a shared communications environment entails giving the application some assurance that there will not be any unwanted interference. Yet, the unique requirements of various applications result in a variety of incredibly unique forms of isolation. Since the Next-Generation Internet (NGI) is a continuous evolution of the current Internet rather than a single, properly specified technical project, it is challenging to make a general statement about its safety. As the NGI is being developed, some of these security shortcomings and vulnerabilities may remain in the current Internet infrastructure. Throughout their work on the NGI, developers and researchers must give security priority and take measures to eliminate any security holes that are discovered. The success of these efforts in addressing security issues will determine how safe the NGI becomes.

1.9 Distributed Systems

1.9.1 Concept of Distributed System

A wide variety of nodes, including high-performance computers, small plug computers, and other small devices, make up contemporary distributed systems. Although these nodes work independently of one another, they are intended to exchange messages to accomplish shared objectives. Each node examines incoming messages, reacts, and continues communication by passing messages to other nodes. Distributed systems are frequently organized using overlay networks. Each node in an overlay network is a software process that has a list of other processes to which it can send messages directly. TCP/IP or UDP channels are used for message passing, and higher-level communication tools might be offered. Two essential characteristics make up the definition of a distributed system [37]. It is first and foremost a group of computational units or nodes, which can function independently of one another. These nodes could be hardware components or computer programs. Second, despite being made up of distinct nodes, users, whether they be humans or applications, perceive the distributed system as a single entity. Because of this, the establishment of collaboration between the nodes is essential to the creation of distributed systems to accomplish the required behavior. Queries for distributed services may be routed to a single node that implements the service or, in the case of replicated services, to a set of nodes. A distribution function is responsible for mapping requests to the nodes that can handle them [38].

1.9.2 Terminologies

A group of machines are linked together in a distributed system interconnected through a network and function collaboratively to offer a fundamental set of services. Within the system, a *node* refers to a single computer, and a site refers to a group of interconnected nodes that form a subset of the system. A *service* is categorized as distributed if it can be offered by numerous nodes, each having the capability to manage a part of the requests for that service [39].

1.9.2.1 Grouping of Autonomous Computer Components

A contemporary distributed system is made up of several computing components that work independently, with nodes ranging from little gadgets to powerful computers. Notwithstanding their independence, nodes are programmed to work together to accomplish shared objectives, which calls for message exchange. A node processes messages it receives and uses message passing to start new conversations. The assumption of a single global clock is impossible because each node has its time reference and operates independently. Due to the absence of a common time reference, synchronization and coordination within a distributed system become difficult [39].

1.9.2.2 Single Coherent System

The objective of a distributed system is for its users to see it as a single cohesive system. While some researchers favor a single-system perspective in which users are fully unaware that processes, data, and control are dispersed, this is sometimes too challenging to implement. The system behaves as expected by the user, which is a weaker definition of coherence. A design objective known as distribution transparency seeks to conceal information about where a process is running or where data are being stored. A distributed system's various nodes make it inevitable for partial failures to cause unanticipated behavior; therefore, this quest for coherence does come with a cost. It can be difficult to provide a uniform and consistent perspective of a distributed system. To achieve this, it must be made sure that end-users are uninformed of a process's precise location or whether a job has been assigned to another process that is operating on a different node. It is crucial that end-users are not concerned about how data are kept or whether they are replicated for better performance. One of the most important design objectives for distributed systems is something called distribution transparency. It follows the same strategy as Unix-type operating systems, which access different resources including a unified file system interface, networks, main memory, storage devices, and files that can all communicate while hiding their differences [40].

1.10 Distributed System Design

Four key goals must be met for the development of a distributed system to be worthwhile. These goals include ensuring that resources are easily accessible, hiding their scattered nature, upholding openness, and facilitating scalability. Meeting objectives is crucial to the success of distributed systems because simply being able to build them does not ensure their viability.

1.10.1 Significance of a Certain Design

There is no global clock since each node has its sense of time as a result of the independence of the nodes. As a result, the distributed system's synchronization and coordination face difficulties. It might be difficult to control group membership in a distributed system. Groups can be classified as either open or closed. Each node can join and communicate with any other node in an open group. Only members of a closed group can communicate with one another, and joining and leaving the group requires a different process. Managing group membership can also present challenges with authentication and confidentiality.

1.10.2 Scale and Scalability

The scale has grown into an increasingly key component in the design of distributed systems. Number, geography, and administration are the three dimensions that make up the scale. The system's user base, object population, and quantity of services make up the numerical dimension. The system's geographic dispersion is referred to as the geographical dimension. The quantity of organizations in charge of various system components is part of the administrative dimension. The three scale dimensions have an impact on several distributed system elements, including naming, authentication, authorization, accounting, communication, using remote resources, and user interaction. Size also has an impact on dependability since it increases the risk of host failure as the system scales numerically and the risk of communication breakdowns as the system scales geographically. The scale has an impact on performance as well because it increases the server load and communication volume with a numerical scale while increasing communication latency with a geographic scale. The scale affects administrative complexity as well since it necessitates interaction between numerous administrative entities with potentially at-odds policies as changes occur more frequently. Finally, the scale has an impact on heterogeneity since larger systems are less likely to have identical components.

1.10.2.1 Definition of Scale

Several factors can affect a system's burden as it scales. The amount of data that network services must manage as the system grows also grows the total number of service requests. Replication, dispersion, and caching strategies are used to reduce the workload by limiting the number of requests that each server must handle. While caching reduces the need for repeated queries, replication and dispersion allow requests to be split across numerous servers. Using many servers, each of which is in charge of keeping a distinct set of files, is one example of distribution. On the other hand, replication calls for the availability of the same system binaries across several servers. With this strategy, servers can be chosen depending on things like load and proximity. Addressing problems with a system's numerical, geographical, and administrative components is a necessary step in scaling it. Each of these elements poses distinct problems, and many systems have used different approaches to solve them.

1.10.2.2 Scalability Analysis

Several scalability metrics have been created to assess an algorithm's performance on platforms of diverse sizes. The scalability of various methods is also evaluated using these criteria. The metrics presuppose that the program executes on a collection of k processors, each of which has a certain architecture, and that the completion time "T" serves as a performance indicator. Strong scalability and weak scalability are two categories into which these metrics might be subdivided. A parallel algorithm's effectiveness is assessed using strong scalability as the number of processors used rises while the size of the task remains constant. On the other hand, weak scalability gauges a parallel algorithm's effectiveness as the size of the problem and the number of processors employed grow correspondingly.

1.11 Distributed System Monitoring

For testing, debugging, and rating the effectiveness of computer programs, monitoring is crucial. Yet, with dispersed systems, monitoring is more difficult.

It is important to dynamically gather information about the interactions between processes to monitor a distributed program. Distributed systems development and monitoring present certain difficulties that need to be carefully considered. Due to the multiplicity of control points, typical monitoring and debugging methods that are effective in sequential systems must be modified for distributed contexts. Also, because distributed, asynchronous systems are inherently non-deterministic, replicating failures or testing low-probability scenarios can be tricky. Communication delays between nodes make it difficult to correctly ascertain the system's state at any one time. In contrast to sequential systems, where a process interruption does not affect later execution, monitoring a distributed system can also change its behavior. Last but not least, interactions between the system and developers can be difficult because it may be essential to move around during several terminals to start processes, create breakpoints, and check traces [41].

1.11.1 Idea of Monitoring System

There are many benefits to using the same IPC mechanism for monitoring data distribution and collection that is utilized by application processes. When an application is relocated to a different network of machines, there is no longer a need to port a separate IPC mechanism, which simplifies the design, debugging, and maintenance of the monitoring system. It is also advised to keep the monitoring data detection and collection from its analysis and presentation. This makes it possible to develop an integrated set of tools that can operate as a unit and have a shared implementation [41]. Additionally, since tool creators do not need to have in-depth knowledge of how monitoring information is acquired at a low level, this technique enables the efficient development of a wide and effective, including textual traces and deadlock detection. A monitoring system must comprehend the semantics of the application system it is watching to be effective. It is not sufficient to merely collect data at the IPC level. The data collection and presentation capabilities of monitoring tools must reflect the structure and behavior of the application system. This necessitates the mapping of lower-level interaction patterns to higher-level logical activities within the application system by the monitoring system. By doing this, the monitoring system can offer insightful information about how the application system behaves and support the detection of performance problems or mistakes that might otherwise be challenging to spot.

1.12 Security in Distributed Systems

Three key components make up distributed system security: authentication, which involves confirming digital credentials; authorization, which helps determine whether a user has the permissions required to carry out a demanded operation; and accounting, which happens to involve documenting and keeping track of a user's actions to prevent irrational resource use. Several systems use a distributed authentication technique while leaving server-by-server permission. Few distributed systems, though, have accounting procedures. Accounting may also have functions that let users be charged for using certain resources.

1.12.1 Reasons for Not Using Encryption

Authentication based on encryption avoids the issues associated with transmitting passwords over a network. Instead, a unique encryption key is assigned to each user and used to authenticate their identity. However, this method has its drawbacks. In larger systems, each principal (user or server) must maintain a key for every other principal they may communicate with, which can become impractical. Specifically, a total of (n*m), if the number n corresponds to the number of clients and m is the frequency of servers, keys are needed [42].

1.12.2 Authorization

Distributed systems can approach permission in several ways. In one, anytime a server must make an access control decision, an authorization service receives a request. The authorization service makes the choice and replies to the server with its decision. This approach enables the decision regarding access control to include factors such as recent usage of certain other servers, global constraints, etc. The drawback is that the process may be challenging and that the authorization service might clog up [42].

1.13 Blockchain and Distributed Systems

Blockchain is a Design Pattern for decentralized consensus concerning public ledgers in distributed systems where the trustworthiness of the agents cannot be considered. Applications of blockchain are typically justified if it is necessary to reach a consensus across separate system actors.

To comprehend blockchain, understanding distributed systems is crucial since blockchain is a decentralized distributed system; it is vital to understand distributed systems to understand blockchain. In a distributed system, numerous nodes work together to accomplish a common objective in a coordinated way while still presenting to the end-user as a single logical platform. Blockchain's distributed consensus does not directly support a transactive grid's decentralized management, but it is compatible either. In reality, there have been initiatives to create decentralized blockchain models to increase the process's transparency. For instance, in one method, players perform off-chain optimization of their local goals and exchange the outcomes with a central smart contract that serves as a data aggregator. A global update is completed by the aggregator, and the modified values are then sent back to the participants for additional optimization. Up till convergence is attained, this process is repeated iteratively. Nodes are distinct entities that have their memory and CPU, are capable of sending and receiving messages, and may be trustworthy, imperfect, or malignant. Any unexpected or harmful activity is included in the definition of a Byzantine node [43], which is a node that can display arbitrary behavior. Coordination between nodes and fault tolerance present the biggest design challenges for distributed systems. Even if some nodes develop issues or network connections fail, the distributed system must keep running flawlessly.

1.13.1 Decentralization

Decentralization is a mechanism to prevent the concentration of power by avoiding a single point of control. It promotes a distributed network where various actors hold different degrees of influence and power and ensures that no actor can gain too much control. This approach offers numerous advantages such as increased efficiency and democracy compared to centralized systems. In a decentralized system, resource allocation decisions are made by resource owners, leading to the possibility of conflicts if there is an imbalance in resource ownership. By distributing control and decision-making across multiple nodes, decentralization enhances fault tolerance and scalability. It reduces single points of failure, as the failure of one node does not disrupt the entire system. Data redundancy and availability are improved through data replication across distributed nodes. Decentralized systems also provide autonomy and local decision-making capabilities, allowing nodes to make independent decisions based on their resources and needs. This enhances efficiency and adaptability. Additionally, decentralization improves security by reducing the vulnerability to attacks and unauthorized access. It enables democratic participation and inclusivity, ensuring that various actors can contribute to the decision-making process. However, achieving effective coordination and resolving conflicts stemming from resource imbalances require the implementation of appropriate governance models, consensus algorithms, and resource allocation mechanisms.

1.13.2 Peer-to-Peer Networks (P2P)

Peer-to-peer (P2P) [44] networks, which offer a decentralized architecture that enables users to connect and share data directly with each other without relying on centralized servers or intermediaries, are a crucial part of distributed systems. Each node in a P2P network function as both a client and a server, and the network as a whole self-organizes; thus, there is no hierarchy or centralized management. For purposes including file sharing, content delivery, and communication, P2P networks are frequently employed in distributed systems. They can provide numerous advantages, including improved scalability, fault tolerance, and attack resistance. These can, however, also pose difficulties, such as guaranteeing data security and network resistance to errors and intrusions.

P2P networks employ a range of strategies to deal with these issues, including distributed algorithms for data replication and routing, cryptographic procedures for data protection, and redundancy and fault tolerance measures to make sure the network can keep running even in the event of failures. P2P networks [45], in general, are a crucial part of distributed systems, offering a decentralized architecture that can have several advantages for data and communication.

1.14 Blockchain-Based Distributed Control Systems

A distributed protocol called the blockchain system makes use of a decentralized network to track and confirm asset transfers. Due to its capacity to keep a transparent and immutable record of transactions, this system is frequently referred to as a distributed ledger. Nodes in the network create transactions, which are then traded across nodes. Special nodes called miners organize these transactions into blocks and put them forward to the other nodes. After then, they compete to decide which block to add to the chain. The blockchain's present status reflects the inter-account transfers of digital assets. Blockchain's distributed consensus does not directly support a transactive grid's decentralized management, but it is compatible either. In reality, there have been initiatives to create decentralized blockchain models to increase the process's transparency. For instance, in one method, players perform off-chain optimization of their local goals and exchange the outcomes with a central smart contract that serves as a data aggregator. A global update is completed by the aggregator, and the modified values are then sent back to the participants for additional optimization. Up till convergence is attained, this process is repeated iteratively.

1.14.1 Control System

To manage complicated, sizable, and widely dispersed industrial units, distributed control systems (DCS), which are autonomous and automated, are used. In a DCS, each group has its dedicated processor, which is connected to the others over a communication channel, in contrast to centralized systems, where the control function is handled by a single central unit. Several field devices, like actuators and sensors, are connected to controllers in a DCS so that they may continuously receive data from them. The data are subsequently transmitted by the controllers across a communication bus to other controllers in the hierarchy. This is accomplished by using a variety of communication methods, including Profibus, HART, arc net, and Modbus [46]. Instead of utilizing a centralized system, the fundamental structure of DCS operates hierarchically from the top to the bottom and conversely. A few of the key benefits of the DCS include the management of complicated processes, pre-defined function blocks, an expandable platform, and more advanced HMI [47]. They differ from centralized systems due to these characteristics.

1.14.1.1 To Handle Complicated Processes

PLC controllers are used in industries to manage and monitor various industrial plant operations and parameters at high speeds. Hence, these systems are chosen for sophisticated computations.

1.14.1.2 Pre-Defined Function Blocks

DCS provides pre-defined programmers and algorithms, making it easier to utilize for applications and problems with complicated requirements. It supports a wide range of programming languages, including function block, ladder, sequential, and others, allowing a user to create a program specifically tailored to his needs.

1.14.1.3 *Scalable Platform*

The capacity of DCS systems depends on the number of I/O that are linked to them. Increase the number of input–output devices and the server to accommodate more users [48].

1.15 Blockchain for Distributed System Security

A key aspect of blockchain technology is decentralization, which reduces the need for outside interference in distributed control systems. The blockchain ensures that security issues like interception and disruption can be avoided by replicating data across all system peers. Furthermore, there is no such thing as an unlawful duplicate of data because blockchain is an open-source system. PLC controllers [49] are used in a variety of industrial settings because they can manage and monitor intricate industrial processes and parameters quickly, making them an excellent choice for complex calculations. The pre-defined programming and algorithms that DCS systems offer make it easier to use the system for applications and problems with complex needs. It supports a variety of programming languages, such as function block, ladder, sequential, and more, enabling users to design a unique program that is tailored to their unique requirements. A blockchain is a powerful tool for stopping data manipulation and falsification. A chain is formed by the hash of each block in the blockchain, which each block includes. This series acts as a security system for the database, preventing any attempts to change the data from succeeding. The blockchain's security system would safeguard the data and keep it safe and secure in the event of a DCS assault.

Double spending is a significant security concern in a blockchain system, which arises due to its decentralized nature. It allows a malevolent user to use the same digital assets in two conflicting transactions, resulting in the potential loss of funds and trust in the system. One common approach to achieving this is by creating divergent forks in the blockchain, which can deceive merchants and other network participants. To mitigate the risk of such a fork scenario, it is necessary to establish a consensus mechanism that ensures all nodes on the network agree on a unique block for each index in the blockchain. Consensus algorithms, such as Proof of Work (PoW) [50] or Proof of Stake (PoS), are used to achieve this by ensuring that all nodes on the network confirm the validity of new transactions and blocks before they are added to the blockchain. This guarantees the integrity and immutability of the blockchain, preventing double spending and other malicious activities.

1.15.1 Fault-Tolerant Consensus in a Distributed System

In a distributed system, despite their physical separation, numerous parts cooperate to accomplish a single objective. The consensus that these elements have come to regarding particular data values is referred to as such. Yet, hostile assaults and unforeseen malfunctions might occur in the system's parts and communication pathways. The consensus problem in message-passing systems is that component smiles can malfunction as a result of crashes or Byzantine faults.

1.15.2 Complications while Using Blockchain for Distributed Computing

Blockchain may be susceptible to multiple cyberattacks if distributed computing is used, such as Distributed Denial of Service (DDoS), which can deplete computing resources by flooding the system with numerous fictitious requests. Tiny blockchain networks with few nodes are especially vulnerable to such attacks, which can allow hackers to take over a large number of nodes to engage in fraudulent activity. Large blockchains like Bitcoin and Ethereum incorporate a "Proof of Work" protocol to address these problems. The requirement to perform backups on each network node complicates the use of distributed computing for blockchain. A robust network of nodes and a lot of computing power are needed for this procedure. Additionally, since there is no central authority to oversee the system as a whole, security measures must be implemented separately on each node. The distributed system's peer-to-peer architecture makes it possible to access any node at any time, necessitating the adoption of strong security measures to stop malicious or illegal activity.

1.16 Conclusion and Future Scope

In conclusion, distributed systems and the next-generation internet are something that computers and communication will look like in the future. These innovations seek to deliver an infrastructure that is more dependable, secure, and scalable to meet the rising demand for data-intensive applications and services. The usage of decentralized architectures based on peer-to-peer networks and blockchain technology is one of the fundamental characteristics of the future generation of Internet and distributed systems. Furthermore, distributed systems and the next-generation internet will be capable of supporting more data-intensive applications and services. The use of edge computing, which involves processing data at the network's edge, will provide faster response times and reduced latency, enabling real-time applications such as autonomous vehicles, telemedicine, and augmented reality. In terms of future enhancements, researchers are currently exploring various approaches to enhance the scalability, performance, and security of these technologies. One promising area of research is the development of new consensus mechanisms for distributed ledger technology. Consensus mechanisms are responsible for ensuring the integrity of the blockchain and are critical to the security and scalability of decentralized systems. These architectures make it possible for users to communicate directly with one another without the use of intermediaries like centralized servers or cloud service providers. As a result, the system becomes more fault-tolerant and resilient, making it less vulnerable to errors and assaults. Although next-generation distributed systems and the internet have numerous advantages, there are also big technical and implementation problems that need to be solved; they present substantial technological and deployment hurdles, but these can be overcome with proper design, implementation, and management. To guarantee that these technologies can be implemented successfully and securely in a variety of applications and environments, it will be crucial to keep addressing these difficulties as they arise.

References

[1] Papavassiliou, S. (2020). Software-defined networking (SDN) and network function virtualization (NFV). *Future Internet, 12*(1), 7.

[2] Cath, C., & Floridi, L. (2017). The design of the internet's architecture by the Internet Engineering Task Force (IETF) and human rights. *Sci. Eng. Ethics, 23*, 449-468.

[3] Cooper, M., & Yen, D. C. (2005). IPv6: business applications and implementation concerns. *Comput. Stand. Interfaces, 28*(1), 27-41.

[4] Atlam, H. F., Azad, M. A., Alzahrani, A. G., & Wills, G. (2020). A Review of Blockchain in Internet of Things and AI. *Big Data Cogn. Comput., 4* (4), 28.

[5] Haji, S. H., Zeebaree, S. R., Saeed, R. H., Ameen, S. Y., Shukur, H. M., Omar, N., ... & Yasin, H. M. (2021). Comparison of software defined networking with traditional networking. *Asian J. Res. Comput. Sci., 9* (2), 1-18.

[6] Chica, J. C. C., Imbachi, J. C., & Vega, J. F. B. (2020). Security in SDN: A comprehensive survey. *J. Netw. Comput. Appl.*, *159*, 102595.
[7] Singh, A., Payal, A., & Bharti, S. (2019). A walkthrough of the emerging IoT paradigm: Visualizing inside functionalities, key features, and open issues. *J. Netw. Comput. Appl.*, *143*, 111-151.
[8] Pianini, D., Casadei, R., Viroli, M., & Natali, A. (2021). Partitioned integration and coordination via the self-organising coordination regions pattern. *Future Gener. Comput. Syst.*, *114*, 44-68.
[9] Abbasi, M., Shahraki, A., & Taherkordi, A. (2021). Deep learning for network traffic monitoring and analysis (NTMA): A survey. *Comput. Commun.*, *170*, 19-41.
[10] Priyadarsini, M., & Bera, P. (2021). Software defined networking architecture, traffic management, security, and placement: A survey. *Comput. Netw.*, *192*, 108047.
[11] Kreutz, D., Ramos, F. M., Verissimo, P. E., Rothenberg, C. E., Azodolmolky, S., & Uhlig, S. (2014). Software-defined networking: A comprehensive survey. *Proc. IEEE*, *103*(1), 14-76.
[12] Rawat, D. B., & Reddy, S. R. (2016). Software defined networking architecture, security and energy efficiency: A survey. *IEEE Commun. Surv. Tutor.*, *19* (1), 325-346.
[13] Singh, S., & Jha, R. K. (2017). A survey on software defined networking: Architecture for next generation network. *J. Netw. Syst. Manag.*, *25*, 321-374.
[14] Software-defined networking (SDN) has emerged as a solution to address the challenges posed by traditional network architectures. It separates the data plane, which is in charge of forwarding network traffic, from the control plane, which is in charge of managing and making decisions about the network.
[15] Parsons, N., & Calabretta, N. (2020). Optical switching for data center networks. *Springer Handbook of Optical Networks*, 795-825.
[16] Keshari, S. K., Kansal, V., & Kumar, S. (2021). A systematic review of quality of services (QoS) in software defined networking (SDN). *Wirel. Pers. Commun.*, *116*, 2593-2614.
[17] Fernandes, S., & Karmouch, A. (2010). Vertical mobility management architectures in wireless networks: A comprehensive survey and future directions. *IEEE Commun. Surv. Tutor.*, *14*(1), 45-63.
[18] Tsirtsis, G., & Srisuresh, P. (2000). *Network address translation-protocol translation (NAT-PT)* (No. rfc2766).
[19] Droms, R. (1997). RFC2131: Dynamic Host Configuration Protocol.
[20] Braun, H. W., Ford, P. S., & Rekhter, Y. (1993, August). CIDR and the Evolution of the Internet Protocol. In *Proceedings of INET* (Vol. 93).
[21] Islam, S., & Grégoire, J. C. (2010). Network edge intelligence for the emerging next-generation internet. *Future Internet*, *2*(4), 603-623.
[22] Gramaglia, M., Sciancalepore, V., Fernandez-Maestro, F. J., Perez, R., Serrano, P., & Banchs, A. (2020, September). Experimenting with SRv6:

A Tunneling Protocol supporting Network Slicing in 5G and beyond. In *2020 IEEE 25th International Workshop on Computer Aided Modeling and Design of Communication Links and Networks (CAMAD)* (pp. 1-6). IEEE.

[23] Kreutz, D. (2020). *Logically Centralized Security for Software-Defined Networking* (Doctoral dissertation, University of Luxembourg, Luxembourg).

[24] McCauley, J., Harchol, Y., Panda, A., Raghavan, B., & Shenker, S. (2019). Enabling a permanent revolution in internet architecture. In *Proceedings of the ACM Special Interest Group on Data Communication* (pp. 1-14).

[25] Qin, Z., Denker, G., Giannelli, C., Bellavista, P., & Venkatasubramanian, N. (2014, May). A software defined networking architecture for the internet-of-things. In *2014 IEEE network operations and management symposium (NOMS)* (pp. 1-9). IEEE.

[26] Krafzig, D., Banke, K., & Slama, D. (2005). *Enterprise SOA: service-oriented architecture best practices*. Prentice Hall Professional.

[27] Li, J. (2008). On peer-to-peer (P2P) content delivery. *Peer Peer Netw. Appl.*, *1*, 45-63.

[28] Paul, S., Pan, J., & Jain, R. (2011). Architectures for the future networks and the next generation Internet: A survey. *Comput. Commun.*, *34*(1), 2-42.

[29] Zhani, M. F., & ElBakoury, H. (2020). FlexNGIA: A flexible Internet architecture for the next-generation tactile Internet. *J. Netw. Syst. Manag.*, *28*, 751-795.

[30] Tourani, R., Misra, S., Mick, T., & Panwar, G. (2017). Security, privacy, and access control in information-centric networking: A survey. *IEEE Commun. Surv. Tutor.*, *20* (1), 566-600.

[31] Wong, W., Villaça, R., de Paula, L. B., Pasquini, R., Verdi, F. L., & Magalhães, M. F. (2008, March). An Architecture for Mobility Support in a Next-Generation Internet. In *22nd International Conference on Advanced Information Networking and Applications (aina 2008)* (pp. 818-825). IEEE.

[32] Koponen, T., Chawla, M., Chun, B. G., Ermolinskiy, A., Kim, K. H., Shenker, S., & Stoica, I. (2007, August). A data-oriented (and beyond) network architecture. In *Proceedings of the 2007 conference on Applications, technologies, architectures, and protocols for computer communications* (pp. 181-192).

[33] Pan, J., Paul, S., Jain, R., & Xu, X. (2009, November). Hybrid transition mechanism for milsa architecture for the next generation internet. In *2009 IEEE Globecom Workshops* (pp. 1-6). IEEE.

[34] Fall, K., & Farrell, S. (2008). DTN: an architectural retrospective. *IEEE J. Sel. Areas Commun.*, *26* (5), 828-836.

[35] Zhang, Z., & Zhang, Q. (2007). Delay/disruption tolerant mobile ad hoc networks: latest developments. *Wirel. Commun. Mob. Comput.*, *7*(10), 1219-1232.

[36] Zhang, S. (2019). An overview of network slicing for 5G. *IEEE Wirel. Commun.*, *26* (3), 111-117.

[37] Romanov, O., Mankivskyi, V., & Saychenko, I. (2020, October). Analysis of Influence of UDP Parameters on QoS in IP Network. In *2020*

IEEE International Conference on Problems of Infocommunications. Science and Technology (PIC S&T) (pp. 551-556). IEEE

[38] Sonbol, K., Özkasap, Ö., Al-Oqily, I., & Aloqaily, M. (2020). EdgeKV: Decentralized, scalable, and consistent storage for the edge. *J. Parallel Distrib. Comput.*, *144*, 28-40.

[39] Kahanwal, D., & Singh, D. T. (2013). The distributed computing paradigms: are P2P, grid, cluster, cloud, and jungle. *arXiv preprint arXiv:1311.3070.*

[40] Van Steen, M., & Tanenbaum, A. (2002). Distributed systems principles and paradigms. *Network*, *2* (28), 1.

[41] Joyce, J., Lomow, G., Slind, K., & Unger, B. (1987). Monitoring distributed systems. *ACM Trans. Comput. Syst. (TOCS)*, *5* (2), 121-150.

[42] Neuman, B. C. (1993, May). Proxy-based authorization and accounting for distributed systems. In *[1993] Proceedings. The 13th International Conference on Distributed Computing Systems* (pp. 283-291). IEEE.

[43] Gao, S., Yu, T., Zhu, J., & Cai, W. (2019). T-PBFT: An EigenTrust-based practical Byzantine fault tolerance consensus algorithm. *China Commun.*, *16* (12), 111-123.

[44] Androutsellis-Theotokis, S., & Spinellis, D. (2004). A survey of peer-to-peer content distribution technologies. *ACM Comput. Surv. (CSUR)*, *36* (4), 335-371.

[45] Koren, I., & Krishna, C. M. (2020). *Fault-tolerant systems*. Morgan Kaufmann.

[46] Sauter, T. (2010). The three generations of field-level networks—Evolution and compatibility issues. *IEEE Trans. Ind. Electron.*, *57* (11), 3585-3595.

[47] Khaled, A. B. A. D. A. (2015). *Distributed Control Systems-DCS-An application on YOKOGAWA CENTUM VP* (Doctoral dissertation, Ministry of Higher Education).

[48] Khan, A. T., Cao, X., Li, S., & Milosevic, Z. (2019). Blockchain technology with applications to distributed control and cooperative robotics: A survey. *Int. J. Robot. Control*, *2* (1), 36-48.

[49] Alphonsus, E. R., & Abdullah, M. O. (2016). A review on the applications of programmable logic controllers (PLCs). *Renew. Sustain. Energy Rev.*, *60*, 1185-1205.

[50] Sriman, B., Ganesh Kumar, S., & Shamili, P. (2021). Blockchain technology: Consensus protocol proof of work and proof of stake. In *Intelligent Computing and Applications: Proceedings of ICICA 2019* (pp. 395-406). Springer Singapore.

2

Decentralized System in Education and Research

Prashant Verma[1]*, Deepaks M. Ratnani[2], Shikha Verma[3] and Sandhya Avasthi[4]

[1]WSI, MeitY, New Delhi, India
[2]WSI, MeitY CDAC, New Delhi, India
[3]MCA Department, ABES Engineering College, Ghaziabad, India
[4]Department of CSE, ABES Engineering College, Ghaziabad, India

Abstract

Decentralization refers to the transfer of authority and responsibility to local communities and educational institutions. Then, they are free to decide for themselves regarding many areas of practice and policy. The majority of nations are testing out or thinking about decentralizing education in some way. Decision-making authority is transferred through the process from central ministries of education to regional governments, local governments, communities, and educational institutions. Yet, the scope of the transfer might range from administrative deconcentration to a considerably wider transfer of financial responsibility to the local or regional level. The degree to which national education systems are decentralized clearly affects inclusive education policy and practice. At all levels of the education system, decentralization can be seen as a crucial variable in context analysis and change management. The chapter intends to identify some developing trends in relation to decentralization in educational and research systems generally, and systems for inclusive education particularly, in order to analyze some of the key challenges. It discusses the level of flexibility and authority required for future aspects. It examines the key advantages and difficulties of various nations' strategies. It begins by introducing the topic of decentralization and the related challenges and opportunities. It then presents examples of approaches that countries are taking and ends with the future direction in building a decentralized system in the education system and research by defining and developing capacity building across levels in the education system. The chapter also concludes with the examination of different scopes and recommendations for decentralization of the education system.

**Corresponding author*: vermaprashant1@gmail.com

Sandhya Avasthi, Suman Lata Tripathi, Namrata Dhanda and Satya Bhushan Verma (eds.)
Decentralized Systems and Distributed Computing, (35–52) © 2024 Scrivener Publishing LLC

***Keywords*:** Decentralization, delegation, deconcentration, devolution, education system, school governance, redistributive model, learning culture

2.1 Introduction

Decentralization is the word used when authority is transferred to local governments and educational institutions [1]. Then, they are free to decide for themselves regarding many areas of practice and policy. In a centralized system, a central authority may administer policy, curriculum, and assessment in addition to managing finances, employees, and resources. Several studies [2–4] demonstrate that resources can be used more effectively in decentralized systems. This, in turn, may result in better student achievement, greater parental satisfaction, and stronger accountability with support from the community.

In the current political climate, decentralization of educational systems is a matter that is very important. This is especially relevant when considering the improvements that will probably be required as nations move towards more inclusive systems. Finding the ideal balance between centralization and decentralization is a key concern in every country's reform attempt [5]. The formulation and implementation of a policy that determines where authority, responsibility, and accountability should belong is the main source of conflict and the topic of dispute. The notion of delegating authority and making decisions to local organizations seems to mesh well with more democratic and participatory approaches. They are essential to inclusive practice and policy. Yet, complicated decentralized networks can make ensuring transparent funding more challenging. Deconcentration, delegation, devolution, and privatization are the four degrees of transfer of power that are frequently used to define decentralization [6]. These words are an excellent place to start even though there is a lot of variety in actual practices. As indicated before, deconcentration measures distributed central authority without passing it to other entities. For instance, a national government can lessen the concentration of power in the national capital by opening offices for test administration in provincial capitals. Reforms deconcentrate power transfer responsibility for enforcing regulations but not for creating them.

Deconcentration may seem to be a step in the direction of a more democratic model since it transfers some power and accountability to "lower" system levels. This architecture typically keeps highly centralized processes, though. Local units may represent the federal government and be tasked with enforcing laws, but not with formulating decisions or policies. For example, a central office may form district offices to carry out central responsibilities on its behalf, but overall power remains with the central government. Figure 2.1 describes this process of transfer of power.

Delegation is the transfer of duties and administrative responsibilities relating to particular tasks, which are often established by a central authority. Because local agents just have to carry out decisions made at the center, there is no change in the balance of power as a result of this process. A national examination board that prepares, conducts, and grades national tests is one instance of how a ministry of education delegated jurisdiction. Devolution is the transfer of power and accountability from national to local entities. The devolution of power in crucial decision-making areas, such funding and staffing, has the potential to strengthen local school communities and enhance student results [7]. Another option is the "redistributive model," which deals with top-down power distribution, the "effectiveness model," which concentrates on the financial benefits and cost-effectiveness of decentralization, and the "learning culture" model, which addresses cultural diversity and the ability of curricula to adapt to local needs [8].

Decentralization refers to the dispersal of something that has been gathered or concentrated around a particular place. This dispersion is frequently described using the metaphor of a pyramid. The majority of public and commercial organizations rely on "leaders," or specifically chosen individuals, to make decisions on behalf of the organization's other members. Large companies typically have a hierarchical structure, or numerous levels of power, in both the public and commercial sectors. Choices made at the "top" layer affect more individuals, and those made at the bottom affect fewer people. The traditional organizational chart delivers the illustration below (Figure 2.2).

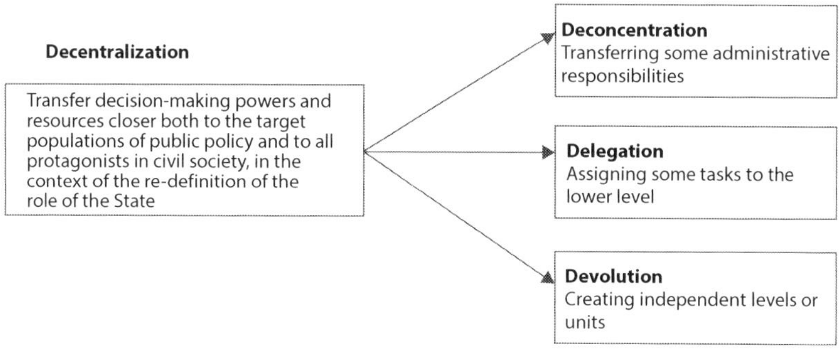

Figure 2.1 Transfer powers in decentralized approach.

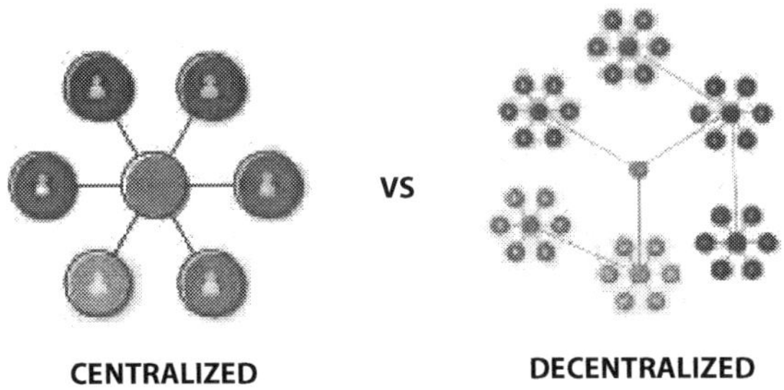

Figure 2.2 Centralized vs. decentralized.

The degree to which power is concentrated at the top of the hierarchy or in the center differs between public and private organizations. Some large firms delegate responsibility to branch managers while others make all decisions at the home office. The degree of authority concentration in one position varies widely across nations as well. In certain nations, the national ministry makes all important choices pertaining to education, whereas in others, the majority of decisions are made by states and districts. The curriculum of schools in centralized countries tends to be more uniform, at least on paper, than that of schools in more decentralized nations.

2.2 History of Decentralization in the Education System

Decentralization in the educational sector may have an impact on all areas, from the concentration of administrative power to more stringent regulatory and financial control [9]. The term decentralization in education refers to the process that transfers financial and decision-making power from higher to lower levels of government and organizational units. This process affects how educational systems decide how to generate and spend resources, organize instruction (curricula, textbooks, teaching methods, and schedule), manage staff, and plan. According to a number of academics, the quantity and quality of education have a significant impact on economic growth and per capita income.

Education has the power to empower people, as well as to increase their ability to make an income. Participation in local and federal government is made possible by education, in order to enhance their quality of life and increase their productivity [10].

While the topic of education continues to be a hot topic in policy discussions, restructuring education to ensure adequate access, equity, and quality education necessitates redefining the role of the government in managing and funding education. International financial institutions, particularly the World Bank and the UN, have supported decentralization reform initiatives in developing countries as a result of this apparent positive correlation between educational decentralization and educational production.

Decentralization has risen to the top of policy deliberations in development and transitional nations over the past 20 years. The provision of education has frequently been considered as one of the first responsibilities to be outsourced to lower levels of government as part of overall decentralization. On occasion, the wrong factors lead to the decision to decentralize education. For instance, central governments may impose unfunded mandates, which are expenditure requirements on sub-national governments, in order to meet deficit targets at the central level or in the absence of sufficient funding [11, 12]. These unfunded mandates put the effective delivery of educational services at risk. However, there is a dearth of comparative studies on education decentralization across countries throughout time, if any at all, despite the literature's abundance of case studies on the effects of decentralization on education output in particular countries. Understanding the components of education and examining education decentralization influence the education system.

> "*Several models, as well as fresh methods and procedures, have been employed by nations all over the world to structure and organize their educational systems. The central ministry of education's organizational frameworks, level of administrative decentralization, and institutional autonomy differ from nation to nation*" [13].

2.3 Advantages and Challenges of Decentralization of the Education System

2.3.1 Advantages

- Due to improved knowledge of local needs, decentralized systems can enhance quality and meet local demand. Such methods may also provide families with greater democratic control, engagement, and choice. Some studies have noted that improved efficiency may result from more local rivalry [14, 15].

- The inertia and bureaucracy of bigger systems may be addressed in terms of reform through the creation of smaller, more adaptable entities. There is not much proof that decentralization by itself improves education quality. It is more likely that decentralization, when paired with other necessary inputs, will focus more on quality issues that raise education quality and bring decentralization efforts closer towards the school/community [16]. Figure 2.3 depicts the centralized and decentralized approaches in education system.
- According to recent finance agency mechanisms, more decentralized systems tend to give a larger chance for the development of novel forms of inclusive education, for inclusive education [17]. These solutions have improved school governance, more flexible learning, and support [18, 19].
- Increased accountability and efficiency may emerge from giving local or provincial governments control and power over education services. It does so through bridging the gap between parents and policymakers or between policymakers and schools. It might also increase managers' capacity to carry out programs and/or increase parental desire for higher quality [20].
- According to [20, 21], centralized governance and management systems in education may overlook key stakeholder groups when making decisions. This limits the ability of the stakeholders to carry out decisions and observes that under centralized administrations, these flaws might compound to create a "legitimacy crisis".

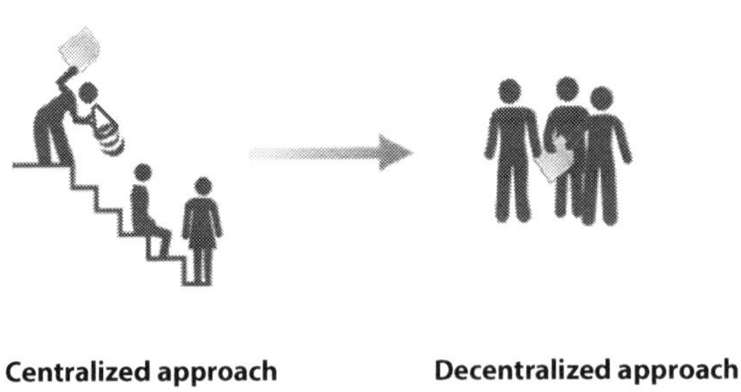

Figure 2.3 Centralized and decentralized approach. (Source- shardeum.org)

- Devolution, as opposed to delegation of authority, will be more effective because it ensures consistency throughout the transition process.
- When power sharing agreements between several levels of government are negotiated rather than enforced, they are more effective.
- The best decentralization strategy is one that transfers power in phases once communities have successfully completed certain readiness assessments. Big Bang strategies are extremely challenging to carry out effectively [22].

2.3.2 Challenges

- Decentralized systems may experience problems with efficacy, equity, and accountability if poorly coordinated provision causes a systemic fragmentation. Some of these issues are brought on by the fact that central ministries do not perform monitoring and training duties to empower subordinate levels with true decision-making and management authority. Inadequate funding from the decision-makers may potentially cause problems with local decision-making and execution.
- When collaborating with education ministries, Bernbaum [23] underlines the necessity to make the roles and duties of the applicable staff clear. Supporting capacity-building is crucial for empowering employees to carry out their duties and responsibilities. Ensuring that there is no overlap in the allocation of duties across various system levels, that funding is commensurate with obligations, and that steps are taken to boost local area capability is particularly crucial. Bernbaum continues by saying that it is crucial to communicate with influential members of the finance ministry, as well as other pertinent ministries and/or autonomous or semi-autonomous entities, in addition to working inside the ministries of education. Ultimately, effective monitoring and accountability procedures are needed to make sure those resources are allocated fairly and that policy objectives are carried out.

- In the most recent work on governance, the significance of matching accountability demands inside and across governance levels, across programs, and stakeholders is discussed [24]. This is done with a focus on conversation and transparency. More school autonomy and competition result from decentralization. In the long run, societal inequities and variations in educational achievements between schools may be exacerbated as a result. By implementing national standards for education services, several nations strive to mitigate the danger of input inequality and service quality disparities. In an effort to counteract the consequences of uneven local taxes, nations may also redistribute resources [25].
- According to [26], decentralization may result in administrative overburden at the local level and more bureaucracy. Performance requirements set at the far-off national level can lead to "gaming" of the system and local institutions being taken over by special interests. When there is a collaborative culture, benefits are seen in school communities, but the challenge is to provide accurate data transmission. Actors at all levels require accurate and pertinent information regarding the academic and financial standing of their institutions in comparison to other institutions. This covers costs, resource usage, and academic results. Also, it is important to think about how education-related data will be used by larger parties.
- As decentralization is frequently associated with a particular center of authority, it is difficult to garner widespread and enduring support. Although it presents a new opportunity, achieving these objectives calls for leadership and cooperation. The decentralization of education must be balanced between federal policymaking and state and local implementation [27].

2.4 Impact of Decentralization in Education and Research

In recent years, there has been substantial focus on the concept of decentralization in education and research. Decentralization is the transfer of authority and power from a central institution to local or regional authorities. Decentralization in the context of education and research can refer to the transfer of authority, control, and obligations from a centralized organization to particular schools or research institutes. Decentralization in research and education has complex and varied relationships. On the one hand, decentralization can give certain universities or research institutions more autonomy and flexibility, allowing them to customize their courses and projects to the particular requirements of their regional populations [28]. Figure 2.4 summarizes the different distribution networks.

Decentralization can also result in more creativity because schools and research institutions are able to try out novel methods of instruction or research without being restricted by centralized rules or regulations [29]. On the other hand, decentralization can also result in wider gaps in the caliber of research and education across different locations or institutions. Certain research centers or institutions may fall behind or lack the resources required to perform cutting-edge research if resources and decision-making authority are not dispersed fairly. The ability to adapt to local needs and promote a sense of ownership and accountability among students and researchers is improved when specific schools or research institutes are given more autonomy and authority.

Decentralization and education and research have a complicated and diverse interaction. Decentralization can have many advantages, but it must be implemented in a way that is fair and sustainable, with enough funding and supervision to guarantee that all students and researchers have access to opportunities for high-quality education and research.

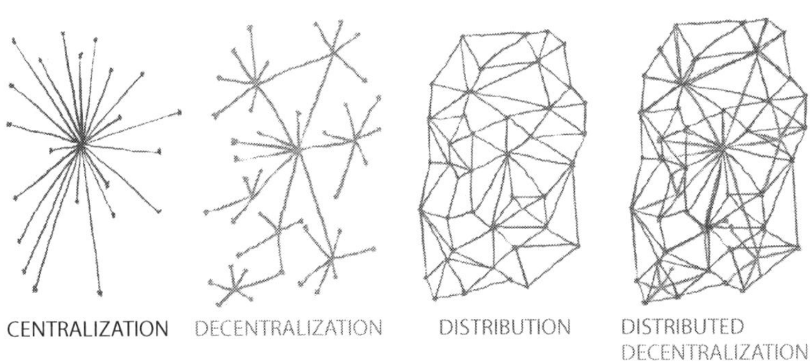

Figure 2.4 Distribution: centralized and decentralized approach (Source: by Lina Etzkorn – own work, CC BY-SA 4.0).

2.5 Current Approaches

2.5.1 World Scenario

Given below are some examples of the current approaches that illustrate the positive impact of decentralization in education and research:

2.5.1.1 Finland Education System

One of the most decentralized educational systems in the world is found in Finland, where individual schools have significant autonomy over their respective curriculum and instructional strategies. This has made it possible for schools to test out cutting-edge educational strategies like project-based learning and interdisciplinary courses [30].

2.5.1.2 Brazil, Federal Universities

The Brazilian federal government started decentralizing the nation's university system in the early 2000s, allowing individual colleges more control over their funding and management. The quantity and caliber of research output significantly increased as a result [31].

2.5.1.3 The United States, National Science Foundation

A federal organization, National Science Foundation (NSF), in the United States provides funding for scientific research. The NSF has a decentralized organization at the program level despite being centralized at the federal level, giving individual programs a significant range of discretion over funding choices [32].

2.5.1.4 Mexico, CONACYT

The funding and promotion of scientific research in Mexico is the responsibility of CONACYT, the nation's council for science and technology. With a more decentralized approach to financing decisions recently implemented by CONACYT, research institutions and universities now have more freedom in how they divide their funds. As a result, the research ecosystem in Mexico is now more varied and active, with a focus on multidisciplinary research and institution-to-institution cooperation.

2.5.2 Indian Scenario

2.5.2.1 Kerala Model of Decentralization Implementation in Education and Research

Kerala adopted a decentralized system of education in the late 1990s that offered local communities and schools more autonomy over the curriculum, the hiring and training of teachers, and the distribution of resources. The state government further provided funds and technical assistance to assist communities and schools in creating and carrying out their own educational plans. In Kerala, the quality of education has significantly improved as a result of this strategy. For instance, the state currently has one of the highest rates of literacy in India, at over 96% [33].

In Kerala, decentralization has also proved effective in fostering innovation in both research and education. For instance, the state government has started a number of programs to encourage innovation in science and technology, such as the development of a network of innovation centers and incubators that offer resources and assistance to startups and business owners. By establishing a variety of research institutes and centers all around the state, Kerala has also been effective in decentralizing research. The promotion of interdisciplinary research and teamwork between academics from many universities and fields has greatly benefited from these centers [34]. Figure 2.5 illustrates the decentralized model implemented in Kerala state, India.

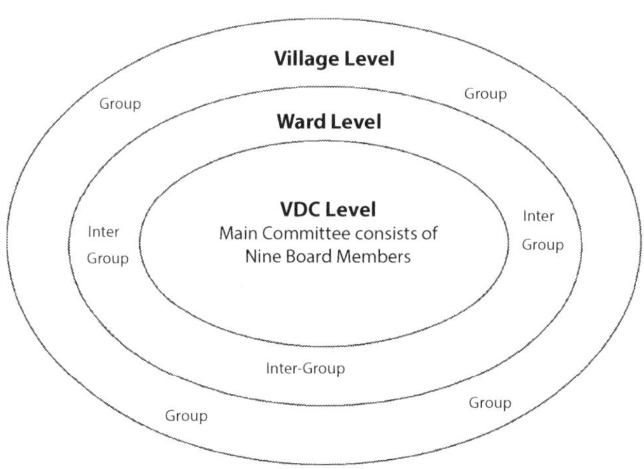

Figure 2.5 Kerala decentralized model.

Ultimately, a lot of elements, such as strong government backing, community involvement, and a focus on innovation and collaboration, are responsible for Kerala's decentralization's success.

2.6 Development Trend Towards Education and Research

Decentralization proponents contend that putting decision-making closer to the beneficiaries will enhance the standard of instruction and the results of children's learning. Yet, it can also result in a number of issues, such as a rise in inequality between cities, districts, states, and educational institutions. Decentralization policies' development and execution must be carefully considered in light of achieving the goals [35].

Decentralization has been viewed as a better alternative to the central government because it is thought to be more responsive and adaptable to local needs. According to [36], a positive outcome is possible if national and local criteria are met and the roles and authority that are shared between the federal and local governments are clearly defined. Govinda and Diwan have outlined many options to create a system of local administration and community involvement in education after evaluating the situation in various states of the nation [37]. The fundamental idea is to provide democratically elected local self-government bodies control over educational governance. Although the majority of governments have embraced the RTE Act's guidelines, the decentralization process varies greatly among states. Countries have devolved authority and transferred management of educational governance to democratically elected local self-government bodies [38].

The majority of nations are testing out or thinking about decentralizing education in some way. Decision-making authority is transferred through the process from central ministries of education to regional governments, local governments, communities, and educational institutions. Yet, the scope of the transfer might range from administrative deconcentration to a considerably wider transfer of financial responsibility to the local or regional level. Although decentralizing education systems has strong theoretical underpinnings, the process needs significant political commitment and leadership to be successful. The reasons for the changes, the initial country and sector conditions, and the interplay of several significant coalitions within the sector all have an impact on the direction, scope, and final success of decentralization reforms. The potential benefits of decentralization have grown increasingly alluring in a world where most governments have encountered the drawbacks of centralized education service provision, namely, opaque decision-making, administrative and fiscal inefficiency, and low quality and access to services. In comparison to centralized systems, the process of decentralization can generally significantly increase the efficiency, transparency, accountability, and responsiveness of service supply. Decentralized education delivery has the potential to be more effective, represent local objectives better, promote involvement, and, ultimately, increase coverage and quality [39].

It is well recognized that the two important aspects of human development are education and research. The significance of education and research in fostering innovation and growth has only increased as our world grows more complicated and interconnected. More investments in education and research, as well as a move towards more decentralized and creative approaches to both, have been trends in recent years. The emphasis on the acquisition of skills necessary for success in the contemporary economy has been one major trend in education. This involves a stronger focus on STEM (Science, Technology, Engineering, and Math) education as well as abilities like critical thinking, problem-solving, and digital literacy. All around the world, governments and educational institutions have acknowledged the significance of these skills [40].

The increasing use of various technologies to support learning is another trend in education. A change towards higher expenditure in research has also occurred, in conjunction to these trends in education. This covers both basic research—which aims to comprehend underlying concepts and phenomena—and applied research—which aims to create workable answers to problems encountered in the actual world [41–43].

2.7 Future Scope and Recommendations

With reference to the current study, the future scope and recommendations are given below in the implementation of decentralization approach:

- The various schools or research centers can modify their programs and research to meet the particular requirements of their communities that helps to encourage diversity and inclusivity.
- It creates opportunities for more egalitarian participation in decision-making based on trust rather than authority,

allowing us to develop educational and research institutions that are truly responsive, inventive, and successful.
- It creates opportunities for fairer participation in decision-making that is guided by trust rather than authority.
- Instead of depending on centralized organizations to issue and verify credentials, the development of decentralized credential systems might result in a particular ecosystem that enables people to own and control their own credentials, making them more portable and flexible.
- Decentralized platforms might make it easier for academics to collaborate and share research data, allowing them to transparently and securely contribute to a single project. This might hasten scientific creativity and discovery, resulting in advancements in technologies of different domains.
- By utilizing technology advancements in conjunction with the growth of decentralized systems, a positive prospective system may be created that has the potential to bring about a number of key advantages, such as improved accessibility, transparency, and security across the board.
- Decentralized systems can only be successful if there is widespread acceptance and standardization. As a result, stakeholders in education and research ought to cooperate to create uniform standards and procedures for distributed systems.
- Security must come first when developing decentralized systems. This entails putting in place reliable data protection safeguards, secure authentication protocols, and strong encryption.
- Infrastructure investment is required to enable decentralized systems. This entails creating decentralized networks and creating the required hardware and software.

2.8 Conclusion

The chapter presents some of the current approaches used in the implementation of decentralization system in education and research. It also discusses the current trends in the development of a decentralization system in education. It also talked about the core challenges and benefits while implementing the current approaches. It concluded that the development trend towards education and research is characterized by a growing focus on the development of skills for the modern economy, the use of technology to support learning, greater investment in research to address global challenges, more collaborative and interdisciplinary approaches to research, and a shift towards greater decentralization in both education and research. By continuing to invest in education and research, and by adopting new and more innovative approaches, we can ensure that we are prepared to meet the challenges and opportunities of the 21st century. The crux of discussion is that decentralized systems in education and research have the potential to transform industries, creating a world-level education system that can lead to greater efficiency and transparency.

References

1. Eurpoean Agency, Decentralisation in Education Systems, Seminar report, 2017.
2. Androniceanu, A. and Ristea, B., 2014. 'Decision Making Process in the Decentralized Educational System' *Procedia – Soc. Behav. Sci.*, 149, 37–42.
3. Busemeyer, M., 2012. Two decades of decentralization in education governance: Lessons learned and future outlook for local stakeholders. Presentation delivered at the *OECD Conference 'Effective local governance in education'*, in Warsaw, 16 April 2012.
4. Urbanovič, J. and Patapas, A., 2012. 'Decentralisation Reforms of Education Management: Theoretical and Practical Issues' *Vieš. Polit. Adm./Public Policy Adm.*, 2012, 11 (4), 659–671.
5. Caldwell, B. J., 2009. 'Centralisation and Decentralisation in Education: A New Dimension to Policy', in J. Zajda and D.T. Gamage (eds.), *Decentralisation, SchoolBased Management, and Quality*. Dordrecht: Springer.
6. Rondinelli, D.A.; Nelson, J.R.; Cheema, G.S. 1984. *Decentralization in developing countries: a review of recent experience*. Washington, D.C.: World Bank, Staff Working Paper No. 581.
7. UNESCO, 2014. *Decentralization as an Education System Reform* (L. Benete and W. Ible, eds.). Bangkok: UNESCO Bangkok.
8. Weiler, H., 1993. 'Control versus Legitimation: The Politics of Ambivalence' in J. Hannaway and M. Carnoy (eds.), *Decentralization and School Improvement: Can We Fulfil the Promise?* San Francisco: Jossey-Bass
9. Heredia-Ortiz Eunice, The Impact of Education Decentralization on Education Output: A Cross-Country Study, Dissertation, Georgia State University, 2007.
10. World Bank. 1995a. *Development in Practice: Priorities and Strategies for Education*. Washington D.C.: World Bank.
11. Prud'homme, Remy. 1995. The Dangers of Decentralization. *World Bank Res. Obs.* 10, no. 2: 201-20.

12. Tanzi, Vito. 1996. Fiscal Federalism and Decentralization: A Review of Some Efficiency and Macroeconomic Aspects. In *Annual World Bank Conferen.*
13. UNESCO International Institute for Educational Planning (IIEP) Advanced Training Programme 2017-2018 course outline, p. 65.
14. Busemeyer, M., 2012. Two decades of decentralization in education governance: Lessons learned and future outlook for local stakeholders. Presentation delivered at the *OECD Conference 'Effective local governance in education'*, in Warsaw, 16 April 2012.
15. Urbanovič, J. and Patapas, A., 2012. 'Decentralisation Reforms of Education Management: Theoretical and Practical Issues' *Vieš. Polit. Adm./Public Policy Adm.*, 2012, 11 (4), 659–671.
16. Bernbaum, M., 2011. *Equip2 Lessons Learned in Education. Decentralization.* Washington, DC: USAID, EQUIP 2 and FHI 360.
17. European Agency for Special Needs and Inclusive Education, 2016a. *Financing of Inclusive Education: Mapping Country Systems for Inclusive Education.* (S. Ebersold, ed.). Odense, Denmark.
18. Stubbs, S., 2008. *Inclusive Education. Where there are few resources.* Oslo: The Atlas Alliance.
19. NESSE Network of Experts in Social Sciences of Education and Training, 2012. Education and Disability/Special Needs. Policies and practices in education, training and employment for students with disabilities and special educational needs in the EU. Report prepared for the European Commission. Brussels: European Commission.
20. Weidman, J.C. and DePietro-Jurand, R., 2011. Equip2 State-of-the-Art Knowledge in Education. Decentralization. A Guide to Project Design Based on a Comprehensive Literature and Project Review. Washington, DC: USAID, EQUIP 2 and FHI 360.
21. Radó, P., 2010. Governing Decentralized Education Systems. *Systemic Change in South Eastern Europe.* Budapest: Local Government and Public Service Reform Initiative, Open Society Foundations
22. Hanson M, *Educational Decentralization: Issues and Challenges*, 1997.
23. Bernbaum, M., 2011. Equip2 Lessons Learned in Education. Decentralization. Washington, DC: USAID, EQUIP 2 and FHI 360.
24. 106 Burns, T., Köster, F. and Fuster, M., 2016. *Education Governance in Action: Lessons from Case Studies.* Paris: OECD Publishing.
25. Herbst, M. and Wojciuk, A., 2014. 'Inequity in a decentralised education system – Evidence from Poland' *Edukacja*, 6 (131), 5–22.
26. Busemeyer, M., 2012. Two decades of decentralization in education governance: Lessons learned and future outlook for local stakeholders. Presentation delivered at the *OECD Conference 'Effective local governance in education'*, in Warsaw, 16 April 2012.
27. Busemeyer, M., 2012. Two decades of decentralization in education governance: Lessons learned and future outlook for local stakeholders. Presentation

delivered at the *OECD Conference 'Effective local governance in education'*, in Warsaw, 16 April 2012.

28. "Decentralization and its Implications for Urban Education" by James R. Fairweather and Betty E. Steffy, in *Handbook of Research in Education Finance and Policy*, edited by Helen F. Ladd and Edward B. Fiske (2008).
29. "*Decentralization in Education: National Policies and Local Practices*" edited by Robert F. Arnove and Carlos Alberto Torres (1999).
30. https://www.oph.fi/en/education-system
31. Fischmann, A. A., & Guedes, M. J. (2015). The decentralisation of the Brazilian university system: Effects on research output. *Stud. High. Educ.*, 40(3), 483-500.
32. National Science Foundation. (2021). https://www.nsf.gov/about/
33. Decentralization in Education: The Kerala Experience, " by K. Anvar Sadath and John Kurien, *J. Educ. Pract.*, Vol. 7, No. 23, 2016.
34. *Science, Technology and Innovation Policy 2015"* by the Government of Kerala, Department of Science, Technology and Environment.
35. https://learningportal.iiep.unesco.org/en/blog/a-conversation-on-the-decentralization-of-education-pros-cons-risks-and-benefits
36. Kullipossa, F. (2004). Decentralisation and democracy in developing countries: An overview. *Dev. Pract.*, 14 (6), 768-779.
37. Govinda, R. & Bandyopadhyay, M. (2010). Changing framework of local governance and community participation in elementary education in India. CREATE Research Monograph No 35. New Delhi: NUEPA.
38. Sinha Rashmi, Kumar Khagendra, Emerging Trends in Decentralization and People's Participation in the Educational Governance, *IOSR J. Humanit. Soc. Sci. (IOSR-JHSS)* Volume 26, Issue 2, Series 9 (February. 2021) 49-55.
39. Education and Decentralization, Decentralization Thematic Team, World Bank.
40. Organisation for Economic Co-operation and Development (OECD). (2021). *Education at a Glance 2021: OECD Indicators*. OECD Publishing.
41. United Nations Educational, Scientific and Cultural Organization (UNESCO). (2021). Futures of Education: Learning to Become. UNESCO.
42. Chauhan, R., Avasthi, S., Alankar, B., & Kaur, H. (2021). Smart IoT systems: Data analytics, secure smart home, and challenges. In *Transforming the Internet of Things for Next-Generation Smart Systems* (pp. 100-119). IGI global.
43. Avasthi, S., Chauhan, R., & Acharjya, D. P. (2023). Extracting information and inferences from a large text corpus. *Int. J. Inf. Technol.*, *15*(1), 435-445.

3

Architecture of Blockchain-Enabled Decentralized Systems

Rohit Kumar[1], Sandhya Avasthi[1*] and Suman Lata Tripathi[2]

[1]Department of CSE, ABES Engineering College, Ghaziabad, UP, India
[2]Lovely Professional University, Jalandhar, India

Abstract
With the expansion in Internet infrastructure, demands for internet technology increased and there is a need for a secure system that is decentralized in nature, combined with a secure and transparent infrastructure that is fulfilled by one of the trending technologies named blockchain. Blockchain technology has significantly advanced over the past few years, with new applications and use cases being discovered and developed every day. The decentralized nature of blockchain, combined with its secure and transparent infrastructure, has made it a highly sought-after solution for various industries, including finance, healthcare, supply chain management, and government services. The expansion of blockchain technology is a result of its increasing adoption and integration into various sectors such as finance, healthcare, supply chain management, and more. One of the key drivers of the expansion of blockchain is the increasing demand for secure, tamper-proof, and transparent transactions. This has led to the development of various blockchain platforms, such as Ethereum, which helps in the development of decentralized applications or dApps for a variety of purposes from financial records to data storage. In a blockchain network, transactions are grouped into blocks and verified by network participants known as nodes. Once a block is verified and added to the chain, the information it contains cannot be altered, providing a secure and tamper-proof record of transactions. By means of this chapter, a blockchain-based architecture of a system is described here.

***Keywords*:** dApps, nodes, tamper-proof, transactions, P2P, contracts, censorship

**Corresponding author*: sandhya_avasthi@yahoo.com

Sandhya Avasthi, Suman Lata Tripathi, Namrata Dhanda and Satya Bhushan Verma (eds.) Decentralized Systems and Distributed Computing, (53–74) © 2024 Scrivener Publishing LLC

3.1 Introduction

As the digital era emerges, blockchain technology has also gained popularity in recent years as a way to create secure, transparent, and decentralized systems. Blockchain is basically a distributed database or ledger of records; by "distributed", we mean that the copy of records is not in a centralized database, but instead of a centralized database, the database copy is also available to all the nodes (i.e., devices connected to the network), such that no possible changes can be made in the records, i.e., immutable in nature.

Blockchain has basically become popular due to the advancement in Bitcoin, which is an open-source online currency that is immutable in nature. Everyone hears the word Bitcoin instead of Blockchain; it is similar to hearing Colgate instead of toothpaste. The other popular blockchain-based E-currencies are Bitcoin, Dogecoin, etc. There are many [1, 2] blockchain networks available worldwide, with Ethereum being one of the most popular. In this chapter, we will provide an overview of the architecture for blockchain technology, involving its key components that work together to support the development of a decentralized system with its advantages and disadvantages. We will also discuss blockchain and smart contracts as well as how we can integrate blockchain into our web-based application by means of a Todo List Application. We will explore the network, the consensus, the data structure, and smart contract components and the challenges that go with these.

3.2 What is a Decentralized System?

A Decentralized System is basically a system of nodes or devices in which the data or ledger is not stored at a centralized server; instead, data are shared over each node. Each node acts as a separate system with all records (see Figure 3.1). As the data are stored across all nodes (i.e., Devices on Network), then it is not possible to tamper any record, as tampering implies updating data at each node, which is not possible [3].

It refers to the transfer of control and decision-making from a central authority to a decentralized authority.

3.2.1 Centralized System vs. Decentralized System

The major differences between a centralized system and a decentralized system are as follows as shown in Table 3.1:

Figure 3.1 Decentralized architecture.

Table 3.1 Comparison between centralized system and decentralized system.

Parameters	Centralized system	Decentralized system
Resources access	Maintained and controlled by a single entity, hence easy to tamper	Is not maintained or owned by a single entity, hence difficult to tamper
Data	Maintained and controlled through a central system	Data are added after consent from all nodes
Control	Control through central entity	No one owns the data; everyone owns the data
Single Point of Failure	Yes	No
Fault Tolerance	Low	Extremely High
Security	Maintained and controlled through central system	Increase when no of nodes increase
Performance	Maintained and controlled through central system	Decrease when no of networks increase
Example	ERP System	Blockchain

(*Continued*)

Table 3.1 Comparison between centralized system and decentralized system. (*Continued*)

Parameters	**Centralized system**	**Decentralized system**
Diagram	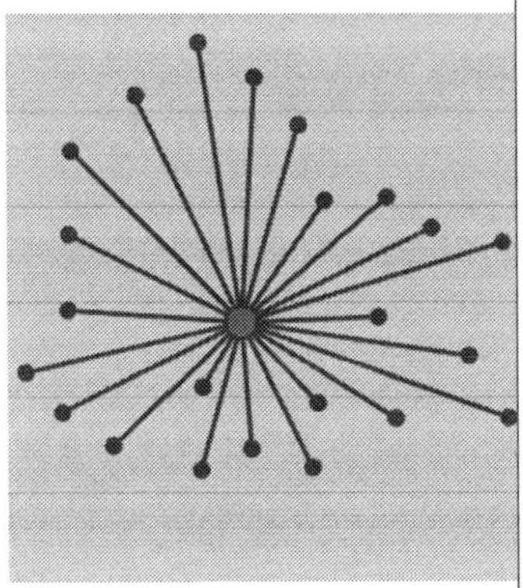	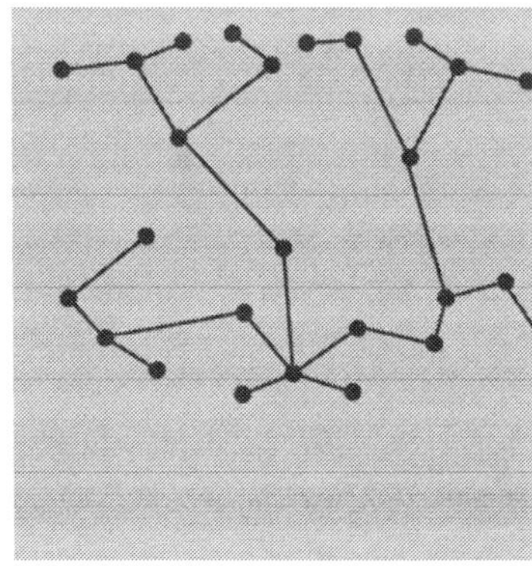

3.2.2 Merits and Demerits of a Decentralized System

Merits of a Decentralized System

- It provides a trustless environment as each one has a copy of the database.
- Improves data reconciliation
- It reduces points of weakness
- It increases or optimizes distribution
- Immutability
- Security
- Transparency and full control by nodes

Demerits of a Decentralized System

- Costly
- Lack of consensus and clarity
- Lack of discipline

3.3 Blockchain

3.3.1 What is Blockchain?

Blockchain is basically a distributed immutable ledger where data are stored in the form of blocks; for each new information or data, a new block is created, and the hash of the previous block is stored in such a way that it creates a chain of records. Hence, blockchain is the composition of various blocks that are linked together where each block consists of data and raw data like timestamp, previous hash, etc. [2–4].

Blockchain has become popular with the advancement of Bitcoin cryptocurrency, which is also the first cryptocurrency in the world.

In blockchain, the records or data are available to all the nodes, as it follows a decentralized fashion, which means each node has the copy of data, but itself, which increases system tolerance as well as decreases single point of failure as it becomes hard to change the data as each node has the data.

In case anyone changes the data in the ledger, then those data are matched with the data stored in the different networks along the blockchain network, and if the majority are in favor that data are tampered, then the data are replaced with the original data.

The structure of a blockchain block is depicted in Figure 3.2.

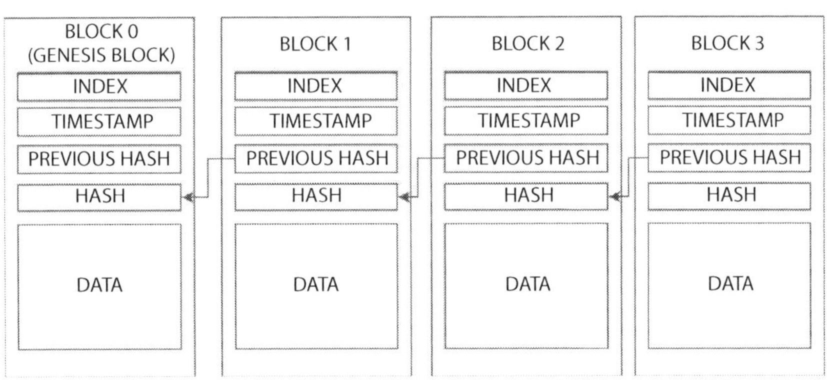

Figure 3.2 Block structure in a blockchain network.

3.3.2 Structure of Blockchain

The structure of blockchain consists of the blocks that are chained together in a regular fashion; each block in the network stores the address of the previous block.

The first block in the blockchain network is called the "Genesis Block" or "Block 0" as the blockchain chain originates from here.

The complete information regarding the structure of blockchain is also depicted in Figure 3.3.

3.3.3 Requirements of Blockchain

As the world transitions towards the digital era, the demand for the database increases. Note that a regular database is easy to modify since all the records are stored in a centralized storage unit, which could be in the same geographical area or in different geographical areas.

Due to data tampering, everyone needs a trusted database that is impossible to tamper and has no centralized storage. Thus, we transition towards decentralized storage where data are not stored in a central server; instead, data are stored on each node, which means a copy of the data is available to each node; thus, in case of any change, the data are matched with the data available to other nodes, which is close to impossible; hence, blockchain prevents data tampering.

The main requirements of blockchain are as follows:

- To develop trusted data that are impossible to tamper.
- To store records like educational and E documents, blockchain-based storage is needed.

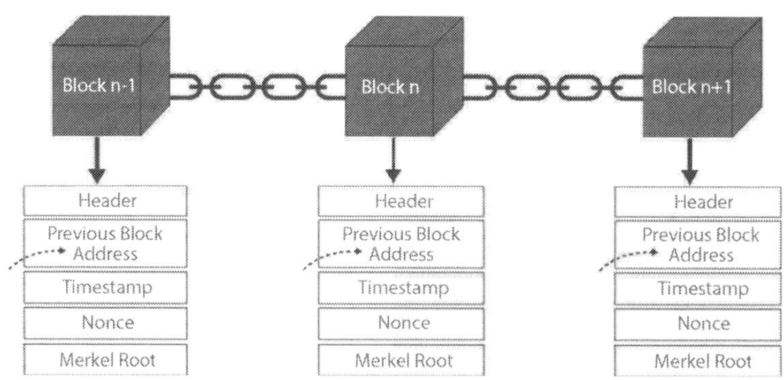

Figure 3.3 Structure of blockchain.

3.3.4 Merits and Demerits of Blockchain

Merits of Blockchain

- **Immutability:** Blockchain supports the concept of immutability, which means the records are impossible to tamper; i.e., blockchain prevents tampering of data within the blockchain network.
- **Transparency:** Blockchain is transparent in nature as the copy of records is available to all nodes available in the network.

Demerits of Blockchain

- **Speed and Performance:** Blockchain is considerably slower than that of a traditional database as while working with records, more operations are carried out, which decreases speed on insertion of records.
  Speed and performance are also proportional to the number of nodes in the network.
- **Costly:** Blockchain is costlier as compared with the traditional database as it requires proper planning and integration with blockchain.
- **Data Modification:** It is not easy to insert or update records in the blockchain; for that, we need to write codes or contracts, which is time-consuming and expensive.

3.4 Smart Contracts and Its Examples

Smart contracts are basically the programs stored in blockchain that automate the task of execution on agreement with predefined conditions, which is set up when developing smart contracts [1, 3].

On the other hand, we can say that smart contracts are basically the authentication system of the blockchain that have predefined conditions, which is first checked when a new block is inserted in smart contracts.

Smart contracts are written through a programming language called "Solidity", which is an object-oriented language and licensed under GNU General Public License v3.0.

Smart contracts also help to automate the task of workflow, triggering the next action when specific conditions are met.

Below is the simple smart contract for Todo List written in solidity:

```
pragma solidity >=0.7.0 <0.9.0;
contract Todos {
    struct Todo {
        string task;
        string timestamp;
        bool status;
    }
    mapping (address => Todo[]) private Users;

    function addTodo(string calldata _task,string calldata _timestamp)
external{
    Users[msg.sender].push(Todo({
        task:task,
        timestamp:timestamp,
        status:false
    }));
}
```

3.5 Blockchain-Enabled Decentralized System

Blockchain is a manner of storing records in the form of a distributed ledger, which uses the concept of a decentralized system. Traditionally, a centralized system was used, which has certain disadvantages and limitations (i.e., no data immutability, single point of failure, and less tolerance), where there is a task of a central authority to control and manage the system. However, in a blockchain-enabled system or decentralized system, there is no central system, which means there is no single authority that manages and maintains the system and there is no central point of control or data.

In a blockchain-powered decentralized system, records are not stored on a central server; instead, records are stored on each node. In blockchain, transactions are validated and recorded on a distributed ledger, which is composed of multiple devices that are connected on the network called nodes. Each node has the copy of the complete ledger and validated by consensus among all the nodes [4]. The process of adding blockchain in shown in Figure 3.4.

The decentralized approach has multiple benefits, namely, it eliminates central authority or intermediaries such as banks and other financial institutions, which reduces running cost as well as increases efficiency; in addition, this type of system increases transparency and security as the records are immutable and tamper-proof and can be audited by any node in the network [5, 6].

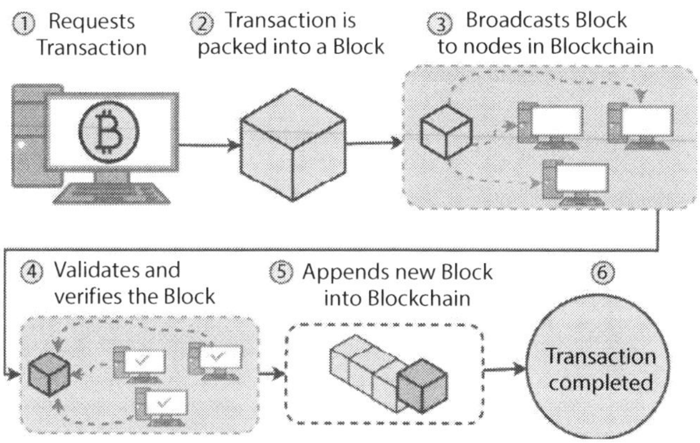

Figure 3.4 Process of block addition in existing blockchain.

The decentralized system powered by blockchain has the potential to transform traditional industries into modern industries by a creating new business model.

Let us take the example of Decentralized Finance (DeFi), a new innovation in the field of blockchain that creates an emerging path for blockchain to take control over financial transactions. DeFi now emerged as the new paradigm in the finance industry, offering a range of financial services involving E-Currency, cryptocurrency, etc. that are accessible to everyone regardless of internet connection, geographical location, and financial status. Other than financial services, blockchain is also involved in supply chain management, e voting, and digital identity verification [5].

3.5.1 Role of Blockchain

In a decentralized system, blockchain helps in creating records that are not on a central server but are on nodes (i.e., devices connected to the blockchain network), which means they are not handled by a central authority and blockchain provides smart contracts that are generally the programs that match the predefined conditions, and on the basis of that condition, the particular action is triggered [1, 2, 7].

Every new record or block on the blockchain is proven to be authentic and digitally signed to ensure its genuineness. Before its addition to blockchain, it should be verified by the majority of nodes among the network.

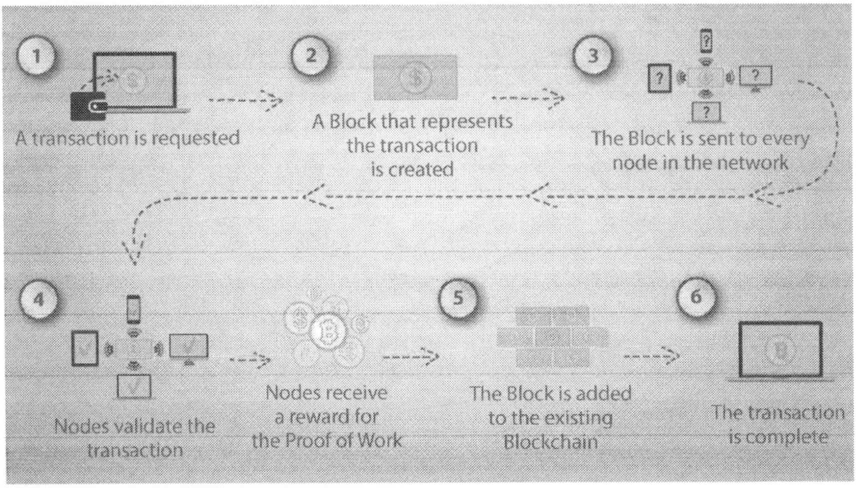

Figure 3.5 Process of addition of block in blockchain.

In blockchain, the new records can go through various phases involving validation:

- A transaction is requested.
- A block that represents the transaction is created.
- The block is sent to every node in the network.
- Nodes validate the transaction.
- Node receives a reward for the proof of work.
- The block is added to the existing blockchain.
- The transaction is complete.

The following is a blockchain architecture diagram that shows how this actually works in the form of a digital wallet.

3.5.2 Types of Decentralization in Blockchain

The blockchain basically follows one of the following levels of decentralization:

1. Fully Centralized
These blockchain systems are entirely controlled and managed by a single central authority.

2. Semi-Decentralized
These blockchain systems are managed and controlled by multiple authorities.

3. Fully Decentralized
These have no central authority for the management.

3.5.3 Types of Blockchain Architecture

Blockchain architecture may be implemented in three ways:

• Public Blockchain Architecture
In the architecture, the data are available to the public who is willing to participate, e.g., Bitcoin and Ethereum follow public blockchain architecture.

• Private Blockchain Architecture
In this architecture, the data are only available to the members of the organization or any other person invited to the organization as a guest [8].

• Consortium Blockchain Architecture
In this architecture, the data are available to the public as well as to the users of an organization. This architecture is also used when there are multiple organizations that can share their data [8].

3.5.4 Requirements

The main reasons for the requirements of the blockchain-enabled decentralized system are as follows:

- **Immutability:** Everyone wants immutable data, which means that data are authentic and cannot be changed by anyone; thus, a blockchain-enabled decentralized system is the solutions for this. It prevents data tampering as all blocks are connected in a chain, and in case a chain is disturbed, another node on the blockchain network has the copy of the data; in case of tampering, the data are recovered through the consensus protocol of blockchain.
- **Transparency:** Blockchain increases transparency in the data as data are available to all the nodes in the network, which means anyone within the network can view the data as per their access. Thus, a blockchain-enabled decentralized system is also required to increase transparency in the data.
- **Single Point of Failure:** The blockchain-enabled decentralized system also reduces the single point of failure as the data are not accessed through a single point or central server.

- **Fault Tolerance:** The blockchain-enabled decentralized system is very fault tolerant as it is not dependent on the central server and there are no loads on the central server; instead, loads are split into the nodes or devices connected to the network.

3.5.5 Pros and Cons

Pros of the Blockchain-Enabled Decentralized System

- This system proves the authenticity of the data, which means that the data stored are authentic in nature and are tamper-proof.
- The system has no single point of failure due to the absence of a central server.
- The system has a maximum fault tolerance as each node on the network has its own hardware and there is no dependence on the infrastructure of the central system.
- The data in the system are transparent in nature as data are available to every node on the network.

Cons of the Blockchain-Enabled Decentralized System

- The system has limited speed and performance, which means speed and performance are inversely proportional to the number of nodes on the network as when the number of nodes increases, the speed and performance decrease, and vice versa.
- The cost of setup of the system is high as compared to a central system.
- Each operation in this system takes time to execute as it has gone through various smart contracts.
- Blockchain has limited scope.

3.6 Architecture for the Blockchain-Enabled Decentralized System

The architecture of the blockchain technology consists of a distributed network of nodes (i.e., network created by connecting multiple devices or nodes together) that work together to validate and record transactions on the blockchain ledger; along with these devices, there are some programs needed for validation of conditions known as smart contracts. This architecture enables decentralized systems to be built, as no single entity or node has access to the whole network.

As per the core fundamentals, blockchain architecture consists of the three main components:

- the network,
- the consensus mechanism or protocol, and
- the data structure needed to store the blocks that consist of records.

3.6.1 The Network

The network component or the setup is generated by connecting multiple devices or nodes to form the decentralized network. These nodes can work together to ensure that the blockchain remains secure, transparent, and tamper-proof.

3.6.2 The Consensus Protocol

The consensus protocol or mechanism ensures that all the nodes on the network have the current state of the blockchain, which is achieved by various algorithms under the consensus protocol, such as proof of work (PoW), proof of stake (PoS), and delegated proof of stake (DPoS) [9].

3.6.3 The Data Structure

The data structure of the blockchain deals with the storage of the transactions records of the transaction that took place on the network. Each block of the blockchain consists of information about the set of transactions, along with a unique cryptographic hash that contains the reference to the previous block that leads to the chain of blocks. On the other hand, we can say that blockchain starts with "Block 0" or the "Genesis Block" up to "Block N", where each block stores the reference to the previous block, which means each n block consists of the reference to n-1 block that creates the chain of records [10–12].

In addition to these core components, i.e., the network, the consensus protocol, and data structure, blockchain architecture also includes smart contracts (self-executing contracts, the type of agreement between parties

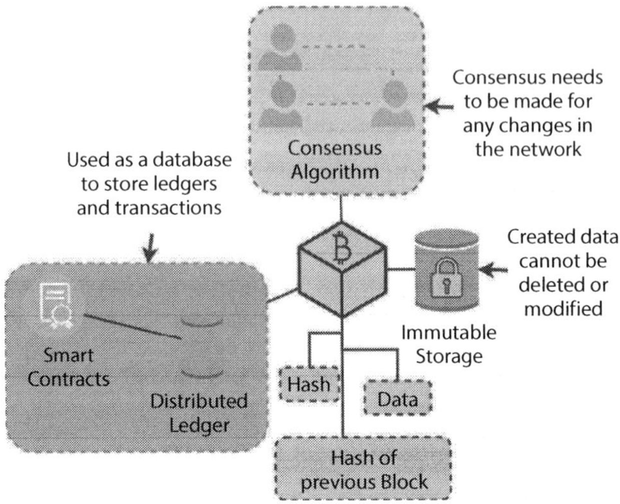

Figure 3.6 The core components of blockchain.

written in the form of code using the solidity programming language). Smart contracts allow the automation of complex transactions and can be programmed to execute triggers based on a specific conditions.

Overall, the architecture of blockchain involves the development of a transparent, tamper-proof decentralized system, by leveraging a distributed network of nodes, consensus mechanism, and a tamper-proof data structure with the potential to create innovations.

For the development of decentralized systems using blockchain, we must discuss the mentioned steps:

1. **Choose the Blockchain Platform:** The first step in the development is to choose a blockchain platform from available blockchain platforms that suits your system requirement; i.e., Ethereum, Hyperledger, and EOS are among those available.
2. **Develop Smart Contracts:** The next step is to develop smart contracts using solidity; here, we have to develop a smart contract that governs our app functionality.
3. **Create Decentralized Application (DApp):** Develop the app that can work with smart contracts and allow users to execute system functionality.
4. **Define User Roles:** Here, we have define the access based on user roles.

5. **Integrate with a cryptocurrency:** To make system functionalities work properly, we need to connect our app with a cryptocurrency such as Ether or Bitcoin.
6. **Publish the DApp:** To make the system accessible to everyone, we need to publish app on the chosen platform.
7. **Data Audit:** Although blockchain is highly secure, we must perform regular auditing of the data and use encryption and other privacy-enhancing technologies to protect user data.

3.7 Decentralized Todo App with Blockchain

To understand in-depth the architecture of the blockchain-based decentralized app, let us create a Todo App where we can store the tasks on the nodes. The blockchain part of this app can be understood as a database.

For creating the application, we first develop the smart contracts and then learn about the functions for doing functionalities.

The components of this app or decentralized application are as follows:

- Ganache: The local blockchain Ethereum that creates the network where we can connect devices.
- Web3 JS: The library used to communicate with the blockchain.
- Solidity: The programming language used to develop smart contracts.
- Web App: The web app developed using front-end technologies like HTML, CSS, Bootstrap, JavaScript, jQuery, etc.

Implementing Smart Contracts

Smart contracts are used to communicate with the blockchain. In our Todo App, smart contracts will allow us to store our task in the blockchain.

To develop smart contracts, we may use Remix IDE for Solidity; for that, we first open Remix IDE on our browser and create a new .sol file.

The first line of smart contracts must declare the SPDX License Identifier and solidity version; after that, we need to declare the structure of our data and then we should map to store our task array with an associated user address.

After the above implementation, we need to create functions like create task, read task, update task, delete task, and get task count.

The smart contract for our todo app is shown below:

```
// SPDX-License-Identifier: GPL-3.0
pragma solidity >=0.8.0 <0.9.0;
contract Bloc{
	struct Task
	{
		string task;
		bool isDone;
	}
	mapping (address => Task[]) private Users;
	function createTask(string calldata _task) external
	{
		Users[msg.sender].push(Task({
			task:_task,
			isDone:false
		}));
	}

	function readTask(uint _taskIndex) external view returns
(Task memory)
	{
		Task storage task = Users[msg.sender][_taskIndex];
		returntask;
	}
	function updateStatus(uint256 _taskIndex,bool _status)
external
	{
		Users[msg.sender][_taskIndex].isDone = _status;
	}
	function deleteTask(uint256 _taskIndex) external
	{
		delete Users[msg.sender][_taskIndex];
	}
	function getTaskCount() external view returns (uint256)
	{
		return Users[msg.sender].length;
	}
}
```

After the above declaration and function, we then compile our contract and then deploy inside Remix IDE; after this, we are done with our smart contract, and in the next part, we have to install ganache and develop front end for our Todo Application.

After all this work, our task now is to connect the front end with the smart contracts; for that, we have to use javascript language; that is, we have to create a javascript file that helps to execute functions that we have developed in our smart contracts.

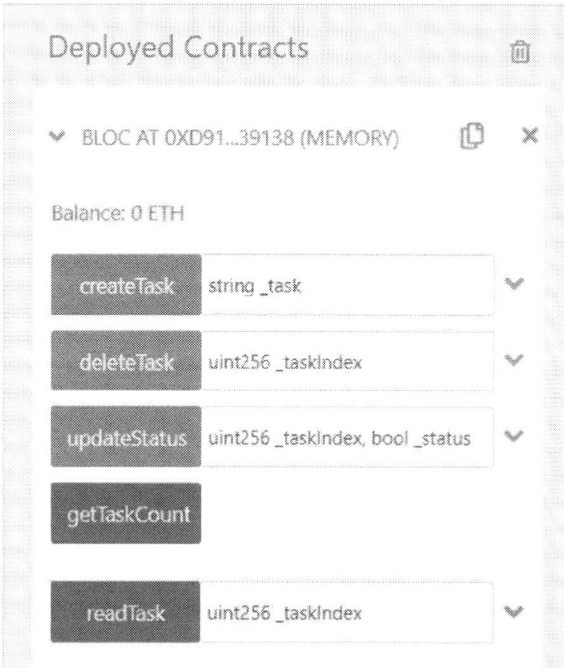

Figure 3.7 Functions available in our Todo Smart Contract.

For integration smart contract with our web application, we need account information that can be retrieved from Web 3 Javascript library like:

```
web3 = new Web3(new Web3.providers.HttpProvider("http://
localhost:7545"));

async function getAccount() {
  let accounts = await web3.eth.getAccounts();
  web3.eth.defaultAccount = accounts[0];
  console.log(web3.eth.defaultAccount + " account detected");
  return web3.eth.defaultAccount;
}
```

After that, we need to write function for the particular task.
For example:
JavaScript function to add task in Blockchain:

```
async function addTask(name) {
  try {
    await contract.methods
    .createTask(name)
    .send({ from: web3.eth.defaultAccount });
  } catch {
    console.log("Error");
  }
}
```

JavaScript function to read tasks from blockchain:

```
async function readTasks() {
try {
  await getAccount();
  contract = new web3.eth.Contract(contractABI, contractAddress);
  try {
    numberOfTask= await contract.methods
      .getTaskCount()
      .call({ from: web3.eth.defaultAccount });
    if (numberOfTask != 0) {
      let taskIterator = 0;
      while (taskIterator < numberOfTask) {
        try {
          let task = await contract.methods
            .getTask(taskIterator)
            .call({ from: web3.eth.defaultAccount });
          if (task[0] != "") {
            addTaskToList(taskIterator, task[0], task[1]);
          }
        } catch {
          console.log("Failed to get Task ");
        }
        taskIterator++;
      }
     }
    } catch {
      console.log("Failed to get task count from blockchain");
  }
  } catch {
    console.log("Failed to get the account");
  }
}
```

JavaScript function to remove specific tasks from blockchain:

```
async function removeTask(taskIndex) {
    try {
        await contract.methods.deleteTask(taskIndex).send({ from:
web3.eth.defaultAccount });
    } catch {
        console.log('Erorr: removeTask');
    }
}
```

JavaScript function to update task status:

```
async function changeTaskStatus(taskIndex) {
    try {
        await contract.methods.updateStatus(taskIndex,
checkbox.checked).send({ from: web3.eth.defaultAccount });
    } catch (error) {
        console.log('Failed to change task status');
    }
}
```

This is the theoretical concept on how we can develop a decentralized application with the blockchain.

3.8 Blockchain-Enabled Decentralized System Development and Challenges

As discussed above in this chapter, the blockchain-enabled decentralized system has the capability to develop a system that provides a way to record and verify transactions securely and transparently. Decentralized mode is achieved by connecting multiple machines on a network that helps in the validation and recording of transactions.

One of the major advantages of a decentralized system based on blockchain technology is that it is free from censorship and control of a single authority or government as nodes ensure that no single node has complete access over the blockchain network. A decentralized system also provides a way to store data in a tamper-proof environment, which can be accessed only with the necessary permissions [13, 14].

However, there are various challenges in this blockchain-based decentralized journey. One of the most common is scalability, as the current blockchain can only handle a limited number of transactions within seconds; thus, a centralized system, which can handle millions of transactions within seconds, takes control of the blockchain decentralized system.

Another challenge associated with the blockchain-based decentralized system is lack of interoperability, which means adoption and integration in various industries and platforms. In addition, there is also the challenge of government regulation as developing regulations also makes adoption slow.

Overall, while blockchain technology helps in the development of the decentralized system, which has its various advantages, there are still several challenges that need to be addressed in order to fully realize the potential of this technology. Further ongoing research on Blockchain is being carried out that helps in creating a backdoor in the adoption of blockchain technology in various fields including E-voting, supply chain management, academic records management, etc.

3.9 Future of the Blockchain-Enabled Decentralized System

The future of blockchain technology, with its ability to develop a decentralized system, is promising. As this continues to advance, it provides better solutions to some of the sectors where it can make a significant impact. One area where blockchain technology can make a significant impact is the financial sector, which is one of the crucial sectors involving money transaction entries, which must be tamper-proof. Decentralized finance (DeFi) applications, which are built on the blockchain technology, have already emerged, enabling P2P lending, i.e., trading without the need of a central authority or financial institutions [4, 15].

Another area where blockchain makes a huge impact is supply chain management, involving management of the records related to the supply of goods. It increases transparency, reduces fraud, and improves efficiency of the supply.

Other areas where blockchain creates a huge impact are online voting, academic records management, etc. Blockchain technology can also be used to develop a decentralized identity system that enables individuals to take control over personal data and protect their privacy online [15].

The future of blockchain includes integration with IoT, artificial intelligence, and machine learning to keep records tamper-proof. These technologies can leverage the decentralized and secure nature of blockchain to develop innovative solutions to various industries. Figure 3.8 depicts the future of the blockchain enabled decentralized system and its impacts.

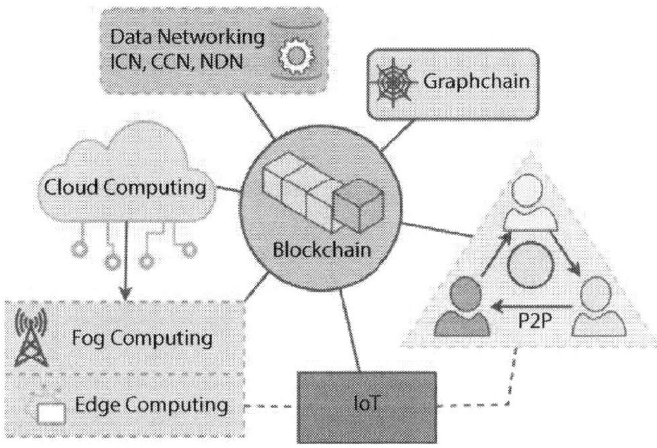

Figure 3.8 Future of the blockchain-enabled decentralized system.

3.10 Conclusion

The blockchain technology has the potential to change the way we build decentralized systems, by providing transparent and tamper-proof management of records and preventing the censorship of records as well as management by a single central authority or government. A decentralized system powered by blockchain has the potential to transform industries involving finance, education, voting, supply chain management, and identity management by promoting transparency and reducing fraud that makes the records authentic and leads us to Web 3.0. There are also some challenges as discussed above that could be reduced in the near future as research on blockchain is ongoing. Overall, the potential benefits linked to blockchain technology and the decentralized system are significant, and it is an exciting time for developers to explore further opportunities and develop world-changing innovations.

References

1. R. Kumar, L. Badwal, S. Avasthi and A. Prakash, "A Secure Decentralized E-Voting with Blockchain & Smart Contracts," *2023 13th International Conference on Cloud Computing, Data Science & Engineering (Confluence)*, Noida, India, 2023, pp. 419-424, doi: 10.1109/Confluence56041.2023.10048871.
2. Yadav, Abhishek. (2020). E-Voting using Blockchain Technology. *Int. J. Eng. Res. V9.*, 10.17577/IJERTV9IS070183.

3. S. Nakamoto, "*Bitcoin: a peer-to-peer electronic cash system*", [Online]. Available: https://bitcoin.org/bitcoin.pdf
4. Nir Kshetri, Jeffrey Voas, "Blockchain-Enabled E-Voting".
5. Gao, S.; Zheng, D.; Guo, R.; Jing, C.; Hu, C. An Anti-Quantum E-Voting Protocol in Blockchain with Audit Function.
6. Kim, T.; Ochoa, J.; Faika, T.; Mantooth, A.; Di, J.; Li, Q.; Lee, Y. An overview of cyber-physical security of battery management systems and adoption of blockchain technology.
7. Avasthi, S., Chauhan, R., & Acharjya, D. P. (2021). Techniques, applications, and issues in mining large-scale text databases. In *Advances in Information Communication Technology and Computing,* (pp. 385-396). Springer, Singapore.
8. Avasthi, S., Chauhan, R., & Acharjya, D. P. (2022). Topic Modeling Techniques for Text Mining Over a Large-Scale Scientific and Biomedical Text Corpus. *Int. J. Ambient Comput. Intell. (IJACI)*, 13(1), 1-18.
9. Pham, H. L., Tran, T. H., Phan, T. D., Le, V. T. D., Lam, D. K., & Nakashima, Y. (2020). Double SHA-256 hardware architecture with compact message expander for bitcoin mining. *IEEE Access*, 8, 139634-139646.
10. Zhang, S., & Lee, J. H. (2020). Analysis of the main consensus protocols of blockchain. *ICT Express*, 6(2), 93-97.
11. Avasthi, S., Sanwal, T., Sharma, S., & Roy, S. (2023). VANETs and the Use of IoT: Approaches, Applications, and Challenges. *Revolutionizing Industrial Automation Through the Convergence of Artificial Intelligence and the Internet of Things*, 1-23.
12. Idelberger, F., Governatori, G., Riveret, R., & Sartor, G. (2016, July). Evaluation of logic-based smart contracts for blockchain systems. In *International symposium on rules and rule markup languages for the semantic web,* (pp. 167-183). Springer, Cham.
13. Wang, S., Ouyang, L., Yuan, Y., Ni, X., Han, X., & Wang, F. Y. (2019). Blockchain-enabled smart contracts: architecture, applications, and future trends. *IEEE Trans. Syst. Man Cybern.: Syst.*, 49(11), 2266-2277.
14. Chauhan, R., Avasthi, S., Alankar, B., & Kaur, H. (2021). Smart IoT Systems: Data Analytics, Secure Smart Home, and Challenges. In *Transforming the Internet of Things for Next-Generation Smart Systems,* (pp. 100-119). IGI Global.
15. Khan, S. N., Loukil, F., Ghedira-Guegan, C., Benkhelifa, E., & Bani-Hani, A. (2021). Blockchain smart contracts: Applications, challenges, and future trends. *Peer Peer Netw. Appl.,* 14(5), 2901-2925.

4

Mobile Edge Computing for Decentralized Systems

Swati Gupta[1] and Puneet Garg[2]*

[1]Vaish College of Engineering, Rohtak, Haryana, India
[2]St. Andrews Institute of Technology and Management, Farrukh Nagar, Delhi NCR, India

Abstract

Decentralized education delivery has the potential to be more effective, represent local objectives better, promote involvement, and, ultimately, increase coverage and quality. Decentralization has the potential to boost efficiency, which appeals to governments with tight budgets in particular. Recently, mobile edge computing (MEC) has gained attention as a viable method for relieving resource-constrained mobile devices of computation-intensive activities. Edge computing technologies have developed as a result of the fact that traditional cloud computing cannot serve the varied data processing needs of today's intelligent society, a fresh approach to computing for performing operations at the edge of the network. It stresses proximity to the consumer and the data source, in contrast to cloud computing. For processing and storing local, small-scale data at the network interface, it is lightweight. By offloading workloads to neighboring MEC servers, devices can increase the quality of their computing experiences. Nevertheless, it is difficult to create and compute offloading policies to reduce the long-term average computation cost in terms of power consumption and buffering delay when a MEC system with numerous mobile users, stochastic task arrivals, and wireless channels is taken into consideration. Decentralization should instead be seen as a continuum that ranges from decentralization to devolution.

***Keywords*:** Resource allocation, mobile edge computing, decentralization, distributed computing, MEC

**Corresponding author*: puneetgarg.research@gmail.com

Sandhya Avasthi, Suman Lata Tripathi, Namrata Dhanda and Satya Bhushan Verma (eds.) Decentralized Systems and Distributed Computing, (75–88) © 2024 Scrivener Publishing LLC

4.1 Introduction

Decentralization is a strategy to boost financing, move power to the level of government with the highest expertise, save money by increasing management efficiency, allow citizens more control over education, and embrace more comprehensive national administrative reforms [1, 2]. These factors, however, can be categorized broadly into three groups: redistribution of power within the educational system, efficiency and effectiveness, and educational financing. Decentralization has been categorized and categorized in numerous research. Decentralization as delegation in the former typically entails the transfer of duties and administrative obligations pertaining to particular activities, which are typically defined by central authorities. The decentralization of tasks does not always equate to a shift in power, especially when it comes to carrying out decisions that have already been made at the central level [3]. Delegation, on the other hand, can imply greater local autonomy just because it is difficult to maintain total central control. Devolution, often known as decentralization, is the transfer of authority and accountability from national to local organizations. The sole kind of decentralization is devolution where there is a discernible rise in local authority and independence [3–5]. To develop effective computation offloading policies separately, a deep deterministic policy gradient DRL method based on continuous action space is used. A series of blocks called a blockchain is made up of data about transactions, a timestamp, and the hash of the preceding block. The peer-to-peer transactions on the bitcoin network are permanently and immutably recorded on the distributed, decentralized ledger known as blockchain. Distributed computing devices can connect to the same file system using the peer-to-peer IPFS distributed file system. IPFS aims to implement off-chain storage, storing hashes of data locations on the blockchain rather than actual files. By averaging the total of locally trained models supplied by users, a global model is created via the blockchain smart contract. Source data are meant to ensure security and privacy in this federated manner. Customers can take the model from the blockchain and use it to train with local data. Decentralization in education necessitates the introduction and adoption of a new management culture, which calls for a fundamental restructuring of educational administration and authority. To create a scalable MEC system with minimal feedback, a decentralized dynamic compute offloading method based on deep reinforcement learning (DRL) is used. Improvements in "efficiency, transparency, accountability, and responsiveness of service supply compared with centralized systems" are promised by decentralizing education.

A centralized vs. decentralized viewpoint is no longer a useful lens through which to examine educational change.

There are various reasons why decentralization is beneficial for development. Both excellent governance and participation in decision-making might be tied to those. People are closer to public services thanks to decentralization, since compared to centralized systems, people have more opportunities to actively participate in local policy and activity decision-making [6]. Decentralization offers individuals the chance to participate in the decision-making process. Because those present at the action are more intimately connected to the issue, this promotes greater flexibility and enables better decision-making [7].

The transfer of power is referred to as decentralization and responsibility to local communities and educational institutions. Then, they are free to decide for themselves regarding many areas of practice and policy. In systems that are centralized, a central authority may administer policy, curriculum, and assessment in addition to managing finances, employees, and resources [8]. Several scholars contend that resources can be used more effectively in decentralized systems [9–11]. This, in turn, may result in better student achievement, greater parental satisfaction, and stronger accountability with support from the community. The only type of decentralization where there is a noticeable rise in local authority and independence is devolution. Decentralization makes it possible for citizens to actively participate in local policy and activity decision-making, which brings public services closer to the people centralized decision-making [12]. Decentralization offers individuals the chance to participate in the decision-making process. Due to the fact that those present at the scene of the activity may make better decisions, this enables more flexibility [13]. Decentralization may encourage greater autonomy from the central authority while simultaneously jeopardizing the integrity of the nation. Regional identities may be promoted more than national identities. Local corruption in the education sector may worsen as a result of decentralization. This might be connected to any potential input they may have on the selection of particular policies and processes [13]. Because different regions in a country may have varying resources, decentralization may encourage the growth of resource-rich areas but may also impede the development of resource-poor areas if top-down management is not well maintained [13].

The blockchain helps the manufacturer, who is the crowd sourcing requester, to check for harmful updates from customers. While our system uses blockchain to track crowd sourcing activities, the old system is maintained by a third party, who charges clients exorbitant service costs. A game-theoretic offloading approach for decentralized algorithms with little overhead was developed [1]. Moreover, the system cost—defined as the weighted sum of energy consumption and execution time—was flexible-regulated with other parameters utilizing dynamic voltage and frequency (DVFS) techniques [2, 3] and was decreased. Moreover, the trade-off between energy and latency has been studied [4]. Under the restrictions of limited energy and delicate latency, communication and processing are simultaneously optimized. Moreover, it has been shown that other state-of-the-art technologies, such as wireless power transfer [5] and non-orthogonal multiple access (NOMA) [6], and MEC's performance may be further enhanced.

MEC adds computational capability to the radio access networks (RANs), which are proximate to mobile users [7]. By allowing mobile devices to transfer compute workloads to the MEC server connected to a base station (BS), it enhances the QoE of mobile applications with much less latency and power consumption [10]. Numerous researchers have been drawn from both academia and industry [8, 9]. However, the effectiveness of wireless data transmission is a key factor in computation offloading; therefore, MEC systems must effectively manage radio resources in addition to compute resources [11].

By bringing decision-making closer to the people, decentralization may provide them more authority to hold service providers accountable and/or influence decisions about their education. Theoretically, schools have the power to determine their own priorities and develop original teaching strategies to improve instruction. Decentralization's advantages are difficult to realize in practice due to a lack of system support, inadequate funding, ill-prepared instructors, and weak managerial skills. Increasing parental involvement in school governance, allowing teachers to select their own textbooks, and giving school directors the authority to hire teachers all have positive effects on educational quality, even though empirical research evidence on education decentralization is inconsistent [16].

In reality, many countries employ decentralized educational programs for reasons unrelated to improving education. For instance, Spain decentralized education to regional governments to accommodate the demands of various ethnic/linguistic communities, while Argentina decentralized education to provincial governments to lower the economic deficit of the federal government [17].

Decentralization refers to the transfer of authority and responsibility to local communities and educational institutions. Then, they are free to decide for themselves regarding many areas of practice and policy. In systems that are centralized, a central authority may administer policy, curriculum, and assessment in addition to managing finances, employees, and resources [18].

4.2 Edge Computing

Edge computing is a new computing paradigm that refers to a range of networks and gadgets that are at or near the user. Near the point of generation, edge focuses on processing data more quickly and in greater volume while delivering real-time, action-driven solutions. Edge enables businesses to combine the physical and digital worlds, incorporating web analytics and data into physical stores to improve the shopping experience developing systems that allow for worker training and environments where humans can learn from machines, and creating intelligent settings that protect our security and comfort. IT and commercial computing are being transformed by edge computing. A thorough examination of edge computing's definition, operation, impact of the cloud, use cases, trade-offs, and implementation issues is necessary. To quickly give in-depth analysis and predictions, edge computing makes use of the growing capacity of in-device computing. The MEC server typically has sufficient processing capabilities, such as a high-frequency multi-core CPU, to gain from edge computing, so it is vital to keep this in mind. Consequently, it may be hypothesized that other applications could be handled concurrently with a very low processing latency, and the feedback delay is disregarded because the calculation output is so minimal [19]. This method will be used to process all of the task data bits that were offloaded to the MEC server via the BS.

With the help of edge computing, you can make the most of the enormous amounts of untapped data that are generated by connected devices (see Figure 4.1, which shows a straightforward illustration of a multiuser

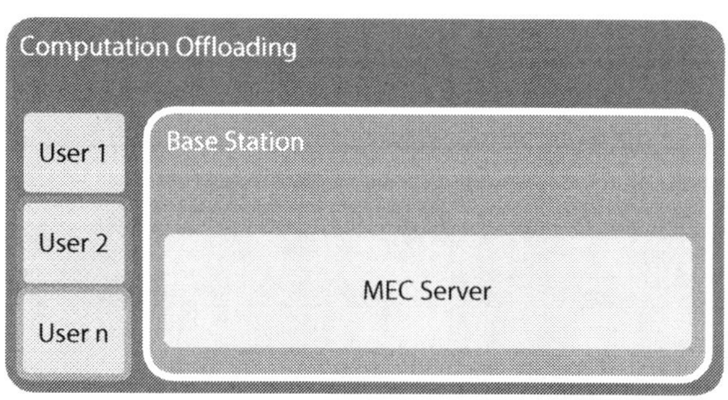

Figure 4.1 Multiuser mobile edge computing system.

mobile edge computing system). Edge computing is a networking method that focuses on bringing processing as close as practical to the source of data in order to reduce latency and bandwidth utilization. The act of moving fewer processes from the cloud to local areas, such as a user's PC, an IoT device, or an edge server, is known as edge computing [20]. Figure 4.2 represents the edge computing architecture. It is divided into three layered architectures: Cloud layer, Edge layer, and Device layer.

The software necessary to complete specific tasks are installed on edge computing devices, which are created to perform specific duties. The employment of sensors to track machine temperature is one instance. An IoT platform or a data center receives temperature data from the machine in this scenario using a sensor that fits in the palm of your hand [21]. Enterprises employ edge devices in a variety of ways. Although their uses in industrial processes are well established, their applications go beyond product production and service delivery. The application of edge computing tools to rescue efforts serves as an illustration of this. In this case, the edge computing device is a rescue robot, which is capable of navigating on its own and gathering visual data in difficult-to-reach places. To aid with decision-making, the gathered data can subsequently be translated to a bigger visual display [23].

Edge computing reduces the requirement for data to be processed in a distant data center by performing computations on-site or close to a specific data source. The architectures needed to do computing tasks must be appropriate for those tasks, and not all computing tasks require the

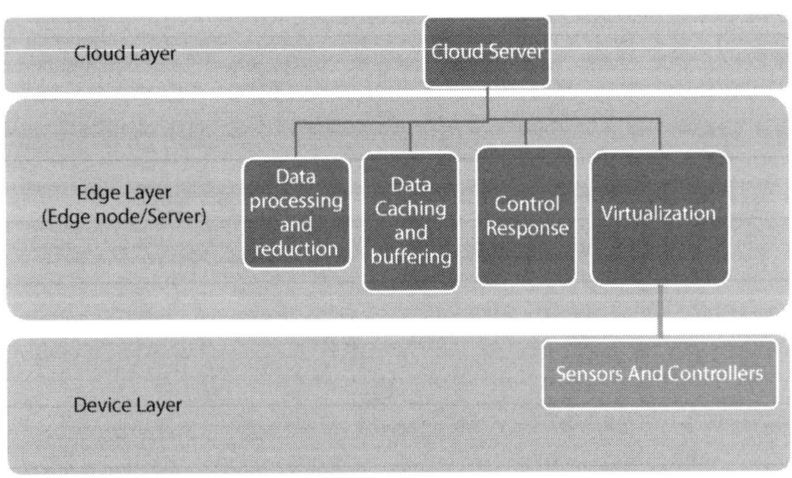

Figure 4.2 Edge computing architecture.

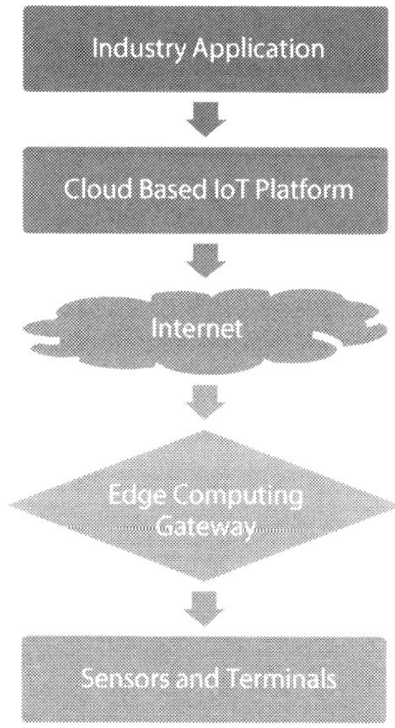

Figure 4.3 Edge computing IoT.

same architecture. Edge computing has evolved as a useful and substantial architecture to bring compute and storage resources closer to—preferably in the same physical location as—the data source [23]. The concepts of remote offices, branch offices, co-location of data centers, and cloud computing are well-established and have a long history. In general, distributed computing models are not very unique. This makes it possible for companies to grow more quickly, introduce new products and services more quickly, and open up possibilities for new revenue sources [24].

MEC servers typically come with enough computing power, such as high-frequency multi-core CPUs. Figure 4.3 clearly defines how edge computing is done in IoT. As a result, it may be predicted that numerous applications can be processed simultaneously with little processing lag. Furthermore, by taking into account applications like video stream analysis and services for connected automobiles, feedback delay is disregarded for the modest computing output. All task data bits offloaded via BS to the MEC server will be processed in this manner [25].

4.3 Benefits of Edge Computing

Edge computing models store and process data locally without transmitting to a cloud platform. It has advantages over on-premise data center models. Putting compute at the edge improves physical asset management and creates new interactive, human experiences. Edge use cases include autonomous automobiles, robots, smart equipment data, and automated shopping.

- **Cost:** Edge computing reduces the need for server resources and bandwidth. Resources in the cloud and bandwidth are limited and expensive [14]. As smart cameras, printers, thermostats, and even toasters become standard in every home and workplace, through statistical projects by 2025, there will be more than 75 billion IoT devices installed globally [26]. It will be necessary to relocate a sizeable portion of computing to the edge in order to support all those devices.
- **Performance:** Processes are relocated to the edge in order to decrease latency. Every time a device needs to communicate with a distant server, there is a lag. When two employees in the same office use an IM platform, for example, there may be a noticeable delay because each message must be sent outside the building, communicate with a server somewhere in the world, and then be brought back before the recipient can see it. If that process was stretched to the extreme and the business's internal router was in charge of sending intra-office messages, there would not be any discernible delay [27]. Similar to this, users of all types of web apps will experience delays when they come across operations that require communication with a remote server. The length of these delays will vary depending on the server's location and available bandwidth, but they can be completely eliminated by moving more tasks to the network edge [28].
- **Security:** In standard cloud computing, all data must be sent to the cloud so that it can be processed there. This method has risks like data loss and data leaks that make it impossible to ensure security and privacy. For example, business secrets, search history, and account passwords are all likely to get out. Edge computing does not need to process data locally because it only does jobs that are in its area of responsibility. This gets rid of the need to send info to the cloud. By doing this, the risks of network communication are removed, making sure that data are safe. When data are attacked, only local data are affected. Not all data are affected.

- **Low cost, low energy consumption**: With edge computing, there is no need to transfer the data to the cloud computing facility for processing. As a result, less network bandwidth is required, which lessens the burden on the system and the energy consumption of smart devices at the network's edge. Edge computing is "small scale," allowing businesses to produce goods at a reduced cost by processing data locally. Additionally, it reduces the amount of energy used by local equipment and improves computing performance.

In summary, edge computing has the following main advantages:

- Reduced latency
- Reduced bandwidth usage and related costs
- Reduction of server resources and related costs
- Additional capabilities

4.4 Classification of Attacks in a 5G IoT System

Attacks are basically classified in 5 layered architectures shown in Figure 4.4.

- The Internet of Things (IoT) and connected devices are distinct data sources that must be protected and recorded in the cloud. These data sources will be nearby or on Edge [29].

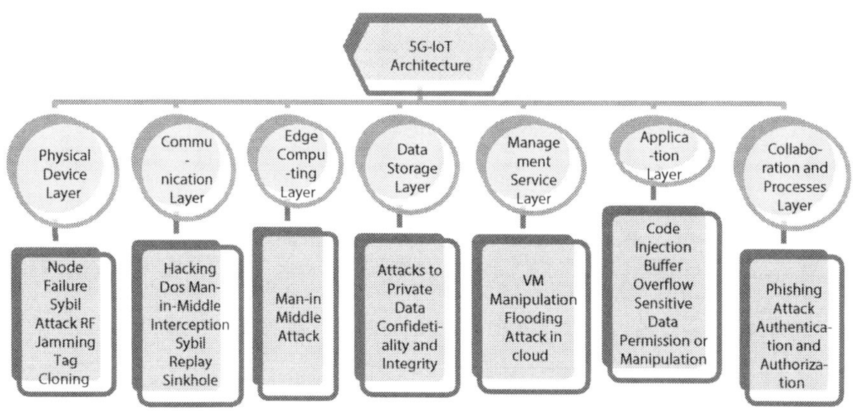

Figure 4.4 Classification of attacks based on 5-G.

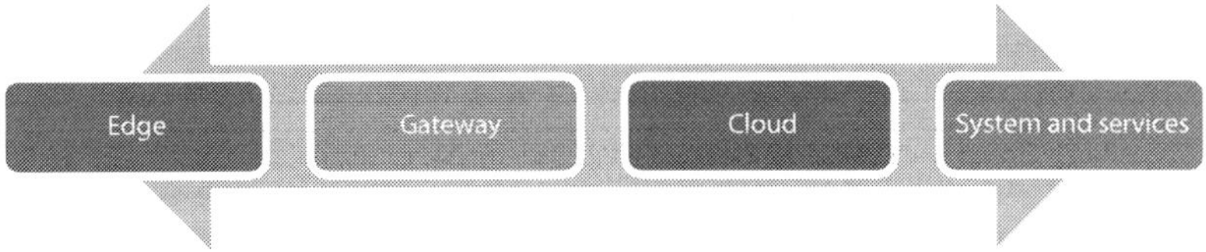

Figure 4.5 Gateways in an IoT system.

- **Containers:** For developers to create and bundle applications, containers offer a standardized deployment environment. Regardless of the device's capabilities, settings, or configurations, containers can be deployed on a variety of hardware [30].
- Service and data mesh offer a mechanism to deploy and query data and services that are dispersed across containers and datastores at the edge. The management and routing of services and data interfaces are abstracted away by these meshes' single interface. Bulk searches for whole populations within the edge are made possible by this essential enabler rather than one query per device [31].
- **Software defined networking:** Users can customize overlay networks thanks to software-defined networking. Also, it makes it simple to adjust bandwidth and routing in order to choose the best way to link edge devices to the cloud [33].
- **Digital twin:** The digital twin, a crucial enabler, makes physical-to-digital and cloud-to-edge organization conceivable. The twin enables data and applications to be described using domain words focused on assets and production lines instead of database tables and message streams. The gateways in an IoT system are shown in Figure 4.5. Digital twins enable domain experts (rather than software engineers) to set programs to perceive, reason, and act in the edge cases [32].

4.5 Decentralized Dynamic Computation Offloading Method

Reducing the decentralized dynamic computation offloading is advised to reduce the long-term average computation cost of each user. Based on local observations of the environment, a continuous action, i.e., powers allotted for local execution and computation offloading in a continuous domain, will be chosen for each slot, and this is specifically possible by leveraging the DDPG [14] algorithm. Since no user has any prior knowledge of the system environment, the information about the number of users M, task arrival data, and wireless channels is all new to each user agent. As a result, there are no models used at all during the online learning process [22]. Following the specification of the state space, action space, and reward function, a DDPG-based framework for decentralized dynamic computation offloading will be built. Additionally, the system incorporates decentralized policy testing and training [15].

4.6 Conclusion

The centralization and decentralization concepts in the educational system each have advantages and disadvantages. A community or a country can decide whether to embrace the system. The negative consequences of the systems may be reduced with effective control and governance.

Restricted capability: One of the benefits that cloud computing offers to edge—or fog—computing is the diversity and reach of the resources and services. It is crucial to define the scope and purpose of the edge deployment before moving further since even a large-scale edge computing deployment serves a specific function at a predetermined size with a minimal amount of resources and services.

Connectivity: Although edge computing circumvents common network limitations, even the most forgiving edge implementation will require at least a minimal amount of connectivity. Planning an edge deployment that considers sporadic or poor connectivity, as well as what happens at the edge if connectivity is lost, is crucial. Success in edge computing hinges on autonomy, AI, and graceful failure planning when dealing with connectivity problems.

Security: It is critical to plan an edge computing deployment that will prioritize proper device management, including policy-driven configuration enforcement, security [14, 34, 35] in the computing and storage resources, including components like software patching and updates, and encryption in the data at rest and in flight. This is necessary because IoT devices are notoriously insecure. Although secure connections are not always present when building an edge site from scratch, they are a feature of IoT services from major cloud providers.

Data lifecycles: Data cycle includes the constant issue with today's data overload and how much of it is superfluous. Consider a medical monitoring gadget. Keeping days of routine patient data is not necessary because only the problem data are important. Real-time analytics typically uses short-term data that are not maintained for an extended period of time. Once studies are completed, a firm must determine which data to preserve and what to discard. Also, the data that are kept must be safeguarded in compliance with corporate and legal guidelines.

References

1. Brennen, Annick M. 2002. Centralization Versus Decentralization. Available online at http://www.soencouragement.org/centralizationvsdecentralization.htm, retrieved on January 3, 2013.
2. C. Maggo and P. Garg, "From linguistic features to their extractions: Understanding the semantics of a concept," *2022 Fifth International Conference on Computational Intelligence and Communication Technologies (CCICT)*, 2022, pp. 427-431, doi: 10.1109/CCiCT56684.2022.00082.
3. Karlsen, G. E. (1999). " Decentralized-Centralism" Governance in Education: Evidence from Norway and British Columbia, Canada. *Can. J. Educ. Adm. Policy*, (13).
4. Chauhan, R., Avasthi, S., Alankar, B., & Kaur, H. (2021). Smart IoT systems: Data analytics, secure smart home, and challenges. In *Transforming the Internet of Things for Next-Generation Smart Systems* (pp. 100-119). IGI global.
5. D. Huang, P. Wang and D. Niyato, "A dynamic offloading algorithm for mobile computing," *IEEE Trans. Wirel. Commun.*, vol. 11, no. 6, pp. 1991–1995, 2012.
6. Garg, P., & Raman, P. K. Broadcasting Protocol & Routing Characteristics With Wireless ad-hoc networks.
7. Garg, P., Arora, N., & Malik, T. Capacity Improvement of WI-MAX In presence of Different Codes WI-MAX: Speed & Scope of future.
8. Garg, P., Dixit, A., & Sethi, P. (2021, May). Link Prediction Techniques for Opportunistic Networks using Machine Learning. In *Proceedings of the International Conference on Innovative Computing & Communication (ICICC)*.
9. Garg, P., Dixit, A., Sethi, P., & Pinheiro, P. R. (2020). Impact of node density on the qos parameters of routing protocols in opportunistic networks for smart spaces. *Mob. Inf. Syst.*, 2020.
10. Garg, P., Saroha, K., &Lochab, R. (2011). Review of wireless sensor networks-architecture and applications. *IJCSMS Int. J. Comput. Sci. Manage. Stud.*, 11(01), 2231-5268.

11. Gupta, A., Garg, P., & Sonal, Y. S. (2020). Edge Detection Based 3D Biometric System for Security of Web-Based Payment and Task Management Application. *Int. J. Grid Distrib. Comput.*, 13(1), 2064-2076.
12. Gupta, S., & Garg, P. (2021). An insight review on multimedia forensics technology. *Cyber Crime and Forensic Computing: Modern Principles, Practices, and Algorithms, 11*, 27.
13. Gupta, M., Garg, P., Gupta, S., & Joon, R. (2020). A Novel Approach for Malicious Node Detection in Cluster-Head Gateway Switching Routing in Mobile Ad Hoc Networks. *Int. J. Future Gener. Commun. Netw.*, 13(4), 99-111.
14. J. Pathak, S. L. Tripathi, "A Novel Model for Resisting Side Channel Attack by Masking of Gates" *J. Eng. Res.* (April 2022) https://doi.org/10.36909/jer.ICMET.17165
15. Varshney, S. Lamba and P. Garg, "A Comprehensive Survey on Event Analysis Using Deep Learning," *2022 Fifth International Conference on Computational Intelligence and Communication Technologies (CCICT)*, 2022, pp. 146-150, doi: 10.1109/CCiCT56684.2022.00037.
16. H. Guo, J. Liu, J. Zhang, W. Sun and N. Kato, "Mobile-edge computation offloading for ultra-dense iot networks," *IEEE Internet Things J.*, 2018.
17. J. Du, L. Zhao, J. Feng and X. Chu, "Computation offloading and resource allocation in mixed fog/cloud computing systems with min-max fairness guarantee," *IEEE Trans. Commun.*, vol. 66, no. 4, 2018.
18. J. Zhang, X. Hu, Z. Ning, E. C. . Ngai, L. Zhou, J. Wei, J. Cheng and B. Hu, "Energy-latency tradeoff for energy-aware offloading in mobile edge computing networks," *IEEE Internet Things J.*, vol. 5, no. 4, pp. 2633–2645, 2018.
19. K. Kumar and Y.-H. Lu, "Cloud computing for mobile users: Can offloading computation save energy?" *Computer*, vol. 43, no. 4, pp. 51–56, 2010.
20. Kanika, Puneet, Swati, " Image augumentaion using gauze filter for photography" *Interdiscip. J. Neurosci. Quantum Phys.*, volume 20 No. 5(2022).
21. Khanna, A., Rani, P., Garg, P., Singh, P. K., &Khamparia, A. (2021). An Enhanced Crow Search Inspired Feature Selection Technique for Intrusion Detection Based Wireless Network System. *Wirel. Pers. Commun.*, 1-18.
22. State Budget Department 2002 *"State Budget: Final Account for 2000 and Plan for 2002"* Financial Publishing House, Hanoi.
23. Kumar, P., Kumar, R., & Garg, P. (2020). Hybrid Crowd Cloud Routing Protocol For Wireless Sensor Networks.
24. L. Saraswat, L. Mohanty, P. Garg and S. Lamba, "Plant Disease Identification Using Plant Images," *2022 Fifth International Conference on Computational Intelligence and Communication Technologies (CCICT)*, 2022, pp. 79-82, doi: 10.1109/CCiCT56684.2022.00026.
25. L. Mohanty, L. Saraswat, P. Garg and S. Lamba, "Recommender Systems in E-Commerce," *2022 Fifth International Conference on Computational Intelligence and Communication Technologies (CCICT)*, 2022, pp. 114-119, doi: 10.1109/CCiCT56684.2022.00032.

26. M. Patel, B. Naughton, C. Chan, N. Sprecher, S. Abeta, A. Neal et al., "Mobile-edge computing introductory technical white paper," *White Paper, Mobile-edge Computing (MEC) industry initiative*, 2014.
27. N. Puri, P. Saggar, A. Kaur and P. Garg, "Application of ensemble Machine Learning models for phishing detection on web networks," *2022 Fifth Int. Conf. Comput. Intell. Commun. Technol. (CCICT)*, 2022, pp. 296-303, doi: 10.1109/CCiCT56684.2022.00062.
28. P. Garg, A. Dixit and P. Sethi, "Ml-fresh: novel routing protocol in opportunistic networks using machine learning," *Comput. Syst. Sci. Eng.*, vol. 40, no. 2, pp. 703–717, 2022.
29. R. Sharma, S. Gupta and P. Garg, "Model for Predicting Cardiac Health using Deep Learning Classifier," *2022 Fifth International Conference on Computational Intelligence and Communication Technologies (CCICT)*, 2022, pp. 25-30, doi: 10.1109/CCiCT56684.2022.00017.
30. S. Bi and Y. J. Zhang, "Computation rate maximization for wireless powered mobile-edge computing with binary computation offloading," *IEEE Trans. Wirel. Commun.*, vol. 17, no. 6, pp. 4177–4190, 2018.
31. S. Kosta, A. Aucinas, P. Hui, R. Mortier and X. Zhang, "Thinkair: Dynamic resource allocation and parallel execution in the cloud for mobile code offloading," in *Proc. IEEE INFOCOM*, 2012, pp. 945–953.
32. Saito, Fumihiko. 2001. Decentralization Theories Revisited: Lessons from Uganda. *Ryukoku RISS Bulletin*, No. 31 March 2001.
33. Sethi, P., Garg, P., Dixit, A., & Singh, Y. (2020, April). Smart number cruncher–a voice based calculator. In *IOP Conf. Ser.: Mater. Sci. Eng.* (Vol. 804, No. 1, p. 012041). IOP Publishing.
34. J. Pathak, S. L. Tripathi, Abhishek Kumar Singh "High Level Transformation techniques for designing Reliable and Secure DSP Architectures" *AI Techniques for Reliability Prediction for Electronic Components* IGI global publishers.
35. J. Pathak and S. L. Tripathi, "Novel Obfuscated Secure Architecture for Baugh Wooley Multiplier," *2023 International Conference on Artificial Intelligence and Applications (ICAIA) Alliance Technology Conference (ATCON-1)*, Bangalore, India, 2023, pp. 1-4, doi: 10.1109/ICAIA57370.2023.10169781

5

Blockchain in Education

Shahnaz Fatima* **and Guru Dev Singh**

Amity Institute of Information Technology, Amity University, Lucknow, Uttar Pradesh, India

Abstract
Blockchain technology has emerged as a disruptive force with the potential to revolutionize various industries, including education. This chapter explores the impact of blockchain in education by examining its benefits, challenges, and use cases. The chapter argues that blockchain can provide secure, transparent, and decentralized solutions for credential verification, learner data privacy, and lifelong learning tracking. Through a comprehensive review of existing literature and examples of successful implementations, the chapter presents the potential of blockchain to transform the education landscape. However, the chapter also acknowledges challenges related to technical complexities, data privacy, and adoption barriers. The chapter concludes by highlighting future prospects and recommending further research to explore the full potential of blockchain in education. Overall, this chapter contributes to the growing body of knowledge on the application of blockchain in education and encourages further exploration of this promising technology.

Keywords: Blockchain, distributed ledger, distributed computing, lifelong learning

5.1 Introduction

Blockchain technology, a distributed ledger technology that enables secure, transparent, and immutable record-keeping, has gained significant attention in recent years for its potential to disrupt various industries [1]. One area where blockchain has the potential to bring about transformative

**Corresponding author*: sfatima1@lko.amity.edu

Sandhya Avasthi, Suman Lata Tripathi, Namrata Dhanda and Satya Bhushan Verma (eds.) Decentralized Systems and Distributed Computing, (89–116) © 2024 Scrivener Publishing LLC

changes is education. The traditional education system has long been plagued by issues such as fraud in credential verification, lack of learner data privacy, and challenges in tracking lifelong learning achievements [2, 3]. However, blockchain technology offers unique features that could address these challenges and revolutionize education as we know it.

Blockchain in education has gained traction as a promising solution to improve the authenticity and security of credentials, provide learners with control over their data, and enable lifelong learning tracking [4, 5]. With blockchain, credentials can be securely stored in a decentralized and transparent manner, eliminating issues of fraud and counterfeit credentials [6, 7]. Additionally, learners can maintain control over their personal information, ensuring data privacy and security. Furthermore, blockchain can enable lifelong learning tracking by creating a comprehensive and immutable record of a learner's achievements and skills, which can be accessed and verified at any point in their career [8].

However, the implementation of blockchain in education is not without challenges. Technical complexities, such as scalability, interoperability, and energy consumption, pose significant hurdles to widespread adoption [9]. Data privacy and security concerns also arise, as the transparency of the blockchain may expose personal information [10]. Adoption barriers, including resistance to change, lack of awareness, and regulatory issues, may also impede the integration of blockchain in the traditional education system [11].

Despite these challenges, there are already successful examples of blockchain implementation in education. For instance, the Massachusetts Institute of Technology (MIT) has developed the "Blockcerts" system, which uses blockchain to issue digital credentials that are verifiable, tamper-proof, and owned by learners [12]. In Malta, the "Learning Machine" platform is being used to issue blockchain-based credentials for formal, non-formal, and informal learning [13]. These examples demonstrate the potential of blockchain in transforming the education landscape.

In light of these developments, this chapter aims to explore the impact of blockchain in education by examining its benefits, challenges, and use cases. By providing a comprehensive review of existing literature and examples of successful implementations, this chapter will shed light on the potential of blockchain to revolutionize education. However, the chapter also acknowledges the challenges that need to be addressed for effective implementation. The remainder of the chapter is structured as follows. First, the benefits of blockchain in education will be discussed in detail, highlighting how it can address existing challenges and improve the current education system. Then, potential challenges of implementing blockchain in education will be examined, including technical complexities, data privacy concerns, and adoption barriers. Next, examples of successful use cases of blockchain in education will be presented to illustrate how this technology has been applied in practice. Finally, the chapter will conclude with a summary of the main arguments, discussion of future prospects, and recommendations for further research in this area.

5.2 Benefits of Blockchain in Education

Blockchain technology has the potential to bring about significant benefits in the field of education. These benefits are outlined below:

- Authenticity and Security of Credentials: In the traditional education system, fraud in credential verification has been a persistent issue [14]. As depicted in Figure 5.1, blockchain can address this challenge by providing a decentralized, transparent, and immutable record-keeping system. Credentials can be securely stored on the blockchain, eliminating the risk of counterfeit credentials and ensuring the authenticity and integrity of the information [15].

```
isChainValid()
{
    for(let i = 1; i < this.chain.length; i++)
    {
        const currBlock = this.chain[i];
        const prevBlock = this.chain[i-1];
        if (currBlock.hash !== currBlock.calcHash())
        {
            return false;
        }
        if (currBlock.prevHash !== prevBlock.hash)
        {
            return false;
        }
    }
```

Figure 5.1 Method to check if the blockchain is valid or not.

For example, the "Blockcerts" system developed by the Massachusetts Institute of Technology (MIT) uses blockchain to issue digital credentials that are verifiable, tamper-proof, and owned by learners [9].

- Learner Data Privacy and Control: Learner data privacy has become a critical concern in the era of digital education. Blockchain can provide a solution by allowing learners to have control over their personal information. Learners can choose what data to share and with whom, and their data can be stored securely on the blockchain, which is resistant to tampering and unauthorized access [16]. This gives learners greater control over their data and ensures privacy and security [17].
- Lifelong Learning Tracking: Blockchain can enable comprehensive and immutable tracking of lifelong learning achievements. A learner's achievements and skills can be recorded on the blockchain in a transparent and verifiable manner, creating a lifelong learning record that can be accessed and verified at any point in the learner's career [5, 18]. This can help learners build a robust and credible profile of their achievements, making it easier for employers and other stakeholders to assess their skills and qualifications.
- Enhanced Trust and Transparency: Blockchain's decentralized nature and transparent record-keeping can enhance trust and transparency in the education system. It can provide a tamper-proof and auditable record of all educational transactions, such as certifications, degrees, and transcripts, which can be verified by multiple parties, including employers, academic institutions, and accreditation agencies [14]. This can help combat issues such as diploma mills and false credentials, leading to a more reliable and trustworthy education ecosystem.
- Improved Efficiency and Cost Savings: Blockchain has the potential to improve the efficiency of administrative processes in education by reducing the need for intermediaries and paperwork. The use of blockchain for record-keeping and verification can streamline processes, reduce bureaucracy, and save costs [2]. For example, the "Learning Machine" platform used in Malta for issuing blockchain-based credentials has been reported to result in significant time and cost savings [10].

- Global Accessibility: Blockchain can enable global accessibility to educational opportunities [19]. Learners from different parts of the world can access credentials and educational records stored on the blockchain, eliminating the need for physical document transfer and simplifying the process of recognition of credentials across borders [1]. This can open up opportunities for learners in underserved regions, providing them with access to quality education and improving their employability prospects.

5.2.1 Blockchain Can Enhance Learner Data Privacy and Security, Allowing Learners to Maintain Control Over Their Personal Information

Blockchain technology can significantly enhance learner data privacy and security by providing learners with greater control over their personal information. Learners can choose what data to share and with whom, and their data can be stored securely on the blockchain, which is resistant to tampering and unauthorized access [20]. This ensures that learners' personal information remains private and secure, and they have the authority to manage and share their data as they see fit.

In a traditional education system, learner data are often stored in centralized databases managed by educational institutions or third-party service providers, which may be susceptible to data breaches or unauthorized access. However, with blockchain, data are stored in a decentralized and distributed manner, making it more resistant to unauthorized access and tampering. Each transaction or data entry on the blockchain is encrypted and linked to the previous data, creating an immutable record that is difficult to alter or manipulate without consensus from the network participants [21].

Blockchain also eliminates the need for a central authority or intermediary to manage learner data, reducing the risk of data misuse or abuse. Learners have control over their data, and they can choose to share it only with trusted parties or revoke access at any time. This puts learners in charge of their personal information, ensuring that their privacy is maintained and their data are used only for authorized purposes.

Furthermore, the transparent nature of blockchain ensures that learners can monitor and audit who has access to their data and how they are being used. Any changes or updates to the data on the blockchain are recorded in a transparent and immutable manner, creating a comprehensive and auditable history of data access and modifications. This enhances the transparency and accountability of data management, giving learners greater confidence in the security and privacy of their personal information [13].

Overall, the use of blockchain in education can significantly enhance learner data privacy and security by providing learners with control over their personal information, creating a tamper-proof and transparent record of data transactions, and eliminating the need for intermediaries in data management [22]. This can help protect learners' privacy and ensure that their data are securely managed in the digital education ecosystem.

5.2.2 Blockchain Can Enable Lifelong Learning Tracking, Creating a Comprehensive and Immutable Record of a Learner's Achievements and Skills

Blockchain technology has the potential to revolutionize the field of education by enabling comprehensive and immutable tracking of lifelong learning achievements. Lifelong learning, also known as continuous learning, is the concept of acquiring knowledge, skills, and credentials throughout one's life to adapt to changing work environments and to enhance personal and professional development. Traditional methods of tracking learning achievements, such as diplomas, transcripts, and certifications, are often paper-based or stored in centralized databases, which can be cumbersome, prone to fraud, and difficult to verify. Blockchain can provide a decentralized, transparent, and secure record-keeping system that can revolutionize the way lifelong learning achievements are tracked, verified, and shared. In this section, we will explore how blockchain can enable lifelong learning tracking, creating a comprehensive and immutable record of a learner's achievements and skills, and the potential benefits of this approach.

5.2.2.1 Blockchain as a Distributed and Immutable Record-Keeping System

At its core, a blockchain is a distributed and immutable ledger that consists of a chain of blocks, where each block contains a list of transactions. Transactions are verified by network participants using consensus algorithms, and once a transaction is recorded on the blockchain, it cannot be altered or deleted without the consensus of the network [23]. This makes blockchain a tamper-proof and transparent record-keeping system, as any changes to the data on the blockchain would require the consensus of the majority of the network participants, making it virtually impossible to alter the data fraudulently as depicted in Figure 5.2, Figure 5.3 and Figure 5.4.

```
console.log("Is Blockchain Valid ??? " + Amity_University.isChainValid());
console.log(JSON.stringify(Amity_University, null, 4));
```

Figure 5.2 Check if the blockchain is valid or not.

```
[Running] node "d:\VS CODE\JS\node_modules\BlockCon.js"
Is Blockchain Valid ???  true
```

Figure 5.3 Output, if the blockchain is valid.

```
[Running] node "d:\VS CODE\JS\node_modules\BlockCon.js"
Is Blockchain Valid ???  false
```

Figure 5.4 Output, if blockchain is tempered.

Blockchain can be used to create a comprehensive and immutable record of a learner's achievements and skills. When a learner completes a course, earns a certification, or acquires a new skill, the information can be recorded as a transaction on the blockchain. This transaction can include details such as the name of the learner, the name of the course or certification, the date of completion, and other relevant information. Once recorded on the blockchain, this transaction becomes a permanent and transparent entry that cannot be altered or deleted. This creates a lifelong learning record that can be accessed and verified at any point in the learner's career, providing a credible and transparent record of the learner's achievements.

5.2.2.2 *Benefits of Blockchain-Enabled Lifelong Learning Tracking*

The use of blockchain for lifelong learning tracking can bring about several benefits for learners, educational institutions, employers, and other stakeholders. Some of the potential benefits of blockchain-enabled lifelong learning tracking are outlined below:

- Comprehensive and Verifiable Learning Record: Blockchain can enable learners to maintain a comprehensive and verifiable record of their lifelong learning achievements. As learners complete courses, earn certifications, and acquire new skills, the information can be recorded on the blockchain in a transparent and immutable manner. This creates a comprehensive learning record that can be accessed and verified by learners, educational institutions, employers, and other stakeholders. This transparent and verifiable record can provide a credible proof of the learner's achievements, making it easier for employers and other stakeholders to assess the learner's skills and qualifications [5].

- Increased Trust and Transparency: Blockchain's decentralized nature and transparent record-keeping can enhance trust and transparency in the field of education. The use of blockchain for lifelong learning tracking can create a tamper-proof and auditable record of all educational transactions, such as certifications, degrees, and transcripts, which can be verified by multiple parties, including employers, academic institutions, and accreditation agencies [24]. This can help combat issues such as diploma mills and false credentials, leading to a more reliable and trustworthy education ecosystem.
- Enhanced Portability and Accessibility: Blockchain can enable learners to have greater portability and accessibility to their learning achievements. Learners can have ownership and control over their learning records stored on the blockchain, which can be accessed and verified from anywhere in the world, eliminating the need for physical document transcripts and certifications that can be lost or damaged. Learners can easily share their blockchain-based learning records with potential employers, educational institutions, or other stakeholders, making it convenient and efficient to demonstrate their skills and qualifications [25]. This can also be particularly beneficial for learners who have taken non-traditional learning paths, such as online courses, workshops, or self-directed learning, as their achievements can be equally recognized and verified on the blockchain.
- Enhanced Privacy and Security: Privacy and security of learner data are critical concerns in the field of education. Traditional methods of storing learner data, such as centralized databases, are vulnerable to data breaches and unauthorized access. Blockchain, on the other hand, can enhance learner data privacy and security as depicted in Figure 5.5 and Figure 5.6. As blockchain uses encryption and consensus algorithms to secure transactions, learner data recorded on the blockchain is highly secure and cannot be easily

tampered with or accessed without proper authorization [1]. Learners can also maintain control over their personal information on the blockchain, as they can choose what information to disclose and to whom, ensuring their privacy rights are respected [26].

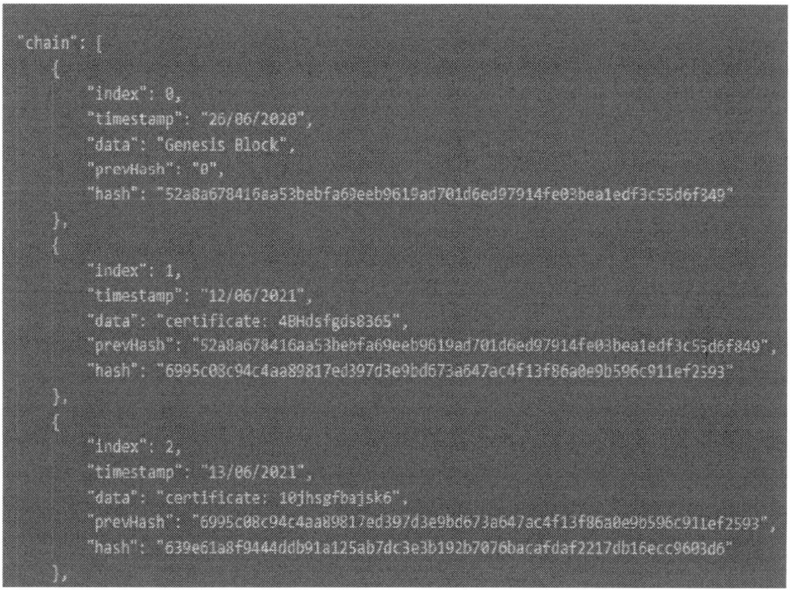

```
"chain": [
    {
        "index": 0,
        "timestamp": "26/06/2020",
        "data": "Genesis Block",
        "prevHash": "0",
        "hash": "52a8a678416aa53bebfa69eeb9619ad701d6ed97914fe03bea1edf3c55d6f849"
    },
    {
        "index": 1,
        "timestamp": "12/06/2021",
        "data": "certificate: 48Hdsfgds8365",
        "prevHash": "52a8a678416aa53bebfa69eeb9619ad701d6ed97914fe03bea1edf3c55d6f849",
        "hash": "6995c08c94c4aa89817ed397d3e9bd673a647ac4f13f86a0e9b596c911ef2593"
    },
    {
        "index": 2,
        "timestamp": "13/06/2021",
        "data": "certificate: 10jhsgfbajsk6",
        "prevHash": "6995c08c94c4aa89817ed397d3e9bd673a647ac4f13f86a0e9b596c911ef2593",
        "hash": "639e61a8f9444ddb91a125ab7dc3e3b192b7076bacafdaf2217db16ecc9603d6"
    },
```

Figure 5.5 Credential verification in blockchain if chain is valid.

```
"chain": [
    {
        "index": 0,
        "timestamp": "26/06/2020",
        "data": "Genesis Block",
        "prevHash": "0",
        "hash": "52a8a678416aa53bebfa69eeb9619ad701d6ed97914fe03bea1edf3c55d6f849"
    },
    {
        "index": 1,
        "timestamp": "12/06/2021",
        "data": "uieygui",
        "prevHash": "52a8a678416aa53bebfa69eeb9619ad701d6ed97914fe03bea1edf3c55d6f849",
        "hash": "f7e092e57b70444e08d3c0f24e34fa590c92f7a16d345da8f523a85aad43b0eb"
    },
    {
        "index": 2,
        "timestamp": "13/06/2021",
        "data": "uihjcsgfvhjsgadfhji",
        "prevHash": "6995c08c94c4aa89817ed397d3e9bd673a647ac4f13f86a0e9b596c911ef2593",
        "hash": "32debbd437a992dc2362536025bf67c20f748113007d4ce3f6066be4aba223c9"
    },
```

Figure 5.6 Credential verification in blockchain if chain is invalid.

- Simplified Credential Verification: Credential verification can be a time-consuming and labor-intensive process for employers, educational institutions, and other stakeholders. With the use of blockchain, credential verification can be simplified and streamlined. As all credentials are recorded on the blockchain in a transparent and immutable manner, stakeholders can easily verify the authenticity of credentials without relying on traditional methods such as contacting educational institutions or manually checking transcripts. This can save time, effort, and costs associated with credential verification, making the process more efficient and reliable [5].

If someone were to tamper with the transactions on the blockchain, it could lead to various consequences. The transaction history stored on the blockchain would be altered, which could mean that some transactions could be removed, added, or modified. The integrity of the data stored on the blockchain would be compromised, which would decrease trust in the blockchain as a secure and transparent system.

Additionally, if someone were to tamper with the cryptographic hashes used to link blocks together, it would disrupt the integrity of the blockchain. The blocks would no longer be linked in the correct order, which would create a fork in the blockchain. This would result in different nodes having different hash values as depicted in the above figures.

- Potential for Micro-credentials and Micro-learning: Blockchain can also enable the issuance and tracking of micro-credentials, which are small units of learning that represent specific skills or achievements. Micro-credentials can be earned through various learning experiences, such as short courses, workshops, or on-the-job training, and can be recorded on the blockchain as a verifiable proof of the learner's skills [27]. This can encourage continuous learning and upskilling, as learners can earn micro-credentials for specific skills they acquire throughout their careers, rather than completing full-length courses or certifications [18, 28]. Micro-credentials can also be easily shared and verified on the blockchain, providing a more granular and comprehensive representation of the learner's skills and achievements.

5.3 Challenges of Blockchain in Education

Some potential challenges of implementing blockchain in education:

- Technical Complexity: Blockchain technology is known for its technical complexity, including different consensus algorithms, encryption methods, and smart contracts, which can pose challenges in understanding and implementing the technology in educational settings [1, 29].
- Cost and Resource Requirements: Implementing blockchain systems can come with significant costs, including infrastructure setup, maintenance, and ongoing management, as well as the potential energy consumption associated with some blockchain networks. Educational institutions may need to invest in new hardware, software, and skilled personnel, which can pose financial challenges [1, 30].
- Interoperability and Standards: Lack of standardized protocols and interoperability among different blockchain platforms can hinder the seamless exchange of data and credentials across different systems, posing challenges in integration with existing IT infrastructure [31].
- Adoption and Acceptance: Widespread adoption of blockchain in education may face resistance from stakeholders due to skepticism about the technology, concerns about privacy and security, and reluctance to change existing systems and processes [21].
- Regulatory and Legal Challenges: The evolving regulatory landscape surrounding blockchain technology can pose challenges in navigating the legal and regulatory requirements associated with implementing blockchain systems in education, such as data privacy, ownership, intellectual property, and jurisdictional issues [32].
- User Experience and Usability: The user experience and usability of blockchain systems can be a challenge for stakeholders, including learners, educators, and employers, due to the complexity of transactions and technical knowledge required [1, 17].
- Scalability and Performance: Some blockchain networks may face scalability and performance issues, such as slow

transaction times and limited transaction throughput, which can pose challenges in accommodating a large volume of transactions and users in educational environments [21].

- Education and Awareness: Educating stakeholders about the benefits and potential of blockchain in education can be a challenge, as many may have limited understanding of the technology and its applications in education [21].
- Ethical Considerations: The use of blockchain in education raises ethical considerations, such as the transparency of student data, consent for data sharing, and the potential for bias in credentialing and certification processes [22].
- Governance and Decision-making: The decentralized nature of blockchain can challenge traditional governance and decision-making structures in educational institutions, as decision-making authority may shift from centralized entities to distributed networks, which may require new policies, procedures, and governance models [20].

In conclusion, while blockchain has the potential to revolutionize the education sector, there are several challenges that need to be addressed, including technical complexity, cost and resource requirements, interoperability and standards, adoption and acceptance, regulatory and legal challenges, user experience and usability, scalability and performance, education and awareness, ethical considerations, and governance and decision-making. Overcoming these challenges will require collaborative efforts from educational institutions, policymakers, and other stakeholders to ensure the successful implementation of blockchain in education.

5.3.1 Concerns Related to Data Privacy, Security, and Compliance with Data Protection Regulations

Addressing concerns related to data privacy, security, and compliance with data protection regulations is a crucial aspect of implementing blockchain technology in education. While blockchain has the potential to bring transparency, efficiency, and security to educational processes, it also raises unique challenges in terms of protecting sensitive student data, ensuring privacy, and complying with data protection regulations. In this article, we will explore these challenges in detail and discuss potential solutions to address them, supported by relevant references.

5.3.1.1 *Data Privacy Concerns*

One of the primary concerns related to the use of blockchain in education is data privacy. As blockchain technology relies on a distributed and immutable ledger, all transactions are visible to all participants, which can raise concerns about the exposure of sensitive student information. Personally identifiable information (PII) such as student names, addresses, and grades could potentially be accessed by unauthorized parties if proper security measures are not in place [21].

To address data privacy concerns, educational institutions must implement robust security measures to protect student data on the blockchain. This may include encryption techniques that encrypt data at rest and in transit to ensure that it is only accessible by authorized parties. Additionally, access controls should be in place to restrict data access to only those who require it for legitimate educational purposes. Multi-factor authentication can also be implemented to add an extra layer of security to prevent unauthorized access to sensitive student data.

Another approach to protecting data privacy on the blockchain is through the use of zero-knowledge proofs (ZKPs). ZKPs allow for verification of data without revealing the actual data, thus maintaining privacy while ensuring the integrity of the information. For example, ZKPs can be used to prove that a student's age is above a certain threshold without revealing the exact age or other personal details [21]. This can be particularly useful in cases where sensitive information needs to be verified, such as verifying student credentials or qualifications.

5.3.1.2 *Compliance with Data Protection Regulations*

Complying with data protection regulations is a critical consideration when implementing blockchain in education. Regulations such as the European Union's General Data Protection Regulation (GDPR) and other relevant local or international regulations impose strict requirements on how personal data should be collected, stored, processed, and shared. The immutability of data on the blockchain can make it challenging to rectify or erase data once they are recorded, which can pose challenges in meeting the requirements of data protection regulations [22].

To ensure compliance with data protection regulations, educational institutions must establish clear policies and procedures for data handling on the blockchain. This may include obtaining explicit consent from students for data sharing and ensuring that data are collected, stored, and processed in accordance with relevant regulations. In cases where rectification or erasure of data is required, educational institutions must have mechanisms in place to address these requests, such as using off-chain solutions or private blockchains that allow for data modification or erasure while maintaining the integrity of the overall blockchain network [21].

It is also essential to educate all stakeholders, including students, faculty, and staff, about their rights and responsibilities regarding data privacy and protection. Training programs, workshops, and information sessions can be conducted to raise awareness and ensure that all parties understand the implications of using blockchain in education and their role in protecting sensitive student data in compliance with data protection regulations.

5.3.1.3 Security Concerns

Security is a critical aspect of any technology implementation, and blockchain in education is no exception. As educational institutions adopt blockchain for various purposes such as student records, credentials, and certificates, it becomes crucial to ensure that the blockchain network is secure from potential attacks or breaches. Security concerns may include the risk of unauthorized access, tampering, or manipulation of data, as well as potential attacks on the consensus algorithm or smart contracts used in the blockchain network [1, 33].

To address security concerns, educational institutions must implement robust security measures at various levels of the blockchain architecture. This includes securing the endpoints, nodes, and the communication channels used within the blockchain network. Encryption techniques, firewalls, and intrusion detection systems can be deployed to protect against unauthorized access and data tampering. Regular security audits and vulnerability assessments should also be conducted to identify and address any potential vulnerabilities in the blockchain network.

Additionally, the consensus algorithm used in the blockchain network should be carefully selected to ensure its security and resilience against potential attacks. Consensus algorithms such as proof of work (PoW) and proof of stake (PoS) are commonly used in blockchain networks, each with its advantages and disadvantages in terms of security, scalability, and energy efficiency [34]. It is essential to thoroughly evaluate the consensus algorithm and choose the one that best aligns with the security requirements of the educational institution.

Another crucial aspect of security in blockchain is smart contract security. Smart contracts are self-executing contracts that run on the blockchain and automate various processes. However, they are susceptible to vulnerabilities, such as coding errors, logic flaws, and security loopholes, which can be exploited by malicious actors. To mitigate smart contract security risks, educational institutions must thoroughly audit and test smart contracts before deployment, follow secure coding practices, and regularly monitor and update them to address any potential vulnerabilities [35].

5.3.1.4 *Collaboration with Industry Experts and Auditors*

Collaboration with industry experts and auditors can play a vital role in addressing concerns related to data privacy, security, and compliance with data protection regulations in blockchain implementation in education. Industry experts and auditors can provide valuable insights, expertise, and best practices in implementing and securing blockchain networks.

Engaging third-party auditors or security firms can help conduct independent security audits and vulnerability assessments of the blockchain network to identify any potential weaknesses or vulnerabilities. Their expertise can help identify and address security risks and ensure compliance with data protection regulations. Additionally, collaboration with industry experts can provide insights into best practices for securing data on the blockchain and addressing privacy concerns, such as the use of encryption, zero-knowledge proofs, and other privacy-enhancing techniques [36].

Educational institutions can also collaborate with industry experts and auditors to stay updated with the latest developments in blockchain technology and its implications for data privacy and security. Participating in industry forums, conferences, and workshops can provide opportunities to learn from experts, share experiences, and gather insights on best practices and emerging trends in securing blockchain in education.

The implementation of blockchain technology in education holds great promise in revolutionizing educational processes, but it also comes with unique challenges related to data privacy, security, and compliance with data protection regulations. Addressing these concerns is crucial to ensure the responsible and ethical use of blockchain in education. Robust security measures, such as encryption, access controls, and multi-factor authentication, should be in place to protect student data on the blockchain. Zero-knowledge proofs can be used to maintain privacy while ensuring data integrity. Compliance with data protection regulations requires clear policies, consent mechanisms, and mechanisms to rectify or erase data when needed. Smart contract security and careful selection of consensus algorithms are also critical aspects of securing the blockchain network. Collaboration with industry experts and auditors can provide valuable insights and expertise in addressing these concerns. It is important for educational institutions to thoroughly evaluate, plan, and implement blockchain technology in a responsible and secure manner, keeping in mind the ethical implications and legal requirements of data privacy and security. By taking appropriate measures to address concerns related to data privacy, security, and compliance with data protection regulations, educational institutions can leverage the benefits of blockchain technology while safeguarding the privacy and security of student data.

5.3.2 Potential Resistance to Change and Adoption Barriers in the Education Sector

The implementation of blockchain technology in the education sector may face potential resistance to change and adoption barriers. As with any technological innovation, there are various factors that can hinder the widespread adoption of blockchain in education. Some of the key challenges related to resistance to change and adoption barriers are discussed below:

- Lack of Awareness and Understanding: One of the significant barriers to the adoption of blockchain in education is the lack of awareness and understanding among stakeholders, including educators, administrators, students, and parents. Many may not fully grasp the concepts, benefits, and implications of blockchain technology, which can result in resistance to change [14]. Education and training programs should be conducted to educate stakeholders about the potential of blockchain in education, its benefits, and its impact on data privacy and security.
- Cost and Infrastructure Requirements: Implementing blockchain technology requires the necessary infrastructure, including hardware, software, and network resources. The costs associated with the setup and maintenance of blockchain infrastructure can be a barrier, particularly for smaller educational institutions with limited budgets [37]. Additionally, integrating existing systems and processes with

blockchain may require additional investments, making it challenging for some institutions to adopt the technology.

- Technical Complexity: Blockchain technology is complex and may require specialized technical knowledge to implement and manage [38]. The technical complexity associated with blockchain, including programming languages, cryptographic algorithms, and consensus mechanisms, can be a barrier for educators and administrators who may lack the necessary skills and expertise. Institutions may need to invest in training and development programs to equip their staff with the required technical skills.
- Legal and Regulatory Concerns: Blockchain in education raises legal and regulatory concerns related to data privacy, ownership, and intellectual property rights [39]. Compliance with data protection regulations such as the General Data Protection Regulation (GDPR) in the European Union and the Family Educational Rights and Privacy Act (FERPA) in the United States can be a challenge. Institutions need to ensure that the implementation of blockchain technology in education aligns with the relevant legal and regulatory requirements and establish appropriate policies and protocols to address these concerns.
- Interoperability and Standards: Interoperability refers to the ability of different blockchain systems to communicate and exchange data with each other. Lack of interoperability and standardization among different blockchain platforms can hinder the seamless exchange of data and transactions, leading to resistance to adoption. Establishing common standards and protocols for blockchain in education can facilitate interoperability and promote wider adoption of the technology.
- Change Management and Cultural Shift: Adopting blockchain technology in education requires a cultural shift in the way educational institutions operate and manage data. This can result in resistance to change and challenges in managing the transition. Institutions need to have effective change management strategies in place, including stakeholder engagement, communication, and training, to overcome resistance and ensure successful adoption of blockchain in education.

- Security Concerns: While blockchain technology is known for its inherent security features, it is not immune to vulnerabilities and risks. Cybersecurity threats, such as hacking, data breaches, and identity theft, can pose challenges to the adoption of blockchain in education. Institutions need to implement robust security measures, including encryption, access controls, and auditing mechanisms, to safeguard against potential security threats and ensure the integrity and confidentiality of data stored on the blockchain [40].

5.4 Use Cases of Blockchain in Education

Blockchain technology has the potential to revolutionize various aspects of the education sector, from enhancing data security and privacy to enabling new forms of credentials and certifications. Here are some notable use cases of blockchain in education:

- Digital Credentials and Certifications: Blockchain can be used to create, verify, and share digital credentials and certifications in a secure and transparent manner. By issuing digital credentials on the blockchain, educational institutions can ensure the integrity and authenticity of the credentials, reducing the risk of fraud and misrepresentation. Blockchain-based digital credentials also allow for easy verification by employers, institutions, and other stakeholders, eliminating the need for time-consuming manual verification processes. For example, the MIT Media Lab has developed the Blockcerts open standard for creating and verifying digital credentials on the blockchain, which has been adopted by several educational institutions worldwide [9]. Similarly, the Open Badges initiative, led by Mozilla, uses blockchain technology to issue and verify digital badges, providing learners with a verifiable record of their skills and achievements [41].
- Learner Profiles and Portfolios: Blockchain can enable the creation of learner profiles and portfolios that capture a comprehensive record of a learner's educational achievements, skills, and experiences. These learner profiles can be owned and controlled by the learners themselves,

allowing them to share their credentials and achievements with employers, educational institutions, and other stakeholders as desired. Blockchain-based learner profiles can also facilitate lifelong learning and the recognition of prior learning, allowing learners to build and showcase their skills and knowledge throughout their careers. For example, Learning Machine, a company that specializes in blockchain-based credentialing solutions, has developed the Learning Machine Wallet, a blockchain-based platform for learners to create and manage their digital credentials and share them with employers [42].

- Secure Storage and Sharing of Educational Records: Blockchain can provide a secure and tamper-proof way to store and share educational records, such as transcripts, diplomas, and other important documents. Blockchain-based storage ensures that the records are protected from unauthorized access, tampering, and data breaches, providing a higher level of security and privacy compared to traditional methods of record-keeping. Blockchain can also simplify the process of sharing educational records between educational institutions, employers, and other stakeholders, reducing administrative overhead and improving efficiency. For example, Sony Global Education has developed a blockchain-based system for storing and sharing educational records, which has been used in pilot projects in Japan to verify the authenticity of educational records and streamline the transfer process [43].
- Micro-credentialing and Stackable Credentials: Blockchain can facilitate micro-credentialing, which involves the issuance of smaller, more focused credentials that represent specific skills or achievements. These micro-credentials can be stacked together to create a more comprehensive credential, allowing learners to build their credentials and demonstrate their skills in a modular and flexible manner. Blockchain-based micro-credentials can provide a decentralized and interoperable way to issue, verify, and share these credentials, making them more portable and recognized across different educational institutions, employers, and industries. For example, the Southern New Hampshire University has implemented a blockchain-based micro-credentialing system called "SNHU Achievements," which allows learners to

earn, verify, and share micro-credentials for specific skills and knowledge areas [44].

- Intellectual Property Rights and Copyright Management: Blockchain can be used to manage intellectual property rights and copyrights related to educational content, such as digital textbooks, research papers, and other educational resources. Blockchain-based systems can provide a transparent and immutable record of ownership and usage rights, ensuring that creators' rights are protected and their work is properly attributed. Blockchain can also enable new business models for content creators, such as micro-payments and royalties for the use of their content, facilitated by smart contracts. For example, the Learning Economy Foundation has developed a blockchain-based platform called "The Learning Economy Protocol" that aims to create a global marketplace for educational content, where creators can publish, license, and monetize their educational resources [45].
- Transparent and Verifiable Peer Review: Blockchain can enhance the transparency and integrity of the peer review process in academic publishing. By recording peer review information on the blockchain, such as reviewer identities, feedback, and timestamps, the process can be made more transparent and accountable. This can help prevent fraud and unethical practices, such as fake peer reviews or biased evaluations, and improve the overall quality and reliability of academic publications. For example, the Peer Community In platform, a blockchain-based peer review platform, has been developed to provide transparent, open, and collaborative peer review for scientific publications [46].
- Academic Transcript Management: Blockchain can streamline the management of academic transcripts, making the process more efficient and secure. Traditional methods of managing transcripts involve manual processes, which can be time-consuming, error-prone, and susceptible to fraud. By leveraging blockchain technology, academic transcripts can be stored in a tamper-proof and transparent manner, ensuring their integrity and authenticity. Blockchain can also enable the secure sharing of academic transcripts between educational institutions, employers, and other stakeholders, eliminating the need for paper-based transcripts and

reducing administrative overhead. For example, the Malta Ministry for Education and Employment has partnered with Learning Machine to implement a blockchain-based system for managing academic transcripts, which aims to improve the efficiency and security of the transcript management process [47].

- Funding and Donations: Blockchain can facilitate transparent and accountable funding and donations in the education sector. Blockchain-based crowdfunding platforms can provide a decentralized and transparent way for educational institutions, researchers, and students to raise funds for educational projects, research initiatives, scholarships, and other purposes. Smart contracts can be used to ensure that funds are used as intended and that donors have visibility into how their funds are being utilized. Blockchain-based fundraising can also enable micro-donations, where donors can contribute small amounts of cryptocurrency to support educational initiatives. For example, the University of California, Berkeley, has launched the Berkeley Blockchain Initiative, which aims to explore the use of blockchain for fundraising and donations to support academic research [48].
- Identity Verification and Fraud Prevention: Blockchain can enhance identity verification and fraud prevention in the education sector. By recording student identities and credentials on the blockchain, educational institutions can create a tamper-proof and verifiable record of students' identities and achievements, reducing the risk of identity fraud and misrepresentation. Blockchain-based identity verification can also help prevent cheating and plagiarism, as it can provide a transparent and immutable record of students' work, allowing for easy verification and detection of academic misconduct. For example, the Holochain project has developed a blockchain-based system for identity verification and fraud prevention in education, which aims to create a secure and transparent ecosystem for student credentials [49].
- Data Security and Privacy: Blockchain can enhance data security and privacy in the education sector. Traditional methods of data storage and management, such as centralized databases, are susceptible to data breaches, hacks, and unauthorized access. Blockchain-based systems, on the other hand, use advanced cryptographic techniques to secure

data, making it tamper-proof and transparent. Additionally, blockchain allows for data to be stored in a decentralized manner, where multiple copies of data are stored across a network of nodes, reducing the risk of a single point of failure. This makes blockchain an ideal solution for securing sensitive educational data, such as student records, research data, and other confidential information. For example, the University of Nicosia in Cyprus has implemented a blockchain-based system called "Block.co" for securing and verifying student records, ensuring the integrity and privacy of student data [50].

Blockchain technology has the potential to revolutionize the field of education by introducing transparency, efficiency, security, and trust in various aspects of the education ecosystem. From credential verification and record-keeping to content sharing and peer review, blockchain can offer numerous benefits to educational institutions, students, educators, and other stakeholders. By leveraging blockchain, educational institutions can create a decentralized, transparent, and trustworthy environment that promotes innovation, collaboration, and lifelong learning. However, it is important to acknowledge that there are challenges and limitations associated with the implementation of blockchain in education. These include regulatory and legal concerns, technological barriers, scalability issues, and the need for standardization and interoperability. Moreover, the adoption of blockchain in education requires careful planning, investment, and collaboration among various stakeholders, including educational institutions, government entities, technology providers, and learners.

As the field of blockchain in education is still in its early stages, further research, experimentation, and pilot projects are needed to fully understand its potential impact and address the challenges associated with its implementation. Nevertheless, with the growing interest and investment in blockchain technology, it is expected that its use cases in education will continue to evolve and expand in the coming years.

5.5 Conclusion

In conclusion, the use cases of blockchain in education are diverse and offer significant potential to transform various aspects of the educational landscape. Through the implementation of blockchain technology, educational institutions can enhance data security, reduce fraud, increase transparency, and streamline administrative processes. The use of blockchain in education has the potential to revolutionize the way credentials are verified, stored, and shared, leading to more reliable and efficient systems for academic credentials management.

The evidence presented in this chapter highlights the numerous advantages of blockchain technology in education. Blockchain can enable the creation of digital credentials that are tamper-proof, verifiable, and transferable, which can help to combat issues related to fraudulent credentials and misrepresentation of qualifications. Additionally, blockchain can facilitate lifelong learning and competency-based education through the creation of digital portfolios that provide a comprehensive and transparent record of an individual's skills and achievements. Furthermore, blockchain can enable secure and transparent data sharing among different stakeholders in the education ecosystem, such as students, educators, employers, and policymakers, which can lead to increased collaboration and innovation in the field of education.

The potential impact of blockchain in education is significant. It can empower learners by giving them control over their own credentials and achievements, enable employers to easily verify the qualifications of job applicants, and create more efficient and transparent systems for educational institutions. Blockchain has the potential to disrupt traditional models of education and provide new opportunities for lifelong learning, micro-credentials, and global mobility of credentials.

However, there are also challenges that need to be addressed for wider adoption of blockchain in education. These challenges include technical complexities, regulatory and legal issues, interoperability among different blockchain platforms, standardization, and the need for skilled workforce to implement and manage blockchain systems. Moreover, there may be concerns related to data privacy, security, and ethical considerations, such as the ownership and control of data stored on the blockchain, the potential for bias and discrimination, and the need for informed consent.

Future research in the area of blockchain in education should focus on addressing these challenges and exploring potential solutions. Further studies can investigate the technical aspects of blockchain, such as scalability, interoperability, and security, to enhance the usability and reliability of blockchain-based educational systems. Research can also examine the legal, regulatory, and ethical implications of using blockchain in education to ensure compliance with existing laws and regulations and to protect the privacy and rights of stakeholders. Additionally, more research can be conducted on the social and economic impact of blockchain in education, including the potential for reducing costs, increasing access to education, and improving employability outcomes.

As blockchain technology continues to evolve and gain momentum, it has the potential to revolutionize the field of education, offering new opportunities for stakeholders in the educational ecosystem. With proper planning, implementation, and collaboration among stakeholders, blockchain has the potential to reshape the way credentials are managed, shared, and verified in the education sector, leading to more transparent, efficient, and learner-centric systems.

In conclusion, the use cases of blockchain in education are diverse and offer significant potential to transform various aspects of the educational landscape. While there are challenges to be addressed, the adoption of blockchain in education holds great promise and warrants further research and exploration to unlock its full potential in revolutionizing the field of education.

References

1. Swan, M. (2015). *Blockchain: blueprint for a new economy.* O'Reilly Media.
2. Chen, M., Hao, Y., Huang, L., & Zhao, H. (2020). Blockchain-Based Credential Verification and Fraud Detection in Massive Open Online Courses. *IEEE Trans. Learn. Technol.*, 13(3), 517-528.
3. Ba-Lam Do, Van-Thanh Nguyen, Hoang-Nam Dinh, Thanh-Chung Dao and BinhMinh Nguyen (2022). Blockchain for Education: Verification and Management of Lifelong Learning Data. *Comput. Syst. Sci. Eng.* DOI:10.32604/csse.2022.023508
4. Rawski, P., Uszok, A., & Górski, A. Z. (2019). Blockchain-Based Solutions for E-Learning Systems. In *2019 IEEE Global Engineering Education Conference (EDUCON)* (pp. 1624-1631). IEEE.
5. Barker, T., & DeFreese, J. D. (2019). How Blockchain Technology Can Enhance E-Learning. *Int. J. E-Learn. Distance Educ.*, 34(2), 1-12.
6. Balasubramanian, V., Srivastava, G., & Kundu, A. (2018). Blockchain Technology in Education: A Systematic Review. In *2018 18th IEEE/ACM International Symposium on Cluster, Cloud and Grid Computing (CCGRID)* (pp. 614-621). IEEE.
7. Belfer, K., & Marciano, R. (2019). Blockchain Technology in Education: A Systematic Literature Review. In *2019 IEEE Global Engineering Education Conference (EDUCON)* (pp. 1550-1557). IEEE.
8. Tapscott, D., & Tapscott, A. (2016). *Blockchain Revolution: How the Technology Behind Bitcoin is Changing Money, Business, and the World.* Penguin.

9. Zheng, Z., Xie, S., Dai, H. N., Chen, X., & Wang, H. (2017). An overview of blockchain technology: Architecture, consensus, and future trends. In *2017 IEEE International Congress on Big Data (BigData Congress)* (pp. 557-564). IEEE.
10. Dai, H. N., Zheng, Z., Xie, S., Chen, X., & Wang, H. (2018). Blockchain for Internet of Things: A survey. *IEEE Internet Things J.*, 5(2), 1270-1281.
11. Johnson, A. C., Adams Becker, S., Estrada, V., & Freeman, A. (2019). NMC/CoSN Horizon Report: 2019 K-12 Edition. New Media Consortium/Consortium for School Networking.
12. MIT Media Lab (2021). Blockcerts. Retrieved from: https://www.blockcerts.org/
13. Government of Malta (2021). The National Blockchain Strategy. Retrieved from: https://strategy.mt/en/Pages/National-Blockchain-Strategy.aspx
14. Chen, C., Wu, F., Zhang, Y., & Xu, L. (2020). Blockchain-based academic certificate verification: A case study in China. *Int. J. Inf. Manage.*, 53, 102126.
15. Balasubramanian, V., Asokan, N., & Zohar, A. (2018). Certcoin: A practical blockchain-based certification authority. *Proceedings of the 27th USENIX Security Symposium (USENIX Security 18)*, 1357-1374.
16. Khan, I. A., Shakil, K. A., Yaqoob, I., & Gani, A. (2020). Blockchain for privacy preserving data sharing in e-learning systems. *Comput. Mater. Contin.*, 64(3), 1071-1090.
17. Rawski, M., Juraszek, C., Stachowiak, A., & Podsiadła, P. (2019). Blockchain technology in education: Applications, challenges, and future perspectives. In C. Wang (Ed.), *Blockchain Technology Applications in Education* (pp. 1-30). IGI Global.
18. Mikroyannidis, Alexander; Third, Allan and Domingue, John (2020). A case study on the decentralisation of lifelong learning using blockchain technology. *J. Interact. Media Educ.*, 2020(1) pp. 1–10.
19. Bratina, B., & Bevk, M. (2020). Blockchain Technology in Education: Opportunities and Challenges. *CEUR Workshop Proceedings*, 2706, 15-20.
20. Khan, N., Ahmed, E., Yaqoob, I., & Al-Fuqaha, A. (2020). A Systematic Review of the Use of Blockchain in Education: Challenges and Opportunities. *IEEE Access*, 8, 109094-109107.
21. Rawski, M., Kopeć, K., Rybarczyk, P., & Wilimowska, Z. (2019). The Use of Blockchain in Education: A Systematic Review. *Sustainability*, 11(13), 3646.
22. Badillo-Urquiola, K., Martinez-Lopez, A., & Herrera-Viedma, E. (2019). Blockchain in Education: A Systematic Review. *Sustainability*, 11(7), 2041.
23. R. Pandey, M.S. Faiyaz, G.D. Singh, Z. Uddin (2023). Functional analysis of blockchain consensus algorithms. *Distributed Computing to Blockchain Architecture, Technology, and Applications*, (pp. 207-233).
24. Rawski, P., Chen, X., Lang, C., & Varshney, L. (2019). Blockchain and lifelong learning: Opportunities for the developing world. In *IEEE International Conference on Advanced Learning Technologies (ICALT)* (pp. 344-345). IEEE.

25. Dinh, T. T. A., Liu, R., Zhang, M., Chen, G., Ooi, B. C., & Wang, J. (2018). Untangling blockchain: A data processing view of blockchain systems. *IEEE Trans. Knowl. Data Eng.*, 30(7), 1366-1385.
26. Koulu, R., Kärkkäinen, T., & Smedlund, A. (2017). Blockchain technology as an enabler of service innovation in the healthcare industry. In *Proceedings of the 50th Hawaii International Conference on System Sciences (HICSS)* (pp. 2360-2369). IEEE.
27. Hada A. Alsobhi, Rayed A. Alakhtar, Ayesha Ubaid, Omar K. Hussain, Farookh Khadeer Hussain (2023) Blockchain-based micro-credentialing system in higher education institutions: Systematic literature review. *Knowledge-Based Syst.* Volume 265, 110238.
28. Varshinee Chukowry, Geeaneswari Nanuck, Roopesh Kevin Sungkur (November 2021). The future of continuous learning–Digital badge and microcredential system using blockchain.
29. Anh Nguyen, H., Kim, J., & Kim, J. (2019). Blockchain for Education: State of the Art and Challenges. *Computers*, 8(2), 26.
30. Educause. (2020). 7 Things You Should Know About Blockchain. https://library.educause.edu/resources/2020/8/7-things-you-should-know-about-blockchain
31. Buchanan, W., *et al.* (2019). Blockchain in Education. In *Handbook of Blockchain, Digital Finance, and Inclusion*, Volume 2: ChinaTech, Mobile Security, and Distributed Ledger (pp. 617-652). Springer.
32. CSES. (2018). *Blockchain in Education.* European Parliament, Policy Department for Structural and Cohesion Policies, Brussels.
33. Accredible. (n.d.). Blockchain Certificate Verification. Retrieved from https://www.accredible.com/blockchain-certificate-verification/
34. Yli-Huumo, J., Ko, D., Choi, S., Park, S., & Smolander, K. (2016). Where is current research on blockchain technology? A systematic review. *PLoS One*, 11(10), e0163477.
35. Christidis, K., & Devetsikiotis, M. (2016). Blockchains and smart contracts for the internet of things. *IEEE Access*, 4, 2292-2303.
36. Buchanan, W., Esfandiari, B., & Liggins, L. (2019). Blockchain: Challenges and opportunities for critical social research. *Big Data Soc.*, 6(1), 2053951719842540.
37. Wang, X., Chen, Y., & Wang, L. (2019). Application of Blockchain in Education: A Systematic Review. *Front. Psychol.*, 10, 2689.
38. Gronlund, A., Al-Saad, S., & Hatakka, M. (2020). Blockchain in Higher Education: A Systematic Review of Challenges and Opportunities. *Comput. Hum. Behav.*, 110, 106385.
39. Zhu, R., Wu, D., Huang, X., & Wang, L. (2019). A Blockchain-Based Platform for Authenticating Digital Certificates. In *2019 18th IEEE International Conference On Trust, Security And Privacy In Computing And Communications/13th IEEE International Conference On Big Data Science And Engineering (TrustCom/BigDataSE)* (pp. 657-664). IEEE.

40. Sintema, E. J., & Fallahkhair, S. (2019). Blockchain-Based Authentic ation of Academic Credentials. In *2019 IEEE Global Engineering Education Conference (EDUCON)* (pp. 1421-1426). IEEE.
41. Open Badges, n.d. https://openbadges.org/
42. Learning Machine, n.d. https://www.hylandcredentials.com/
43. Sony Global Education, 2018 https://www.sony.com/en/SonyInfo/News/Press/201708/17-071E/
44. Southern New Hampshire University, n.d. https://www.snhu.edu/
45. The Learning Economy Foundation, n.d. https://www.learningeconomy.io/
46. Peer Community In, n.d. https://rr.peercommunityin.org/
47. Malta Ministry for Education and Employment, 2017 https://spectrum.ieee.org/malta-pilots-blockchainbased-credentials-program
48. University of California, Berkeley, 2019 https://blockchain.berkeley.edu/
49. Holochain, n.d. https://www.holochain.org/
50. University of Nicosia, n.d. https://www.mastersportal.com/universities/545/university-of-nicosia.html

6

Pattern Recognition Applications in Distributed Systems and Distributed Machine Learning

Kalyani S.[1*], Swarup K. S.[2], Sandhya Avasthi[3] and Tanushree Sanwal[4]

[1]*Subject Matter Expert (Electrical), L&TEduTech, Larsen & Toubro Limited, Chennai, Tamilnadu, India*
[2]*Department of Electrical Engineering, Indian Institute of Technology Madras, Chennai, India*
[3]*Department of CSE, ABES Engineering College, Ghaziabad, Uttar Pradesh, India*
[4]*KIET Group of Institutions Delhi-NCR, Ghaziabad, India*

Abstract
Identifying patterns and structures within data constitutes a fundamental problem across various disciplines, facilitating informed decision-making processes. Distributed computing has emerged as a crucial enabling technology to effectively solve these challenges. The reason for this phenomenon can be attributed to the rapid growth of data and the intricate nature of modern pattern recognition algorithms. Distributed computing harnesses the collective computational capabilities of numerous interconnected nodes to facilitate the parallel processing and analysis of large datasets. This methodology provides scalable, efficient, and prompt resolutions for issues pertaining to pattern recognition. As the data grow for pattern recognition application, efficient machine learning techniques are need to process such huge data in a distributed manner. Distributed system, a powerful technology, is formed by multi-computers connected via a network and communicate with each other and share their hardware or software components. The performance efficiency of distributed systems is improved by machines that remotely work on a specific task. Some of the prominent features of a distributed computing approach includes fault tolerance, agility, cost convenience, and resource sharing. Distributed computing finds applications in a variety of fields that include Internet of Things, real-time process control, speech recognition, GPS, etc. The chapter

**Corresponding author*: skalyani.mdu@gmail.com

Sandhya Avasthi, Suman Lata Tripathi, Namrata Dhanda and Satya Bhushan Verma (eds.) Decentralized Systems and Distributed Computing, (117–144) © 2024 Scrivener Publishing LLC

focuses on elucidating the advantages of distributed computing in effectively managing large datasets and building intricate models, particularly machine learning methods employed for jobs involving pattern recognition.

***Keywords*:** Pattern recognition, feature selection, classifier, distributed computing, distributed machine learning

6.1 Introduction

The rate of data collection is increasing due to rapid growth in new internet technologies, Internet of Things, use of mobile devices, and applications. All these applications are data-intensive applications; the data generated can be used in applications such as pattern recognition and classification. Machine learning (ML) methods can be employed to perform these tasks and evaluate the results. The results are useful in various decision-making activities. Controlling self-driving cars [1], recognizing speech [2], and forecasting customer behavior [3] are a few examples. Due to the extended training period, solution designers are sometimes forced to create distributed systems in order to maximize parallelization and overall I/O bandwidth. A centralized solution is often unfeasible because the data are too vast or organically distributed to fit on a single computer. Processing transactions involving data kept at different locales in larger businesses [4] and astronomical data that are too massive to transmit and organize [5] are two examples.

6.1.1 What is Pattern Recognition?

Pattern recognition is a method that possesses the capacity to classify objects. The objects under consideration are frequently referred to as the all-encompassing phrase "patterns." Pattern recognition, often known as PR, encompasses the cognitive mechanism through which raw data are received and subsequently utilized to make informed decisions or initiate actions depending on the categorization of said data. The main aim here is to use prior knowledge or statistical information to obtain patterns and then classifying them as per certain parameters. Most of the research works in the field of pattern recognition focus on PR as a classification scheme. The classification or description scheme can be either statistical (decision theoretic) or structural (synaptic). There exist two types of Pattern Recognition Techniques: the first is Statistical and the second is Structural. The Statistical method recognizes classes utilizing statistical characterization of patterns, whereas the Structural method recognizes structural relationship among input features. This chapter provides conceptual details about statistical pattern recognition methods and process.

6.1.2 Distributed Computing Environment

For all types of application such as self-driving cars or smart transportation systems, massive amount of data need to be processed efficiently in real time, which depends on commodity servers. The commodity servers are relatively small as compared to an expensive centralized server and have a small storage size. This is a better strategy where a large number of limited capacity servers are used offering a higher bandwidth. This is beneficial for data-intensive applications due to use of data ingestion. Pattern recognition is one of the widely recommended tools in the field of engineering and research. Some applications that utilize pattern recognition as solutions are ethnology, cognitive science, engineering, forensics, computer vision, and statistics. Applications such as speech recognition are capable of classifying speech text into different classes or groups, email classification into spam or non-spam classes, recognizing handwritten postal codes, and recognition of human faces automatically [2–5].

6.1.3 Basic Model of the Pattern Recognition System

The basic model of a complete pattern recognition system is depicted in Figure 6.1. It consists of a data capturing system based on in-built sensors that observes behavior and gathers information to be used for classification task, a feature selection or extraction mechanism for selecting most dominant attributes from the observations, and uses a classification technique for the purpose of classifying in a binary or multi-class mode.

To perform a pattern recognition task, three basic steps are required: preprocessing, feature selection, and classification algorithm design [3]. In the first stage, depending on the environment within which the objects are to be classified, data are gathered via a set of sensors and preprocessed. The role of preprocessing is to define a compact representation of the pattern. Preprocessing techniques include noise filtering, smoothing, and normalization. In the second stage, a feature space is constituted in order to reduce the space dimensionality. The feature selection/extraction stage selects the optimal feature subsets from large-dimension patterns. Finally, at the third stage, a classifier is constructed based on the selected feature set for the discrimination process. Preprocessing techniques capture a compact representation of pattern acquired from data. The learning strategy for a

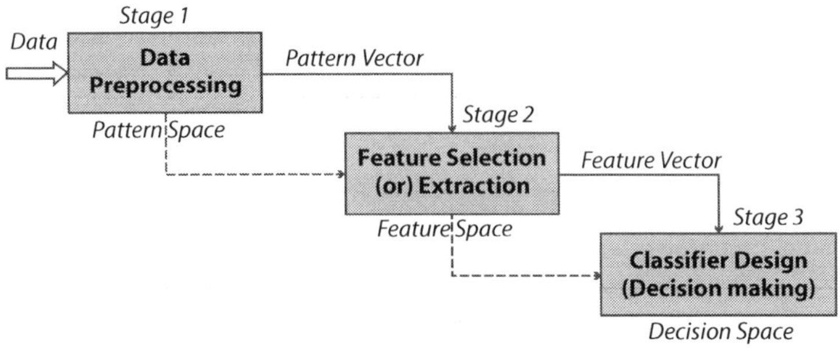

Figure 6.1 Basic model of a pattern recognition system.

classifier design can be either supervised or unsupervised learning. A wide range of learning algorithms can be applied for pattern recognition, from simple Naive Bayes classifiers and KNN decision rules to powerful neural networks and machine learning techniques [4, 5].

6.2 Data Generation and Preprocessing

Any pattern recognition system utilizes data to perform machine learning; the system is mainly based on two steps, the first is training and then building a classifier. The training set generation is performed offline, where a great deal of operating points is considered and their class label is analyzed for classification.

- **Data (Pattern) Generation:** Choosing primary attributes that form a pattern vector is the primary task in any pattern recognition system. The system's overall accuracy depends on the quality of the training set or training data. The training set comprises an input vector and a target vector for learning on a knowledge database. All variables describing an operating condition constitute the components of the pattern vector represented as $X = \{x_1, x_2, \ldots.., x_n\}$. The data patterns needed for training and testing purposes can be obtained online from real-time occurrences or can be synthesized from offline simulations [5]. The training set should contain enough data points representing various classes that will maximize its ability to generalize classes.
- **Data Preprocessing:** Preprocessing is done on the train set and test set before model training in the PR system.

Any existing noise or missing values or any outlier is removed, making the training set good for modeling purposes. Data preprocessing includes cleaning, normalization, filtering the noisy or unreliable data, and data transformation.

- **Data Normalization:** In many real-world problems, the attribute variables have values in different ranges. The attributes having a high value influence cost function more than small values but do not affect the overall significance of the design of the classifier. This problem is overcome by normalizing the attribute (pattern) variables so that their values lie within similar ranges [6]. Data Normalization is a form of data transformation that involves the rescaling of attributes. Data Normalization is otherwise called *data scaling,* where the data samples are scaled to fall within a small specified range. Data scaling is employed to address the challenge posed by large numerical amounts, ensuring that no single property overwhelms the others. Additionally, it serves to mitigate any numerical complications during calculations.

6.3 Feature Selection

6.3.1 What is Feature Selection?

To reduce dimensions from training data, feature selection methods are utilized, and building a good classification model is required [7]. Generally, the feature selection step simplifies the need for required resources to represent a large dataset. For pattern recognition application, the data are huge with a large number of attributes and so handling the data becomes quite difficult. Overall memory requirement and computational power are high in such data, often leading to overfitting issues and failure to generalize unseen test samples [33]. In machine learning and statistics, the feature selection process chooses a subset of original features removing the redundant and irrelevant features for building robust learning models [8]. A feature vector represented as $Z = \{z_1, z_2, \ldots.., z_m\}$ is constructed using selected features. It is important to acknowledge that the value of m is smaller than n, where m represents the number of components in the feature vector and n represents the number of components in the pattern vector [33]. The PR system classification accuracy and processing times is superior if the feature set used for modeling purposes is good. A good feature set should be insensitive to disturbances, robust, and easily processed. The systematic method for feature selection are mainly of two types: wrapper methods and filter methods; these methods are described in the following sections [9].

6.3.2 Filter Models of Feature Selection

Some filter methods for the feature selection process are Chi-Squared Test, Pearson Correlation Coefficient, T-test, and Relief Algorithm. The procedure of feature selection carried out by each of these filter models is described in this section.

(a) T-Test (T-Statistics)

The most commonly employed strategy for feature selection is the filtering method. This feature selection strategy is more appropriate for binary classification problems. The objective of the Fisher Linear Discriminant approach [10] is to determine the optimal linear separation for two classes of data. The methodology employed in this approach is characterized by its simplicity and lack of reliance on data mining methods. Instead, it leverages the inherent characteristics of the data. The attributes within the pattern vector are sorted and assigned scores based on statistical estimations of the mean and standard deviation. For each pattern attribute x_j, the mean $\mu_j^{(1)}$ ($\mu_j^{(0)}$) and standard deviation $\sigma_j^{(1)}$ ($\sigma_j^{(0)}$) are calculated using the samples belonging to binary 1 (binary 0) classes. Then, a figure of merit called the T-score or Fisher's score is obtained for each attribute as given by Equation 6.1. This score provides an easy and straightforward measure of information content in each attribute [11].

$$T(x_j) = \frac{\left|\mu_j^{(1)} - \mu_j^{(0)}\right|}{\sigma_j^{(1)} + \sigma_j^{(0)}} \tag{6.1}$$

While making feature selection, we can simply evaluate T-score for all attributes and extract those attributes with high scores as discriminatory features. However, this mode of selection does not yield an optimal feature subset. Hence, it is required to follow an iterative procedure to obtain a better-discriminating feature set. One such iterative procedure is called the "Single Ranking" Method. The Single Ranking algorithm is described here; the method is an iterative process and is designed based on the most dominant pattern attribute.

SR ALGORITHM

Step 1) Compute T-score for all attributes in pattern vector in training set using Eq. (5).

Step 2) Start with an empty feature vector Z = {} and set feature count, k=1.

Step 3) Select the attribute with highest T-score and store it as the first feature Z_1 in the feature vector Z.

Step 4) Determine the correlation coefficients between the selected feature and other attributes in the train set pattern vector. (ρ_{kj}, j=1,2...n ; j≠k)

Step 5) Discard all variables highly correlated to Z_k (say with $\rho_{kj} \geq 0.8$) and hence modify the pattern vector.

Step 6) Increment the feature count, k=k+1

Step 7) Now, from remaining pattern attributes, select the one with largest T-score and assign to the feature Z_k.

Step 8) Repeat from Step 4 until required number of features has been extracted.

(b) Chi-Squared Test (χ^2-Statistics)

The relationship between class and input properties can be evaluated by employing the chi-squared statistic. Before applying this strategy, it is necessary to discretize the range of each characteristic into a specific number of intervals. One option to achieve this is by employing a method such as the entropy-based discretization approach [12]. The χ^2 value of an attribute is defined by Equation 6.2.

$$\chi^2(x_j) = \sum_{i=1}^{m}\sum_{j=1}^{k}\frac{(A_{ij}-E_{ij})^2}{E_{ij}} \quad where \quad E_{ij} = \frac{R_i \times C_j}{N} \tag{6.2}$$

In the above expression, m is the number of intervals, k is the number of the distinctive classes, A_{ij} is the number of samples in the ith interval and belonging to jth class, R_i is the number of samples in the ith interval, C_j is the number of samples belonging to the jth class, E_{ij} is the expected frequency of A_{ij}, and N is the total number of data samples. After calculating the chi-squared (χ^2) value of all attributes in the pattern vector, they are sorted in descending order. The significance and discriminatory power of a certain feature increase as the chi-squared value of the attribute becomes larger. Within the hierarchical arrangement of attributes, a subset comprising 20% to 30% of the total attributes is chosen to form an ideal feature set.

(c) Pearson's Correlation Coefficient (PCC)

The approach devised by Karl Pearson offers insights into both the magnitude and the orientation of the correlation. Pearson's correlation coefficient is computed by dividing the covariance of two variables by the sum of their standard deviations [13]. The Pearson correlation coefficient (PCC) quantifies the degree of correlation between the attributes in the pattern vector and the class label (category) vector. Pearson's linear correlation coefficient is defined by Equation 6.3.

$$\rho(X,C) = \frac{E(X,C) - E(X)E(C)}{\sqrt{\sigma^2(X)\sigma^2(C)}} = \frac{\sum_i (x_i - \overline{x_i})(c_i - \overline{c_i})}{\sqrt{\sum_i (x_i - \overline{x_i})^2 (c_i - \overline{c_i})^2}} \tag{6.3}$$

$\rho(X,C) = \pm 1$ if X and C are linearly dependent and zero if they are completely uncorrelated. The error function is given by Equation 6.4.

$$P(X) = erf\left(\rho(X,C) \times \sqrt{N/2}\right) \tag{6.4}$$

The pattern characteristics X are organized in descending order based on their respective values of P(X), which serves as a mechanism for ranking the features. For categorization purposes, a subset of variables comprising the first 20%–30% of the organized list is selected as discriminating qualities.

(d) Relief Algorithm

The utilization of a statistical approach for the identification of pertinent features is employed in the relief algorithm [14]. The technique is built on the concept of feature weighting, drawing inspiration from instance-based learning algorithms. The relief method identifies occurrences classified as "Near Hit" or "Near Miss" using a Euclidean distance measure. The term "Near Hit" refers to the specific instance inside a class that has the smallest Euclidean distance when compared to all other examples within the same class. On the other hand, "Near Miss" refers to the instance that has the smallest Euclidean distance when compared to all instances belonging to a different class. The weights of characteristics, which are initially set to zero, are adjusted according to a logical notion that an attribute holds greater significance if it effectively differentiates between an instance and its Near Miss, while it holds lesser significance if it differentiates between an instance and its Near Hit.

RELIEF ALGORITHM

Relief (TrainData, NoSamples)

(1) Set T = φ (empty feature set)

(2) Initialize Weights of all features, W_k (k = 1,2 .... m) to zero

(3) For i = 1 to NoSamples

Find the "*Near Hit*" and "*Near Miss*" of the ith instance

For j = 1 to n

$$[W_j]=[W_j]-\frac{\left([X_j]-[Near\,Hit_j]\right)^2}{j}+\frac{\left([X_j]-[Near\,Miss_j]\right)^2}{j} \tag{6.5}$$

end jth and ith loop

(4) Collect first 20%–30% of the attributes with the highest weights and append to T.

Return T

Advantages: Filter models are commonly characterized by a non-iterative calculation of the dataset, which renders them both straightforward and efficient in terms of execution. The scalability of high-dimensional datasets is a straightforward task, whereas the significance of the classification process is negligible. In contrast to alternative methodologies for feature selection, this particular methodology has a reduced computational complexity. The focus of filters is primarily on the intrinsic attributes of data rather than their interaction with a particular classifier. Consequently, the outcomes derived from filters possess a higher degree of universality, thereby rendering them advantageous for a wide range of classifiers within a given family.

Disadvantages: Although filter techniques are highly computationally efficient, they show little resistance to feature interactions and redundant data. An important problem that needs attention is finding a suitable ranking threshold that effectively separates important features from unimportant noise. The dependency of features and the interaction of the classifier were neglected.

6.3.3 Wrapper Models of Feature Selection

In the context of feature selection, a wrapper model involves the utilization of a search algorithm to explore the space of potential features. Each subset of features is afterwards assessed by executing a model on it. A search approach is employed within the domain of potential feature subsets, wherein several feature subsets are constructed and evaluated. This approach is tailored to a certain classification algorithm since it employs a distinct classification model that has undergone training and testing to assess a specific group of attributes. The classification model is subsequently included by a search strategy to explore the entire space of feature subsets [15].

The wrapper method requires (a) a search strategy to select candidate subsets and (b) an objective function to evaluate these candidates as shown in Figure 6.2. The objective function is a pattern classifier, which evaluates the feature subsets by their predictive accuracy and returns a measure of their "goodness", a feedback signal, used by the search strategy to select new candidates. The search methodology used in this work may be divided into two categories: a stochastic strategy-based method and a greedy strategy-based method, which both use sequential algorithms and decision trees, respectively.

(a) Sequential Search Algorithms

The sequential search algorithm-based wrapper model employs a greedy strategy approach to add or delete features from a candidate subset, guided

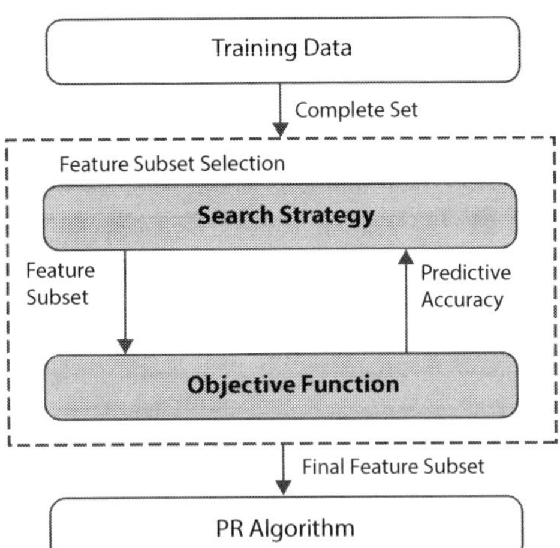

Figure 6.2 Feature selection process of the wrapper model.

by the evaluation of a specific criterion. The criteria function of this model adheres to a standard loss measure, which encompasses the summation of squared errors (mean-squared error) for regression models and the count of misclassified data (misclassification rate) for classification models. Sequential Forward Selection (SFS) and Sequential Backward Selection (SBS) are often utilized as sequential search methods (Reference [9]). The container model that employs the sequential search algorithm employs a greedy strategy. It includes or excludes characteristics from a subset of candidates based on the evaluation of a particular criterion. The Sequential Backward Selection (SBS) algorithm begins with a set of all features and iteratively eliminates the least promising ones until there is no further improvement in the criterion. When the efficacy of the learning algorithms falls below a specified threshold, the Sequential Backward Selection (SBS) procedure is terminated. The Sequential Backward Selection (SBS) method has difficulty accurately determining the impact of each variable on the objective variable when dealing with high-dimensional data. This could lead to the untimely elimination of potentially useful predictors. Sequential Feature Selection (SFS) is an alternative method that begins with an initial vacant set of features and iteratively selects one feature at a time. This procedure begins with the most promising feature and continues until there is no further improvement in the criterion function. In comparison to the SBS method, the SFS approach to feature selection is regarded as more effective and widely preferred.

SFS Algorithm:

Step 1) Start with an empty set $Z_0 = \{\ \}$

Step 2) Set the feature count, $k = 1$

Step 3) Select the best pattern attribute X^* giving minimum error rate, $X^* = \min_{x \notin Z_k} [\ J\ (Z_{k-1} + X^*)\ ]$.

Step 4) Set $Z_k = Z_{k-1} + X^*$; $k = k + 1$.

Step 5) Move to Step 2 until there is no further decrease in criterion function J.

SBS Algorithm:

Step 1) Start with the full set $Z_m = \{X\}$

Step 2) Set the feature count, $k = m$

Step 3) Select the wort pattern attribute X^* giving maximum error rate $X^* = \max_{x \notin Z_k} [J\ (Z_k - X^*)]$.

Step 4) Set $Z_{k-1} = Z_k - X^*$; $k = k - 1$.

Step 5) Move to Step 2 until there is no further increase in criterion function J.

(b) Genetic Algorithm Based Feature Selection

The Genetic Algorithm (GA) can be classified as a stochastic heuristic search technique. In the context of feature selection problems, the process of sampling in feature space is typically carried out using a population-based probabilistic or random approach [16]. An elementary genetic algorithm encompasses the following stages: the formation of an initial population, the utilization of tournament selection for reproduction, and the implementation of crossover and mutation operations with probabilities set by the user. Every individual (chromosome) is a binary string that serves as a representation of the input feature selection. A population of chromosomes is randomly created. Each chromosomal unit, also known as a gene, represents the presence or absence of the jth attribute.

Crossover and mutation, two often employed genetic processes, are utilized to generate offspring for reproduction. The assessment of an individual's fitness relies on the prediction accuracy of a designated learning algorithm. The fitness evaluation function is mostly employed in the

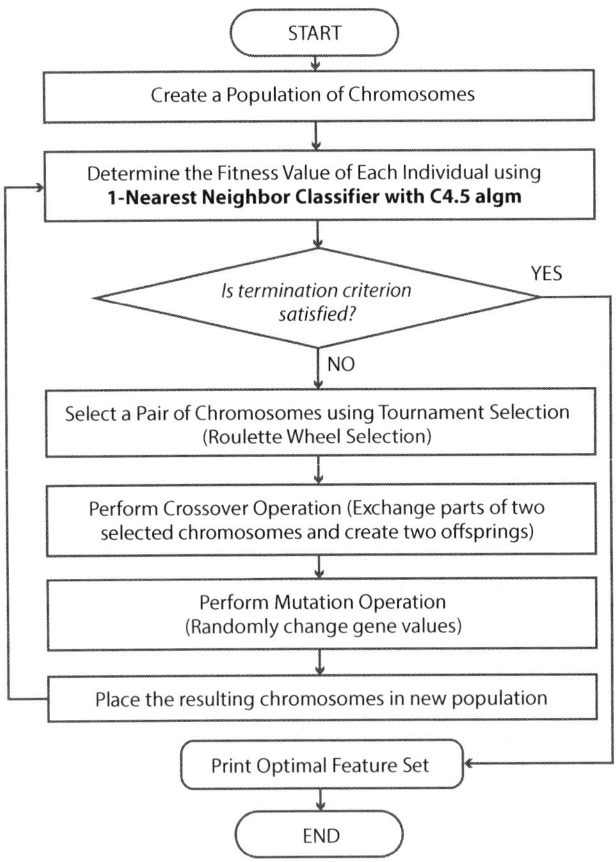

Figure 6.3 Flowchart for genetic-based feature selection algorithm.

K-Nearest Neighbor (KNN) method along with the C4.5 algorithm. This approach is utilized to assess the effectiveness of each particular feature subset in terms of quality. Figure 6.3 depicts the flowchart for genetic algorithm based feature selection process.

(b) Decision Tree Based Feature Selection

The Decision Tree (DT) is a well-known method in the field of machine learning that depicts categorization principles using a knowledge representation approach. The process of feature selection in decision trees entails an intrinsic evaluation of the suitability of characteristics for distinguishing objects belonging to distinct classifications. In the realm of decision trees, each branch is composed of a series of interconnected nodes extending from the root node to a terminal node, also known as a leaf. In addition, it is noteworthy that each node in this structure represents a particular characteristic.

DT Algorithm

Step 1) Construct DT (T) using X_L (Train Data Matrix) and Y_L (Class Label) based on *Gini Splitting* rule and *C4.5 algorithm*.

Step 2) Compute the Balanced Error Rate (BER) for T given by Eq. (10).

$$BER = \frac{1}{k}\sum_{i=1}^{k}\frac{F_i}{T_i + F_i} \tag{10}$$

T_i number of cases actually in class k and classified correctly by T.

F_i number of cases actually in class k and misclassified by T.

k number of classes.

Step 3) For each non-leaf node k, prune the kth node and call the pruned tree as T`. Compute BER for the pruned tree T` and compute the node importance index, $n_{im}(k)$ = BER(T`) - BER(T).

Step 4) N_{tree} is set of all non-leaf nodes in T.

Step 5) Find the node with maximum n_{im} and its corresponding attribute, F_{temp}. Append F_{temp} and its corresponding n_{im} to Feature Vector Z and remove F_{temp} (used nodes) from N_{tree}. Repeat this step until N_{tree}=NULL.

Step 6) Return the list of features in feature vector Z in increasing order of n_{im}( ). From the rearranged Feature Vector Z, select those features with n_{im} greater than a threshold as the optimal feature subset.

The significance of an attribute is indicated by the presence of a particular characteristic within a tree structure. Multiple values are associated with the branches, resulting in the division of a particular property.

This capacity can be utilized promptly for the purpose of conducting feature selection [17]. The provided pseudo-code describes the feature selection algorithm based on a decision tree.

Advantages: Wrapper methods are considered to be more successful than filter methods due to their classifier-specific nature. Filter models are outperformed by it in terms of accuracy, as they are specifically optimized for the interactions between the classifier and dataset. The model exhibits a strong generalization capability, as it commonly employs cross-validation techniques to assess its forecast accuracy. The models incorporate dependencies and engage in interactions with the classifier.
Disadvantages: Complexity and high time complexity are characteristics of wrapper methods, which may contribute to the problem of overfitting. In general, the procedure has limited scalability and a relatively moderate rate. Moreover, it is susceptible to becoming stuck in a locally optimal solution. Due to its dependence on the prejudice of the classifier used in the evaluation function, the solution's generality is limited. This results in the optimal feature set being more specific to the constructed classifier.

6.4 Design of Classifier

Classification is a fundamental method used to divide the feature space into distinct regions, where each region corresponds to a specific category or class designation [18]. Hence, the classifier serves as the delineation between distinct categories. The diagram given in Figure 6.4 illustrates a standard representation of a classifier function. It showcases a prominent boundary line that effectively separates data points belonging to two unique categories, namely, cats and dogs. This depiction is based on a two-dimensional feature space and is referenced from a scholarly source. The primary objective of the classifier is to assign each data point within the feature space to one of the available classes. The feature set obtained from the feature selection stage serves as the input database for the classifier. The foundation of classifier design is in the construction of the design (training) set. This collection has operational components from various unique classes. When assessing unlabeled test set samples, a classifier model's primary criterion for evaluation is achieving a high level of classification accuracy while minimizing the occurrence of misclassifications. An effective pattern classifier should possess several key attributes, including online adaptability, nonlinear separability, the capability to manage overlapping

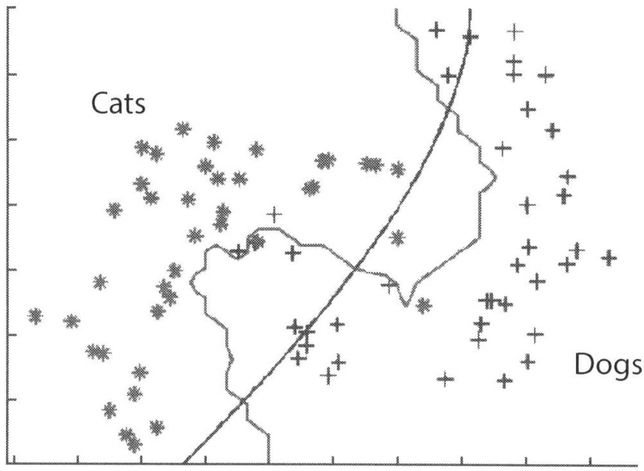

Figure 6.4 Typical example of a binary classification system.

classes, prompt decision-making, and the capacity to create both soft and hard conclusions.

Classifiers are commonly constructed using labeled data, therefore denoting these difficulties as Supervised Classification. There are several frequently employed methods for supervised pattern categorization, such as the closest neighbor rule, Bayes' maximum likelihood classifier, and the perceptron rule. The issues associated with constructing classifiers using unlabeled data are sometimes referred to as unsupervised classification [18]. Unsupervised classification is a methodology that entails the examination of a considerable volume of unannotated data to discern and classify them into discrete categories, relying solely on intrinsic patterns or groups. Various clustering techniques, such as k-means, are commonly employed as preliminary steps in the development of classifiers for unsupervised classification tasks.

6.4.1 Conventional Classifiers in Distributed Environment

(a) Naive Bayes Classifier

In practical problems, most data are noisy, and so classes produced may overlap in feature space. In modeling, the feature values are random values applied in the probabilistic approach and the most common is the Bayes Maximum likelihood classifier that groups utilizing likelihood function values. The Bayes classifier is a fundamental probabilistic classifier that relies on the utilization of Bayes' theorem, derived from Bayesian statistics while assuming strong (naive) independence. Naive Bayes classifiers can handle a wide range of independent variables, including both continuous and categorical data. The predictor variables set, $S = \{S_1, S_2, \ldots, S_d\}$, each of dimension m; we want to construct the posterior probability for the event C_j among a set of possible outcomes $C=\{C_1,C_2,\ldots,C_d\}$. Using Bayes' rule:

$$p(C_j \mid S_1,S_2,\ldots,S_d)\ \alpha\ p(S_1,S_2,\ldots,S_d \mid C_j)p(C_j) \tag{6.6}$$

where $p(C_j \mid S_1,S_2, \ldots, S_d)$ is the posterior probability of class membership, i.e., the probability that S belongs to C_j. Given the assumption of statistical independence among the conditional probabilities of the independent variables, Naive Bayes allows for the decomposition of the likelihood into a product of terms.

$$p(S \mid C_j)\alpha \prod_{k=1}^{d} p(S_k \mid C_j) \tag{6.7}$$

The posterior can be rewritten as:

$$p(C_j \mid S)\alpha\ p(C_j)\prod_{k=1}^{d} p(S_k \mid C_j) \tag{6.8}$$

A new case S_{new} can be assigned to a class level C_j that optimizes the posterior probability by using the Bayes method outlined above. The Naive Bayes Classifier method is particularly effective when the inputs are highly dimensional. Naive Bayes has been seen to routinely beat more complicated classification techniques, even though it appears to be simple. By reducing the task of predicting high-dimensional density to that of estimating one-dimensional kernel density, Naive Bayes efficiently makes it simpler.

The utilization of the Naive Bayes Classifier inside distributed systems necessitates the resolution of several issues, including but not limited to data partitioning, communication, synchronization, and scalability. The Naive Bayes algorithm is a probabilistic method that computes the likelihood of a data point being assigned to a specific class based on its observed properties. This section outlines the process of adapting the Naive Bayes algorithm for distributed systems.

(b) K-Nearest Neighbor Classifier

The KNN classification is a well-known decision rule based on distance functions widely used in pattern classification problems. K-Nearest Neighbor (K-NN) is an instance-based learning algorithm for classifying objects based on closest training examples in the feature space [19]. The proximity of an unknown pattern to a class serves as a measure for its classification. The k-nearest neighbor classification rule consists of determining the k-nearest neighbors to unseen sample x and uses the majority voting principle to assign the class label. It is imperative to maintain the grouping of related data components on a single node when distributing data across multiple distant nodes. Data partitioning can be performed using several methods, such as random partitioning, partitioning based on keys, or other strategies that aim to maintain data proximity.

K-NN ALGORITHM

Step 1) Assume parameter k, the number of nearest neighbors.
Step 2) Calculate the distance between unlabeled (unseen) new test sample X_{new} and all samples in the training set.

$$DM_i = D(X_{new}, X_i) \quad i = 1,2,...N_{train} \tag{6.9}$$

DM can be Euclidean or ***Mahalanobis or any other distance*** measure.
Step 3) Sort the distance array DM and determine K (first) nearest neighbors of new query instance and gather the category (class label) of the k nearest neighbors.
Step 4) The class label of these nearest neighbors with a majority vote is assigned for the new unseen sample.

In a distributed environment, to apply kNN method, the training data are distributed across the network clusters using a hashing algorithm. The objective is to identify k neighbors so that predictions can be made. It is typical for partition-based hashing hashes to comprise multiple elements. Therefore, when seeking a particular partition, it is probable that all the required points can be obtained by matching the query. It is possible to utilize neighboring data points when the required number of k points is not met. As stated previously, this dataset is more suited for partitioning than clustering [20].

(c) Least Squares Classifier

The implementation of the Least Square Classifier, also known as the Linear Regression Classifier, in a distributed environment requires the partitioning of data across multiple nodes, the execution of local calculations, and the aggregation of results to generate the final model. This paper provides a thorough analysis of the adaptation of the Least Square Classifier to a distributed scenario [21, 22]. The classifier function represented by S (z) can be built by assuming a specific form and finding the correct coefficients of this function by minimizing the estimated average risk of misclassification. A simple classifier function may take the form of a linear first-order function of selected features as defined by Equation 6.10.

$$\begin{aligned} S(z) &= w_0 + w_1 z_1 + w_2 z_2 + \ldots + w_m z_m \\ &= w_0 + \sum_{i=1}^{m} w_i z_i \end{aligned} \tag{6.10}$$

The coefficients w_0, w_1, w_2 ......, w_m are called weighting coefficients or simply "*weights*" of discriminating features. The process through which the weights are determined is called training. Least Squares is a simple training algorithm, which determines the weighting coefficients w = {w_0, w_1... w_m} by Multiple Linear Regression analysis. Once the weights are determined, the S (z) value is computed for any new unseen sample using Equation 6.10. The new pattern is labeled appropriately after rounding the S (z) value to the nearest integer.

The linear regression model should be trained on each node using the data that are specifically accessible on that node. The local model will acquire knowledge of a linear function to make predictions about the class labels, based on the observed attributes. To ensure the consistency of the models, it is important to establish regular communication and synchronization of the model parameters across the dispersed nodes. The implementation of this phase is necessary to ensure that all nodes achieve convergence towards a consistent global model.

6.4.2 Neural Network Classifiers in a Distributed Environment

In a distributed environment, neural network classifiers function by distributing the training data and the neural network model across multiple nodes or computational resources. The goal is to leverage parallelism and distributed computational capabilities to accelerate the training process and efficiently manage large datasets [23, 24].

(a) Multilayer Perceptron (MLP) Classifier

The simplest linear classifier is the perceptron, which is a single-layer network. The single-layer perceptron model's main drawback is that it can only resolve issues that can be separated along linear axes. However, most issues in the real world are incredibly nonlinear [25, 26]. It maps input datasets to output datasets using a feed-forward neural network model called MLP, and three or more layers of neurons (nodes) and nonlinear activation functions; it is a variation on the traditional linear perceptron. Therefore, because it can distinguish non-linearly distinct data, it is more effective than the perceptron.

Multilayer Perceptrons (MLPs) are versatile non-linear models that are commonly employed in scenarios where the relationship between the predictor and response variables is not fully comprehended. The fundamental structure of an MLP network is illustrated in Figure 6.5. The network consists of input, output, and hidden layers composed of nodes that exhibit nonlinear activation. Every individual node inside a given layer establishes connections with all other nodes present in the subsequent layer, whereby

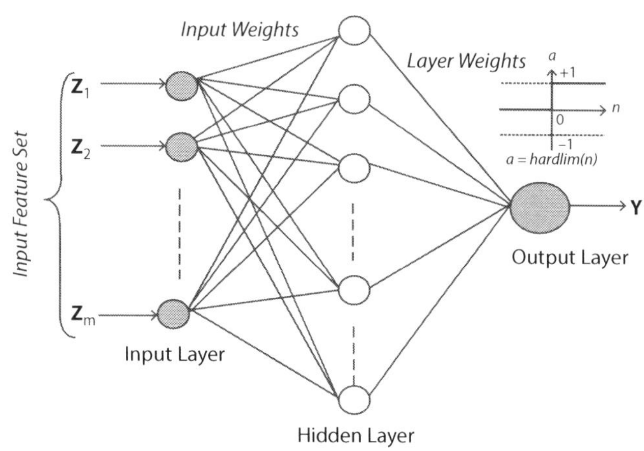

Figure 6.5 Structure of the multilayer perceptron (MLP) network.

each connection is assigned a specific weight denoted as w_{ij}. The quantity of input features selected is equivalent to the number of neurons present in the input layer. Following a series of experimental trials, the determination is made regarding the optimal configuration of hidden layers and the number of neurons within each hidden layer.

The dissemination of network segments across all available devices can accelerate neural network evaluation in a widely deployed distributed environment such as Internet of Things or smart transportation systems. Nonetheless, earlier research on neural network dispersion has indicated that communication overhead quickly becomes the dominant restriction [27–29]. As a result, this approach is primarily used to accelerate the training process on a cluster of nodes connected by a high-speed network. Krizhevsky *et al.* (2012) described a method for training a larger neural network by effectively spreading the network's computational load among two graphics processing units (GPUs). As a result of allowing direct access to each other's memory, the GPUs effectively reduce communication latency.

(b) Learning Vector Quantization (LVQ) Classifier

To discriminate between groups of similar input vectors, several neural network topologies employ a competitive layer. A supervised method of training competitive layer neurons is called Learning Vector Quantization (LVQ) [30, 31]. It is a prototype-based supervised classification approach. An LVQ network's structure is shown in Figure 6.6. It has two layers: a linear layer and a competitive layer. While the linear layer transforms the competitive layer's classes into user-defined target categories, the competitive layer learns to classify the input vectors.

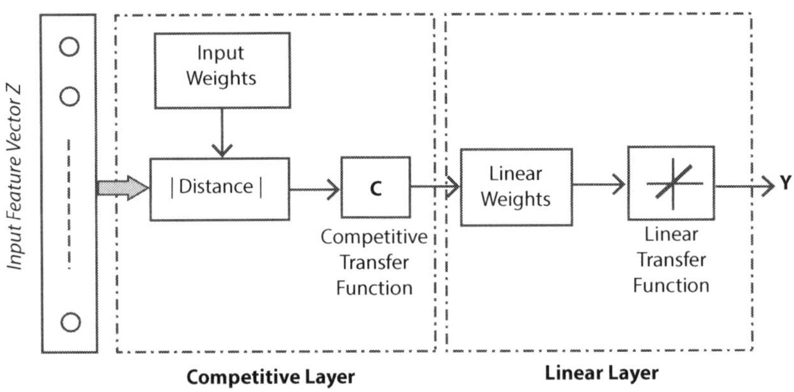

Figure 6.6 Structure of the learning vector quantization (LVQ) network.

The distributed implementation of the Learning Vector Quantization (LVQ) algorithm involves the dividing of data across several nodes. Each node independently trains local models on its respective partitioned data. Subsequently, the results obtained from each node are integrated to obtain the final global model. The training data should be distributed across the different nodes in smaller segments. Each node will utilize its own local data partition for its operations [32]. The prototype vectors are initialized independently on each node, either randomly or through a preset initialization mechanism. Each prototype vector in the neighboring dataset represents a class in the dataset.

(c) Probabilistic Neural Network (PNN) Classifier

Incorporating a Probabilistic Neural Network (PNN) classifier into distributed computing requires partitioning the data, training distinct models on each node, and combining the results to derive the final global model. The PNN classifier is a robust algorithm that uses probability estimation to classify data points, taking into consideration the probability of a data point being assigned to a particular class. On each node, a local PNN classifier is trained using its respective local data. The training procedure for a Probabilistic Neural Network (PNN) involves the construction of a static kernel density estimator for each class, based on training data specific to that class [33, 34].

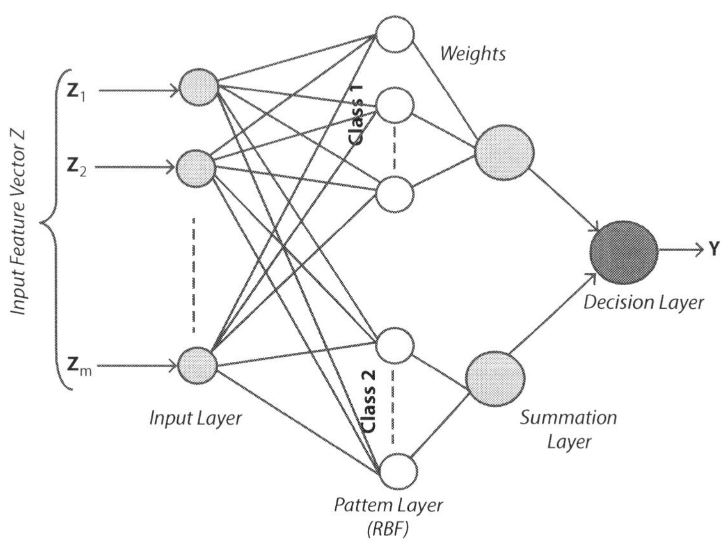

Figure 6.7 Architecture of the probabilistic neural network (PNN).

The probabilistic neural network (PNN) is a multilayer neural network commonly employed in the domain of pattern classification. The probability density function for each data class is determined by the utilization of the PNN's non-linear, nonparametric pattern recognition approach, which relies on the training set and the ideal kernel width value.

The probabilistic neural network's four-layered design is shown in Figure 6.7. The activation function for the buried layer is the Radial Basis Function (RBF), a kernel with an exponential distribution. In the input layer, there is one neuron for each feature variable. A neuron in the buried layer (pattern layer) is used to represent each training sample. Along with the target class, the hidden neurons also store the values of the feature variables. Each target variable category has one neuron in the summation layer. Neurons in this layer determine the weighted vote for every class category. The decision layer, the top layer after the summation layer, evaluates the weighted votes and uses the vote with the highest weight to predict the target category (final output Y).

Limitation of PNN in a Distributed Environment: The process of aggregating probability distributions in PNN classifiers entails the construction of communication connections between distant nodes. Communication overhead may be a bottleneck, particularly in cases involving a high number of nodes or frequent model modifications during training. PNN classifiers may be unable to handle imbalanced datasets effectively in a distributed scenario. Nodes with uneven class distributions may not effectively represent the global data distribution due to the limited scope of operation of each node [35].

6.5 Designing Machine Learning Models for Distributed Environment

When designing a machine learning model for a distributed environment, it is essential to carefully consider data distribution, communication overhead, synchronization, and fault tolerance, among other considerations. In a distributed system, data are frequently distributed across multiple nodes. To assure balanced workloads and reduce transmission costs, develop a method for dividing and distributing data among nodes in an efficient manner. Choose a machine learning model based on the characteristics of your challenge and the distribution of your data. Ensure that the selected model can be effectively distributed and parallelized. Before feeding data into the machine learning model, each node should perform preprocessing. To avoid inconsistencies, the preprocessing techniques should be uniform across all nodes. Communication between dispersed nodes could incur additional overhead [36]. Utilize efficient communication protocols and pattern optimization to reduce the number of data transfers. During distributed training, the model weights between nodes must be routinely synchronized [37]. Choose the synchronization frequency and technique (e.g., parameter server, all-reduce) based on the distributed framework and problem type. Faults are possible in distributed environments. Utilize fault tolerance tools to prevent node failures during training and inference. Checkpointing and redundancy are two techniques that can aid in failure recovery.

6.6 Role of Distributed Computing in Pattern Recognition

Distributed computing is the method of utilizing the combined computational capacity of multiple computers to collaboratively solve a shared problem. Network virtualization entails the consolidation of multiple computer networks into a unified entity that resembles a robust and high-capacity single computer system. This merger facilitates the provision of vast resources capable of tackling complex and demanding tasks. Distributed computing technologies such as Peer-to-Peer (P2P) and Grid computing [38] are among the most significant technological advancements that have had a significant impact on the Internet. One of the numerous methodologies and frameworks available for distributed and grid computing is depicted in Figure 6.8. In the context of distributed systems, especially when large-scale applications are considered, failures or unavailability of individual tasks or nodes are more frequent than not. Therefore, it is essential to implement appropriate failover methods to ensure that distributed calculations are executed without interruption. Pattern recognition is an essential technology in the field of distributed computing because it facilitates the simplification of complex problems and the comprehension of minute particulars.

The utilization of distributed computing is imperative in the field of pattern recognition as it effectively handles and examines extensive datasets through the utilization of the computational capabilities and storage capacity of several devices or nodes. Pattern recognition refers to the process of identifying and categorizing patterns or regularities within data.

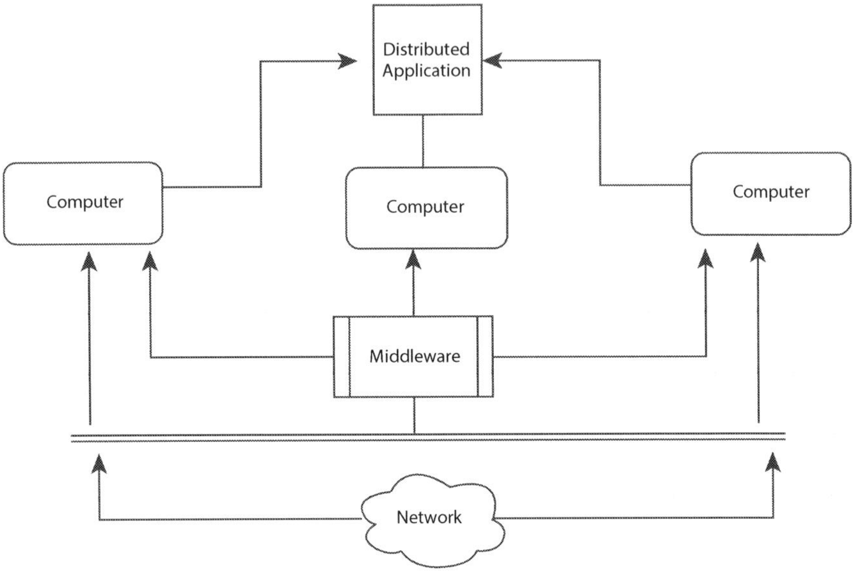

Figure 6.8 A simple concept of distributed computing.

The utilization of distributed computing in this domain offers a multitude of advantages and opportunities [39]. Frequently, pattern recognition tasks require the analysis of large datasets containing high-dimensional features. Through the dissemination of data and computations across multiple nodes, distributed computing facilitates the horizontal scalability of algorithms, thereby enabling the efficient processing of large-scale datasets.

By employing distributed computing methodologies, self-driving vehicles are capable of efficiently handling vast quantities of data, hence enabling instantaneous perception and decision-making capabilities. Enhancing individuals' capacity to maneuver intricate road settings and guaranteeing the safety and reliability of self-driving vehicles are the key benefits of this development. Due to their effectiveness in image identification and object detection tasks, convolutional neural networks (CNNs) have grown significantly in prominence in the field of autonomous cars. The efficiency of learning is increased by distributed computing, which makes it easier to train and infer complex Deep Learning models on distributed GPUs or accelerators [40].

In order to design a voice recognition system within a distributed computing environment, it is necessary to decompose the intricate responsibilities of speech processing into more manageable sub-tasks. These sub-tasks can then be allocated among multiple nodes to enable parallel processing [41, 42].

6.7 Conclusion

With a large selection of solutions that differ in architecture, algorithms, performance, and efficiency, the distributed Machine Learning ecosystem is expanding. Some fundamental challenges need to be overcome before distributed machine learning becomes viable, like finding ways to effectively parallelize data processing while combining the outcomes into a single coherent model. When adding more resources, a common trade-off is to reduce wall-clock time at the expense of total aggregate processing time (i.e., decreased efficiency). When computational resources are sufficiently cheap, rapid training is most useful for many machine learning real-world applications. The fact that this usually necessitates a large increase in total compute resources and associated energy consumption is overlooked as long as a model saves more money than it costs to train. Different techniques available for the feature selection process and classifier design process along with their algorithms have been widely discussed in this chapter. Through distributed computing, a complicated problem can be solved efficiently by utilizing several computers working together by sharing data. This finds wide applications and has a vital role in pattern recognition and machine learning problems. Major features of the distributed system include performance, transparency, scalability, data integrity, high reliability, security, etc. The material presented is intended to benefit the reader in understanding the concepts of pattern recognition in a distributed computing environment and enables them in developing suitable classifier systems for their application domain.

References

1. Badue, C., Guidolini, R., Carneiro, R. V., Azevedo, P., Cardoso, V. B., Forechi, A., ... & De Souza, A. F. (2021). Self-driving cars: A survey. *Expert Syst. Appl.*, *165*, 113816.
2. Ghai, W., & Singh, N. (2012). Literature review on automatic speech recognition. *Int. J. Comput. Appl.*, *41*(8).
3. Wang, X., Zhang, M., & Ren, F. (2018). Learning customer behaviors for effective load forecasting. *IEEE Trans. Knowl. Data Eng.*, *31*(5), 938-951.
4. Verbraeken, J., Wolting, M., Katzy, J., Kloppenburg, J., Verbelen, T., & Rellermeyer, J. S. (2020). A survey on distributed machine learning. *ACM Comput. Surv. (CSUR)*, *53*(2), 1-33.

5. Sen, S., Agarwal, S., Chakraborty, P., & Singh, K. P. (2022). Astronomical big data processing using machine learning: A comprehensive review. *Exp. Astron.*, *53*(1), 1-43.
6. Sergios Theodoridis, Konstantinos Koutroumbas, '*Pattern Recognition*', Academic Press, Fourth Edition, 2009.
7. A.K. Jain, R.P.W. Duin, and J. Mao, 'Statistical Pattern Recognition: A Review', *IEEE Trans. Pattern Anal. Mach. Intell.*, Vol. 22, No. 1, January 2000, pp. 4-37.
8. D. Duttam Ajumdar, 'Trends in Pattern Recognition and Machine Learning', *Def. Sci. J.*, Vol. 35, July 3, 1985, pp. 327-351.
9. Andrew R. Webb, '*Statistical Pattern Recognition*', John Wiley & Sons Ltd, Second Edition, 2004.
10. S.Y. Oh, 'Pattern Recognition and Associative Memory Approach to Power System Security Assessment', *IEEE Trans. Syst. Man Cybern. SMC* - 16, Issue 1, January/February 1986, pp. 62-72.
11. J. Quackenbush, 'Microarray Data Normalization and Transformation', *Nat. Genet.*, Volume 32, 2002, pp. 496-501.
12. I. Guyon, A. Elisseeff, 'An Introduction to Variable and Feature Selection', *J. Mach. Learn. Res.*, Vol. 3, 2003, pp. 1157–1182.
13. H. Liu, Y. Lu, 'Toward Integrating Feature Selection Algorithms for Classification and Clustering', *IEEE Trans. Knowl. Data Eng.*, Vol. 17, Issue 4, April 2005, pp. 491-502.
14. Iffat A. Gheyas, Leslie S.Smith, 'Feature Subset Selection in Large Dimensionality Domains', *Pattern Recognit.*, Vol. 43, Issue 1, 2010, pp. 5-13.
15. C. Jensen, M. El-Sharkawi, Robert J. Marks, 'Power System Security Assessment using Neural Networks: Feature Selection using Fisher Discrimination', *IEEE Trans. Power Syst.*, Vol. 16, Issue 4, 2001, pp. 757-763.
16. C.K. Pang, F.S. Prabhakara, Ahmed H. El-Abiad, A.J. Kovio, 'Security Evaluation in Power Systems using Pattern Recognition', *IEEE Trans. Power Appar. Syst.*, Vol. PAS - 93, May/June 1974, pp. 969-976.
17. Huiqing Liu, Jinyan Li, Limsoon Wang, 'A comparative study on Feature Selection and Classification Methods using Gene Expression Profiles and Proteomic Patterns', *Genome Inform.*, Vol. 13, 2002, pp. 51 - 50.
18. J. Biesiada, W. Duch, 'Feature Selection for High-Dimensional Data: A Pearson Redundancy Based Filter', *Adv. Soft Comput.*, Vol. 45, 2008, pp. 242-249.
19. M. Dash, H. Liu, 'Feature Selection for Classification', *Intell. Data Anal.*, Vol. 1, No. 3, 1997, pp. 131-156.
20. Adhikary, S., & Banerjee, S. (2023). Introduction to distributed nearest hash: On further optimizing cloud based distributed knn variant. *Procedia Comput. Sci.*, *218*, 1571-1580.
21. Xx Dalei, J., & Mohanty, K. B. (2016). Fault classification in SEIG system using Hilbert-Huang transform and least square support vector machine. *Int. J. Electr. Power Energy Syst.*, *76*, 11-22.

22. Ramakrishnan, N. (2010). *Distributed Learning Algorithms for Sensor Networks* (Doctoral dissertation, The Ohio State University).
23. Guo, M. F., Yang, N. C., & Chen, W. F. (2019). Deep-learning-based fault classification using Hilbert–Huang transform and convolutional neural network in power distribution systems. *IEEE Sens. J.*, *19*(16), 6905-6913.
24. Avasthi, S., Chauhan, R., & Acharjya, D. P. (2022). Topic modeling techniques for text mining over a large-scale scientific and biomedical text corpus. *Int. J. Ambient Comput. Intell. (IJACI)*, *13*(1), 1-18.
25. S. Haykin, '*Neural Networks: A Comprehensive Foundation*', Prentice Hall, 2008.
26. Abhisek Ukil, '*Intelligent Systems and Signal Processing in Power Engineering*', Springer-Verlag, 2007.
27. Avasthi, S., Prakash, A., Sanwal, T., Tyagi, M., & Yadav, S. (2023, January). Tourist reviews summarization and sentiment analysis based on aspects. In *2023 13th International Conference on Cloud Computing, Data Science & Engineering (Confluence)* (pp. 452-456). IEEE.
28. De Coninck, E., Verbelen, T., Vankeirsbilck, B., Bohez, S., Simoens, P., Demeester, P., & Dhoedt, B. (2016). Distributed neural networks for internet of things: The big-little approach. In *Internet of Things. IoT Infrastructures: Second International Summit, IoT 360° 2015, Rome, Italy, October 27-29, 2015, Revised Selected Papers, Part II* (pp. 484-492). Springer International Publishing.
29. Leroux, S., Bohez, S., De Coninck, E., Verbelen, T., Vankeirsbilck, B., Simoens, P., & Dhoedt, B. (2017). The cascading neural network: building the internet of smart things. *Knowl. Inf. Syst.*, *52*, 791-814.
30. Diao, C., Kleyko, D., Rabaey, J. M., & Olshausen, B. A. (2021, July). Generalized learning vector quantization for classification in randomized neural networks and hyperdimensional computing. In *2021 International Joint Conference on Neural Networks (IJCNN)* (pp. 1-9). IEEE.
31. Shobha, & Nalini, N. (2022, February). Performance Study of Data Fusion Using Kalman Filter and Learning Vector Quantization. In *Proceedings of Second International Conference on Sustainable Expert Systems: ICSES 2021* (pp. 79-88). Singapore: Springer Nature Singapore.
32. Basaklar, T., Tuncel, Y., Narayana, S. Y., Gumussoy, S., & Ogras, U. Y. (2021). Hypervector design for efficient hyperdimensional computing on edge devices. *arXiv preprint arXiv:2103.06709.*
33. Kusy, M., & Kowalski, P. A. (2018). Weighted probabilistic neural network. *Inf. Sci.*, *430*, 65-76.
34. Perera, N., & Rajapakse, A. D. (2010). Recognition of fault transients using a probabilistic neural-network classifier. *IEEE Trans. Power Deliv.*, *26*(1), 410-419.
35. Liao, L., Lai, Y., Yang, F. and Zeng, W., 2023. Online computation offloading with double reinforcement learning algorithm in mobile edge computing. *J. Parallel Distrib. Comput.*, 171, pp. 28-39.

36. Li, Zhuozhao, Ryan Chard, Logan Ward, Kyle Chard, Tyler J. Skluzacek, Yadu Babuji, Anna Woodard *et al.* "DLHub: Simplifying publication, discovery, and use of machine learning models in science." *J. Parallel Distrib. Comput.*, 147 (2021): 64-76.
37. Yu, Yuan, Martín Abadi, Paul Barham, Eugene Brevdo, Mike Burrows, Andy Davis, Jeff Dean *et al.* "Dynamic control flow in large-scale machine learning." In *Proceedings of the Thirteenth EuroSys Conference*, pp. 1-15. 2018.
38. Davies, E. K., Glick, M., Harrison, K. N., & Richards, W. G. (2002). Pattern recognition and massively distributed computing. *J. Comput. Chem.*, *23*(16), 1544-1550.
39. Exner, P., & Nugues, P. (2014, March). KOSHIK-A large-scale distributed computing framework for NLP. In *International Conference on Pattern Recognition Applications and Methods* (Vol. 2, pp. 463-470). SciTePress.
40. Daily, M., Medasani, S., Behringer, R., & Trivedi, M. (2017). Self-driving cars. *Computer, 50*(12), 18-23.
41. Jo, K., Kim, J., Kim, D., Jang, C., & Sunwoo, M. (2014). Development of autonomous car—Part I: Distributed system architecture and development process. *IEEE Trans. Ind. Electron.*, *61*(12), 7131-7140.
42. Zhang, W., He, L., Chow, Y. L., Yang, R., & Su, Y. (2000, June). The study on distributed speech recognition system. In *2000 IEEE International Conference on Acoustics, Speech, and Signal Processing. Proceedings (Cat. No. 00CH37100)* (Vol. 3, pp. 1431-1434). IEEE.

7

Next-Generation Distributed Computing for Cancer Detection

Kapil Kumar Gupta[1]*, Namrata Dhanda[1] and Neeraj Kumar[2]

[1]*Department of Computer Science & Engineering, Amity University Lucknow, Uttar Pradesh, India*
[2]*Department of Computer Science & Engineering, Shri Ramswaroop Memorial University, Barabanki, Uttar Pradesh, India*

Abstract

Increasing levels of cancer research are used to improve patient outcomes by distributed computing techniques in healthcare. These techniques offer promising applications and major challenges. Medical imaging, natural language processing of medical documents, and genetic information are the three main areas of distributed computing. Many of these fields are diagnosed, detected, and predicted. A broad medical device infrastructure currently generates data but often does not provide a supportive infrastructure for the efficient use of such data. The various forms of medical data exist in which data formatting is also challenging and may cause noise. The ability to capture, share, and deliver data is a high priority, given that digitization disrupts every industry, including healthcare. The challenges of the huge amounts of data may be addressed through Distributed computing, Big Data, and Artificial Intelligence (AI). Distributed computing can also assist health organizations in meeting increasing medical needs, improving operations, and decreasing costs. Innovative Distributed computing can help health practitioners more efficiently and with greater precise and personalized healthcare to detect and treat diseases. The aim of the biological modeling process is to describe data, but ultimately, it aims at how systems may be set and how system targets, algorithms, and mechanisms can be better understood. Distributed computing enables data to be modeled extremely well, without making strong assumptions about the modeling system, thanks to technical applications. Distributed computing usually describes data more clearly than biomedical models, providing an essential benchmark as well as engineering solutions. It can also be an instrument

**Corresponding author*: kapilkumargupta2007@gmail.com

Sandhya Avasthi, Suman Lata Tripathi, Namrata Dhanda and Satya Bhushan Verma (eds.) Decentralized Systems and Distributed Computing, (145–162) © 2024 Scrivener Publishing LLC

to promote understanding. The expansion of translational cancer research has been made possible by improvements in next-generation sequencing (NGS) and mass spectrometry (MS) technology, but they have also posed enormous problems in terms of data storage and interpretation. Traditional computer models can never be applied to the ensuing informatics issue. Recent advancements in scalable computing and related infrastructure, notably distributed computing for Big Data, may hold the answers to solving these challenges.

Keywords: Artificial narrow intelligence (ANI), artificial general intelligence (AGI), mass spectrometry (MS), distributed computing, cancer detection, brain tumor

7.1 Introduction

Data-driven medicine is becoming more and more popular right now, and it has the potential to completely transform the healthcare industry by attracting significant interest in diagnosis, clinical decision-making, and research [1–3]. The effectiveness of treatments and clinical trials depends on the accessibility of sufficiently large and trustworthy datasets, such as patient records or medical imaging, particularly in research [4]. Because AI approaches can learn from past data and make predictions, they are being used more and more in the healthcare industry, which increases the demand for a tremendous amount of data [1, 5, 6]. The increasing success is nevertheless accompanied by recent obstacles. Because patient data are sensitive, complying with patient rights and privacy laws is a significant challenge [7]. For instance, the European GDPR provides extra security for this sensitive patient data. The worry of losing sovereignty over the data instances prevents healthcare facilities from sharing their data, even though these policies improve data privacy [8]. As a result, researchers are constrained by the data that are accessible within institutional boundaries, which lowers the efficacy of therapeutic outcomes [7, 9–11]. A data centralization strategy is typically used to bring data into the analysis from an abstract standpoint. Therefore, information is gathered from many institutions and sent to a hub point. As was already said, this technique is challenging to implement in accordance with current data privacy laws. By bringing the analysis to the data, DA, on the other hand, flips this process. As a result, data never leave their source and always remain in the ownership of the data owner [10]. While DA is a first step towards privacy-preserving analyses and creates linkages between institutions, its highly decentralized structure also raises new research questions, especially throughout the previous few years. Several flaws in DA systems have been exposed by data management and the fairification of digital assets. Despite the fact that a few DA concepts have been conceptually assessed in terms of their contribution to the FAIR data, there has not yet been any real progress made in this approach [10]. In addition to these more ethereal concepts, standard procedures frequently call for significant technical expertise needed to solve the above said problem [11].

7.2 Research Motivation

The development of this research was prompted by the rise in cancer incidence and death cases observed globally [12]. According to the Health Metrics and Evaluation definition of seven superregions, Azamjah *et al.*'s [13–16] research sought to ascertain the mortality rate in these areas. Breast cancer develops in communities, which is higher in industrialized nations, according to a study by Momenimovahed *et al.* [17]. In an analysis published by Haggar *et al.* [18], the prevalence, mortality, and survival rates for colon cancer were demonstrated while taking into account regional variations and long-term changes. Cancer Incidence in Five Continents was used as the source for Zhang *et al.*'s [19] inquiry to acquire data on the frequency of CRCs. The socio-economic progress of a country was found to positively correlate with incidence by Wong *et al.* [20, 21]. According to Lee *et al.* on COVID-19 clinical history, the probability of death was significantly correlated with the aging of the patient population, with 226 (28%) patients dying. The histology trends of skin cancer in Yemen's southern regions were examined by Al-Zhou *et al.* [22] in addition to their evaluation of the demographic traits. One of computer science's outstanding accomplishments, artificial intelligence (AI), was first proposed in the 1940s [23, 24]. The goal of AI is to investigate the links between medical procedures and patient outcomes. Artificial intelligence (AI) has shown promise in the field of cancer research for improving patient outcomes by influencing a variety of aspects of cancer therapy, speeding up and improving the accuracy of cancer diagnosis, and providing more reliable therapeutic recommendations [25, 26]. The level of precision with which AI can predict cancer is not comparable to that of a conventional statistical expert [27, 28]. As a result, the medical professionals in hospitals are allowed to make unhindered medical judgments.

7.3 Cancer Statistics

The International Agency for Research on Cancer (IARC) estimates that there are thirty-six different types of cancer. One in five people will eventually develop cancer, and one in eight men and one in eleven women will pass away from it, according to their research. Eight million people globally lose their lives to cancer each year [29, 30]. The most common types of cancer among both men and women were lung cancer and breast cancer, respectively, according to GLOBOCAN 2020 estimates [31]. Table 7.1 displays the top new cancer cases globally in 2020. A total of 400,000 children are diagnosed with cancer each year, according to WHO. Table 7.2 displays cancer deaths statistics.

Table 7.1 New cancer cases in 2020 all over world (WHO, 2022).

Cancer type	Deaths
Breast cancer	2,260,000
Lung cancer	2,210,000
Colon and Rectum	1,930,000
Skin (non-melanoma)	1,210,000
Prostrate	1,410,000

Table 7.2 Worldwide cancer deaths statistics in 2020 (WHO, 2022)

Cancer type	Deaths
Lung cancer	10,000,000
Colon and Rectum	916,000
Liver cancer	830,000
Stomach	769,000
Breast	685,000

7.4 Modern Cancer Diagnosis and Treatment Methods

A patient who experiences symptoms like hard lumps on their bodies or odd patterns on their skin typically seeks medical assistance from a doctor. In order to swiftly send patients for additional testing and treatment, if necessary, the screening test looks for people who have a specific cancer or pre-cancer but have not yet shown any symptoms. Several scan modalities can be used to do a pre-stage analysis. This is carried out as a preventative measure to avoid the overdiagnosis of cancer in a high-risk group [32, 33]. They compile, synthesize, and analyze the pertinent data while also recommending a diagnosis. The appropriate course of therapy is suggested, and the patient is advised of the current diagnosis and prognosis. With possibilities for speedy diagnosis and the capacity to learn from mistakes, this procedure is advantageous for patients as well as the healthcare system. Nevertheless, this procedure has room for error and is adaptable based on the medical specialty. A skilled physician can diagnose cancer non-invasively by visually examining prospective locations that resemble the symptoms of malignant tumors using a number of medical imaging techniques. The degree of accuracy of the diagnosis is significantly influenced by the clinician's experience. Radiologists frequently disagree while analyzing the same image since correctly reading these scans involves years of training. Due to the enormous amount of medical data, this strategy is also laborious and inconsistent. Inferences from non-standardized data from a variety of modalities may be difficult to make because doctors frequently have little time to dedicate to making a diagnosis. Additionally, the process could take longer than anticipated [34] because a diagnosis needs to be made by a variety of experts from various medical disciplines. Because patient-specific therapeutic combinations must be explored and tested, conventional cancer treatment is difficult. Patients with malignant illnesses are typically treated using mechanical, physical, pharmacological, and biological therapies. Malignant diseases can be treated at any anatomic site in the body with chemotherapy, a type of intravenous systemic therapy. When medicine must be given systemically throughout the body to treat metastatic disease, it is the only option. Chemotherapy kills cancerous cells primarily by using medications that target cells that divide quickly. The drugs that are used to shrink tumors have negative side effects [35, 36]. When a person has breast or prostate cancer, hormones have a significant impact. Adoptive cell transplants and checkpoint inhibitors are two types of immunotherapy. Radiation therapy affects malignant cells' deoxyribose nucleic acid (DNA), either destroying them or preventing their proliferation. Medical experts frequently recommend this therapy as a preventative step to lessen cancer symptoms or shrink tumors. Physical techniques for eliminating tumors include radiation, hyperthermia, heat, and photodynamic therapy. Leukemia and lymphoma are two blood-related diseases that could potentially benefit from stem cell transplantation. Chemotherapy-damaged blood cells must be removed as part of the treatment. Targeted therapy improves immunity and stops cancer from spreading. Target therapies include monoclonal antibodies and small-molecule medications. Tumors can be removed mechanically through surgery. The most common option is surgery, since it offers the best chance of success. The removal, inspection, and repair of tissue are all part of the medical strategy. This local therapy works best on solid tumors that are localized to a particular region. Patients who have cancerous cells in their bodies often undergo surgery. Additionally, eliminating the lymph nodes helps stop the disease's spread. Since it may not always totally destroy every cancerous cell in the patient, the tumor may eventually come back. The methods used to diagnose and treat cancer today have drawbacks, including higher rates of false-positive test results that suggest someone does not genuinely have lung cancer. On CT scans, for instance, some noncancerous lung changes resemble cancer quite closely. Additionally, early identification may prevent some of the malignant nodules from being evident to the human eye. The pain a patient experiences throughout therapy is made worse by an increased number of treatment trials and related costs. Repeated radiation exposure for routine screening comes with substantial danger [37]. The limitations of current cancer diagnosis and therapy have been shown to be minimized by advances in AI, which has the potential to bring about revolutionary breakthroughs in the field of cancer healthcare. The different approaches for cancer research is shown in the Figure 7.1.

A range of data about the patient and the cancer cells can be examined by medical specialists using big data and AI to provide

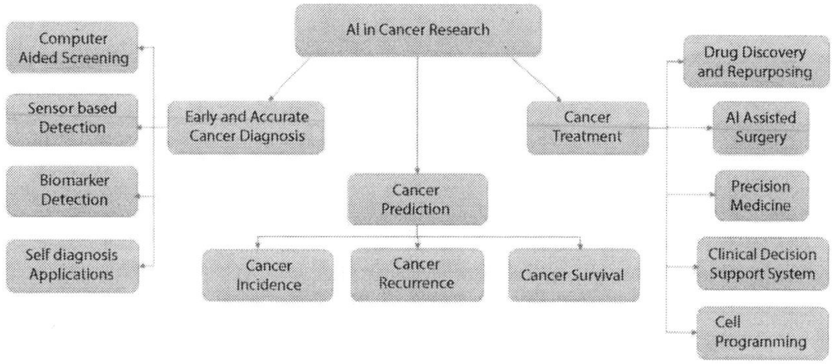

Figure 7.1 Approaches for cancer research using distributed system using AI.

customized treatments. This type of therapy will have milder side effects. Healthy cells will be less negatively impacted, whereas cancer cells will be more negatively impacted. Researchers are putting more effort into understanding how genetics impacts cancer development and treatment. Fisher argues that although cancer "might not go away, once you learn how to control it, it's a problem solved". Finding the cause of the issue and devising a solution come next. Fisher wants to fully comprehend how a cell's programming, which governs behavior, works. When the workings of a healthy cell are known, the workings of a diseased cell can be compared to them. Because of this, researchers may be able to keep track of how the body's millions of genes and proteins interact to cause cancer and swiftly come up with the safest and most effective strategy to treat each particular patient. The system can be used to conduct a variety of experiments, such as contrasting the information provided by a cancer patient [38]. The billions of chemicals, proteins, and genes that interact within the human body present an excessive number of variables, making it impossible to perform these computations using straightforward algorithms. Another important area where AI has made a huge contribution is in the repurposing and creative drug design industries. AI can predict how various medications will impact cancerous cells. The development of novel anticancer drugs and the timing of their administration are made easier by this information.

7.5 Methodology

To categorize tumors and cure brain cancer, Harvard University and University of Pennsylvania researchers developed a DL algorithm. The intrusive procedures that are frequently necessary can be avoided by using MRI images of gliomas to detect and describe the isocitrate dehydrogenase (IDH) mutations. Today's clinical practice makes extensive use of guided needle biopsies and pathology updates since they can reduce the incidence of unnecessary surgical excisions. Rapid healing and a return to normal life are made possible by robotic surgery.

A range of data about the patient and the cancer cells can be examined by medical specialists using big data and AI to provide customized treatments [38, 39]. This type of treatment will have milder side effects. Healthy cells will be less negatively impacted, whereas cancer cells will be more negatively impacted. To understand how genetics affects the development and treatment of cancer, more research is being done. Cancer "might not go away, but once you learn how to control it, it's a problem solved," says Fisher. They claim that employing technology to learn more about cancer in general and the biological processes that give rise to cancerous cells is the best approach to doing this. The next step is to identify the problem's cause and figure out how to fix it. The programming of the cell, which governs a cell's behavior, is Fisher's ultimate objective. Understanding the way a healthy cell works allows us to compare it to a sick one. This can help with locating the root of issues and finding solutions. In order to quickly devise the most efficient and secure course of action for treating each patient personally, the millions of genes and proteins in the human body that affect cancer may be connected, according to studies. Clinicians can enter all of the patient's biological data into the system using BMA. Another important area where artificial intelligence has made a considerable contribution is in the repurposing and creative drug design industries. AI can predict how different medications will impact cancerous cells. The development of novel anticancer drugs and the timing of their administration are both aided by this information. Drug development is a costly and time-consuming procedure, but AI can increase productivity by using neural networks. Combining reinforcement learning with deep learning yields a groundbreaking method for the discovery of novel pharmaceutical compounds by approximating the statistical relationship between prospective actions and outcomes. AI has the ability to change every aspect of the pharmaceutical lifecycle, from regulatory procedures to pharmacovigilance. Numerous computer tools have been proposed for the identification of cancer-related drugs based on various AI approaches.

The T cells in our immune system are constantly on the lookout for indications of cancer and other foreign invaders, according to Tao Wang from the UT Southwestern Medical Centre. When these cells find neoantigens, they connect to one another. Even though some neoantigens go unrecognized, some can lead to cancer.

7.6 Technical Challenges

Even though AI has many benefits, it is limited by a number of issues that limit how well it can be used in cancer research. The new wave of innovation presents a number of challenges that must be overcome if cancer practices are to be drastically altered at various scales. Some of the barriers to the proper implementation of AI include rules, payments, knowledge, practical issues, and rigid healthcare systems. The AI classifier and predictor models must be trained using labeled data. Datasets still require manual annotation or, at the very least, curation, even though raw data can simply be fed into AI models. Involving several subject experts in the data annotation process can help ensure that the data labels are accurately evaluated. One of the main obstacles to the broad use of AI algorithms and decision-support systems in the delivery of cancer care is the lack of inclusive and diverse training datasets. One of the most prevalent issues is a shortage of data needed to train the model. The bulk of sophisticated AI models require a large sample size in order to successfully train the model better than a constrained one. When a dataset has high dimensionality, there are more features than there are medical data records. Although there are feature-selection and dimensionality-reduction techniques for dealing with different dimensions to address this issue, correct implementation of these techniques is crucial for better results. Data classification models have a tendency to favor the class with the bulk of samples. The majority of the techniques now in use manage imbalances in binary classes well but fail in multi-class patterns. Numerous studies have demonstrated excellent AI cancer prediction outcomes. However, to eliminate data cleaning processes and achieve high cancer prediction accuracy, a computationally efficient feature-selection strategy is still required. Research has to be refocused on improving the generalizability of the model. A prediction model that has only been tested on one region has been proposed by the bulk of studies. To make the models more generalizable, they must be tested across a number of sites. When used in various healthcare systems and foreign populations, as was previously indicated, reproducibility of AI-model output is challenging to establish. However, even in the environment in which they were developed, data drift over time can have an effect on the performance of AI algorithms and complex decision-support systems. In the opinion of experts, employing AI and ML continuously in therapy or diagnosis may be harmful because distributional shifts may occur, which means that goal data may differ from the ongoing patient data used to train the model and may produce false conclusions. The link between the data items is probably going to change as a result of changes in technology, the healthcare system, and the population (including the gene pool). Although the models are frequently utilized in cancer research, their practical use in clinics is rarely proactively considered. The predictions of these models frequently need to be validated in a clinical setting to assist the medical professional in validating the diagnosis judgements. Broad deployment in clinics has been hindered by significant issues with data transparency and interpretability brought on by AI's "black box" process, as well as an inherent bias towards minority cohorts that restricts the reproducibility of AI models and sustains healthcare disparities. Deep architectural problems can be resolved by utilizing smart and efficient optimizers, intelligent initialization strategies, and activation functions based on local competition. The stacking of numerous non-linear transformations, however, has left deep network training with a number of issues that still need to be fixed. Though there are methods to find the bias in model training, their efficacy still has to be increased. For cancer prediction using medical imagery, around 90% of the studies have validated DL techniques over other approaches. On the other hand, DL-based methods are very challenging. While the CNN classifier performed brilliantly, it did so at a cost to the trials' computations of roughly 41% [40]. Additionally, the lack of expertise among therapists and medical professionals in computing algorithms and technology may hinder the implementation of AI in many developing countries. Weak AI is the term used to describe the clinical–diagnostic AI algorithms of today. In other words, they are trained to conduct a single activity, lack general intelligence, and are not versatile enough to handle new clinical diagnostic duties. However, utilizing transfer-learning strategies, a fully trained AI algorithm can be changed to do closely similar tasks.

7.7 Future Directions

AI will affect cancer in a variety of ways that go beyond prevention, diagnosis, and treatment [32]. Every current technology that is being presented goes through a hype cycle. The technology trigger phase, the peak of the inflated expectations phase, the trough of the disillusionment phase, the slope of the enlightenment phase, and the plateau of the production phase are the five stages of a hype cycle. The advancement of a technology through these stages creates a barrier to its genuine utility being realized. Moving the focus from diseases with high incidence, such as lung, colon, and breast cancers, to those with slower medical advancement is necessary. Innovations being made across the board will impact cancer treatment in the future. The Internet of Medical Things (IoMT) and telemedicine for virtual consultations are two additional ways that these platforms might be used with other digital technologies to improve referral practices. AI can accurately categorize cancer stages according to risk and determine the most effective course of treatment. When analyzing data, common determinants can be discovered at many different levels. AI models can use these determinants to determine the causal relationships between variables. Having both integrated knowledge and more interpretable AI models can improve the accuracy of AI research on specific cases of cancer-related sickness. Two emerging trends are interpretable DL and explainable AI [41]. For DL models, interpretability is an issue because their great accuracy comes at the expense of excessive abstraction. The area has always been split between AI proponents and opponents in healthcare. The key to the success of these models is that both biologists and computer scientists must use their knowledge of their respective specialties.

BMA talks in a language that biologists can understand and uses tools that visually mimic those that researchers may employ in the lab [36]. Self-diagnostic smartphone applications, like SkinVision, were developed to assist users in looking for skin irregularities indicative of cancer. They can take a picture of a questionable location on their skin with their smartphone's camera and email it for analysis. Neil Dalchau, a biologist, wants to investigate the idea of using cells to code [36]. The author is a member of a team trying to create computing in cells rather than silicon. An in-depth knowledge of a cell's computations is necessary for effective programming. The ability of immune networks to combat defects or bugs that cause cancer in healthy cells is a futuristic idea. Biologists with in-depth knowledge of the subject may contribute hypotheses that instruct computer scientists on what to search for in the data in this kind of partnership. Additionally, computer scientists can do the necessary studies to confirm or deny their hypothesis. Examples of diagnostic technology include imaging, digital pathology, and tailored biopsies. Future therapeutic possibilities will be made possible by the combination of AI and novel therapeutic modalities, such as enhanced antibody design, cell therapies, nucleic acid therapies, and cancer vaccines. Combination immunotherapies continue to be covered in the media. Combination therapy is becoming more popular in oncology, as evidenced by the fact that 81% of immuno-oncology (IO) trials use combinations of two or more medications [42]. Cell therapies like CAR-T are now being used to treat B-cell malignancies that were difficult to treat by other methods. In order to further enhance specificity, safety, and amplification in conjunction with AI, the second and third generations of CAR-T therapies include "activation" and "kill" switches. However, recent failures, spurred on by terrible circumstances, have rekindled skepticism about the viability of important schemes. Cell treatments may not be as effective overall against solid tumors, according to certain research [43–46]. Preclinical research has favored a high-precision gene-editing technique known as clustered regularly interspaced short palindromic repeats (CRISPRs), which has the potential to deliver a long-lasting cure. CRISPR's most innovative clinical applications revolve around how it might be utilized to support cell therapies. CRISPR could be used to delete T-cell receptors or major histocompatibility complexes from the DNA to lessen the immunogenic potential of allogeneic transplantation. Researchers never imagined that it would be possible to quickly and easily change the genetic code of living cells. CRISPR operates like a pair of scissors and can precisely cut out, add to, or change certain DNA sequences inside cells. A significant issue is the CRISPR editing tool's propensity to cause mutations in double-stranded DNA while it is being repaired. The targeted sequence and the guide RNA (gRNA) are believed to have a major impact on the mistakes, among other things, but they also seem to follow a recurrent pattern. A method that employs machine learning to predict cell changes has allegedly been developed by researchers at the Wellcome Sanger Institute. This might improve the effectiveness of CRISPR research and make it easier to turn it into workable and secure treatments [46].

7.8 Conclusion

In this paper, we present the opportunities, trends, challenges, and future directions of AI in cancer research. Cancer is identified as one of the leading causes of death by the WHO. Initiatives like Cancer Moonshot intend to reduce the cancer death rate by half in the next 25 years. Conventional practices for cancer diagnosis and treatment include radiology, chemotherapy, surgery, and radiation. AI models could improve the accuracy and effectiveness of the methods used for cancer prediction, diagnosis, and treatment. To develop these models, we need large datasets for training and validation, and many such datasets are available publicly. Healthcare providers can access AI services conveniently through cloud-based AI platforms. AI-based classifiers help pathologists diagnose cancer accurately and consistently. Predictive AI models establish a person's likelihood of getting a certain type of cancer by identifying the risk factors. Personalized cancer treatment with the help of AI is one of the key methods for improving cure outcomes. AI makes it possible for the patient to have a longer life, less pain, and more cures by appropriately controlling the timing of restaging and surveillance tests, the dosages of systemic-cancer medicines and radiation, and the choice and sequencing of diagnostic and treatment measures. Proper adoption of AI for cancer research is constrained by regulatory controls with respect to data security and privacy, the non-availability of labeled data, data bias, imbalanced data, and so on. It is a fact that AI can and will make faulty decisions in certain situations because its decision-making is predictive and probabilistic. So, if AI makes a faulty decision, we cannot hold it responsible. Explainable AI and interpretable DL are steps towards mitigating this risk and providing rationale for decision-making. Unless a human-medical expert works with the AI system, we cannot have a human-verification element in the process. Therefore, no one anticipates that AI will entirely replace medical professionals in the near future. AI-based precision medicine is going to be the key to cancer treatment in the future. Living databases will power extremely complex models that can customize therapy selection, dose calculation, surveillance modality and schedule, and so on. However, the migration from artificial narrow intelligence (ANI) to artificial general intelligence (AGI) will witness the automation of almost all processes for cancer prediction, diagnosis, and treatment.

References

1. Balicer, R.D.; Cohen-Stavi, C. Advancing Healthcare Through Data-Driven Medicine and Artificial Intelligence. In *Healthcare and Artificial Intelligence*; Nordlinger, B., Villani, C., Rus, D., Eds.; Springer International Publishing: Cham, Switzerland, 2020; pp. 9–15. [CrossRef]
2. Alyass, A.; Turcotte, M.; Meyre, D. From big data analysis to personalized medicine for all: Challenges and opportunities. *BMC Med. Genomics,* 2015, 8, 33. [CrossRef] [PubMed]
3. Deo, R.C. Machine learning in medicine. *Circulation* 2015, 132, 1920–1930. [CrossRef] [PubMed]

4. Geifman, N.; Bollyky, J.; Bhattacharya, S.; Butte, A.J. Opening clinical trial data: Are the voluntary data-sharing portals enough? *BMC Med.*, 2015, 13, 280. [CrossRef] [PubMed]
5. Sidey-Gibbons, J.A.M.; Sidey-Gibbons, C.J. Machine learning in medicine: A practical introduction. *BMC Med. Res. Methodol.*, 2019, 19, 64. [CrossRef] [PubMed]
6. Giger, M.L. Machine Learning in Medical Imaging. *J. Am. Coll. Radiol.*, 2018, 15, 512–520. [CrossRef]
7. Rieke, N.; Hancox, J.; Li, W.; Milletari, F.; Roth, H.; Albarqouni, S.; Bakas, S.; Galtier, M.N.; Landman, B.; Maier-Hein, K.; *et al.* The Future of Digital Health with Federated Learning. *NPJ Digit. Med.*, 2020, 3, 119. [CrossRef]
8. Rosenblatt, M.; Jain, S.H.; Cahill, M. Sharing of Clinical Trial Data: Benefits, Risks, and Uniform Principles. *Ann. Intern. Med.*, 2015, 162, 306–307. [CrossRef]
9. Sheller, M.J.; Reina, G.A.; Edwards, B.; Martin, J.; Bakas, S. Multi-Institutional Deep Learning Modeling Without Sharing Patient Data: A Feasibility Study on Brain Tumor Segmentation. In *Brainlesion: Glioma, Multiple Sclerosis, Stroke and Traumatic Brain Injuries*. BrainLes (Workshop); Springer: Cham, Switzerland, 2019; Volume 11383, pp. 92–104. [CrossRef]
10. Beyan, O.; Choudhury, A.; van Soest, J.; Kohlbacher, O.; Zimmermann, L.; Stenzhorn, H.; Karim, M.R.; Dumontier, M.; Decker,S.; da Silva Santos, L.O.B.; *et al.* Distributed Analytics on Sensitive Medical Data: The Personal Health Train. *Data Intell.*, 2020, 2, 96–107. [CrossRef]
11. Deist, T.M.; Dankers, F.J.W.M.; Ojha, P.; Scott Marshall, M.; Janssen, T.; Faivre-Finn, C.; Masciocchi, C.; Valentini, V.; Wang, J.; Chen, J.; *et al.* Distributed learning on 20,000+ lung cancer patients—The Personal Health Train. *Radiother. Oncol.*, 2020,144, 189–200. [CrossRef] [PubMed]
12. Andriole G, Kramer B, Berg C (2009) Mortality results from a randomized prostate cancer screening trail. *N. Engl. J. Med.*, 360:1310–1319.
13. Rong F, Gong W, Pan J, Wang W (2019) Analysis of mortality and survival rate of liver cancer in Zhejiang. Province in China: A general population-based study. Hindawi, 1–7.
14. Dolatkhah R *et al* (2020) Breast cancer survival and incidence: 10 Years cancer registry data in the northwest, Iran. *Int. J. Breast Cancer*, 1–6.
15. Goodarzi E, Moslem A, *et al.* (2021) Epidemiology, incidence and mortality of thyroid cancer and their relationship with the human development index in the world: An ecology study in 2018.
16. Azamjah N, Zadeh Y, Zayeri F (2018) Global trend of breast cancer mortality rate: A 25-year study. Global Trend Breast Cancer Mortal 20:1–6.
17. Momenimovahed Z, Salehiniya H (2019) Epidemiological characteristics of and risk factors for breast cancer in the world. *Breast Cancer Ther.*, 151–164.
18. Haggar F, Boushey R (2009) Colorectal cancer epidemiology: incidence, mortality, survival, and risk factors. *Clin. Colon*, 22:191–197.

19. Zhang Y, Li M, Gao X, Chen Y, Liu T (2019) Nanotechnology in cancer diagnosis: Progress, challenges and opportunities. *J. Hematol. Oncol.*, 12:1–13. https:// doi. org/ 10. 1186/s13045-019-0833-3
20. Wong M, Goggins W, Fung F *et al.* (2017) Incidence and mortality of kidney cancer: temporal patterns and global trends in 39 countries. *Sci. Rep.*, 7:1–10.
21. Nguyen Q, Lee J, Huang M, Khullar A, Raymond P (2015) Diagnosis and treatment of patient with thyroid cancer. *Clinical* 1:1–40.
22. AlZhou A, Thabit M, Sakkaf K, Basaleem H (2017) Skin Cancer. 17, 3195-3199.
23. Adegun A, Viriri S (2020) Deep learning techniques for skin lesion analysis and melanoma cancer detection: A survey of state-of-the-art. In: *Artif. Intell. Rev.* (Issue 0123456789). Springer Netherlands. https://doi.org/10.1007/s10462-020-09865-y
24. Patil RS, Biradar N (2020) Automated mammogram breast cancer detection using the optimized combination of convolutional and recurrent neural network. *Evol. Intell.* https://doi.org/10.1007/s12065-020-00403-x.
25. Zhang N, Lou W, Ji F, Qiu L, Tsang BK, Di W (2016) Low molecular weight heparin and cancer survival: Clinical trials and experimental mechanisms. *J. Cancer Res. Clin. Oncol.*, 142(8):1807–1816. https://doi.org/10.1007/s00432-016-2131-6
26. Zhang R, Zheng Y, Wing T, Mak TWC, Yu R, Wong SH, Lau JYW, Poon CCY (2016) Automatic detection and classification of colorectal polyps by transferring low-level CNN features from nonmedical domain. *IEEE J. Biomed. Health Inform.*, 21(1):41–47. https://doi.org/10.1109/JBHI.2016.2635662
27. Selvathi D, Poornila A (2017) Breast cancer detection in mammogram images using deep learning technique. *J. Sci. Res.*, 25(2):417–426. https://doi.org/10.5829/idosi.mejsr.2017.417.426
28. Zhang C, Guo H (2018) Smart Software can diagnose prostate cancer as well as pathologist. *Europ. Assoc. Urol.*, 1–3.
29. Cancer. Who. Int. 2022. Available online: https://www.who.int/news-room/fact-sheets/detail/cancer (accessed on 13 October 2022).
30. Available online: https://www.mckinsey.com/industries/life-sciences/our-insights/delivering-innovation-2020-oncology-market-outlook (accessed on 14 October 2022)
31. GLOBOCAN 2020: New Global Cancer Data|UICC. Uicc.Org. 2022. Available online: https://www.uicc.org/news/globocan-2020-new-global-cancer-data (accessed on 13 October 2022)
32. Lee, D.; Yoon, S. Application of Artificial Intelligence-Based Technologies in the Healthcare Industry: Opportunities and Challenges. *Int. J. Environ. Res. Public Health* 2021, 18, 271. [CrossRef]
33. Lee, H.; Chen, Y. Image Based Computer Aided Diagnosis System for Cancer Detection. *Expert Syst. Appl.*, 2015, 42, 5356–5365. Available online: https://www.sciencedirect.com/science/article/abs/pii/S0957417415000986?via%3Dihub (accessed on 13 October 2022).

34. Mirbabaie, M.; Stieglitz, S.; Frick, N. Artificial intelligence in disease diagnostics: A critical review and classification on the current state of research guiding future direction. *Health Technol.,* 2021, 11, 693–731. [CrossRef]
35. National Cancer Institute. Can Artificial Intelligence Help See Cancer in New Ways? National Cancer Institute: Bethesda, MD, USA, 2022. Available online: https://www.cancer.gov/news-events/cancer-currents-blog/2022/artificial-intelligence-cancer-imaging (accessed on 13 October 2022)
36. Ginghina, O.; Hudita, A.; Zamfir, M.; Spanu, A.; Mardare, M.; Bondoc, I.; Buburuzan, L.; Georgescu, S.E.; Costache, M.; Negrei, C.; *et al.* Liquid Biopsy and Artificial Intelligence as Tools to Detect Signatures of Colorectal Malignancies: A Modern Approach in Patient's Stratification. *Front. Oncol.,* 2022, 12, 856575. [CrossRef]
37. Mascalchi, M.; Sali, L. Lung cancer screening with low dose CT and radiation harm from prediction models to cancer incidence data. *Ann. Transl. Med.,* 2017, 5, 360. [CrossRef] [PubMed]
38. News.Microsoft.Com. 2022. Available online: https://news.microsoft.com/stories/computingcancer/ (accessed on 13 October 2022)
39. Available online: https://www.cnbc.com/2017/05/11/from-coding-to-cancer-how-ai-is-changing-medicine.html (accessed on 13 October 2022).
40. Rani, P.; Dutta, K.; Kumar, V. Artificial intelligence techniques for prediction of drug synergy in malignant diseases: Past, present, and future. *Comput. Biol. Med.,* 2022, 144, 105334. [CrossRef]
41. Shao, D.; Dai, Y.; Li, N.; Cao, X.; Zhao, W.; Cheng, L.; Rong, Z.; Huang, L.; Wang, Y.; Zhao, J. Artificial intelligence in clinical research of cancers. *Brief. Bioinform.,* 2022, 23, bbab523. [CrossRef]
41. Available online: https://www.mckinsey.com/~{}/media/McKinsey/Industries/Healthcare%20Systems%20and%20Services/Our%20Insights/The%20next%20wave%20of%20innovation%20in%20oncology/The-next-wave-of-innovation-in-oncology. ashx (accessed on 14 October 2022).
42. Available online: https://www.mckinsey.com/industries/life-sciences/our-insights/playing-to-win-in-oncology-keycapabilities-for-success (accessed on 14 October 2022)
43. Available online: https://www.mckinsey.com/industries/life-sciences/our-insights/delivering-innovation-2020-oncologymarket-outlook (accessed on 14 October 2022)
44. Cancer Moonsho. National Cancer Institute. 2022. Available online: https://www.cancer.gov/research/key-initiatives/moonshot-cancer-initiative (accessed on 14 October 2022)

45. Vadla P.K., Naga Pawan Y.V.R., Kolla B.P., Tripathi S.L. (2021) Accurate Detection and Diagnosis of Breast Cancer Using Scaled Conjugate Gradient Back Propagation Algorithm and Advanced Deep Learning Techniques. In: Sengodan T., Murugappan M., Misra S. (eds) *Advances in Electrical and Computer Technologies*. Lecture Notes in Electrical Engineering, vol. 711. Springer, Singapore. https://doi.org/10.1007/978-981-15-9019-1_9
46. Allen, F.; Crepaldi, L.; Alsinet, C.; Strong, A.; Kleshchevnikov, V.; De Angeli, P.; Páleníková, P.; Khodak, A.; Kiselev, V.; Kosicki, M.; *et al.* Predicting the mutations generated by repair of Cas9-induced double-strand breaks. *Nat. Biotechnol.*, 2018, 37, 64–72. [CrossRef] [PubMed]

8

Benefits and Challenges of Decentralization in Education for Resource Optimization and Improved Performance

Ankita Singh[1], Shobhit Sinha[2], Satya Bhushan Verma[2]* and Bipin Kumar Singh[3]

[1]Graphic Era Hill University, Haldwani, India
[2]Shri Ramswaroop Memorial University, Barabanki, India
[3]Department of Mechanical Engineering, Sri Eshwar College of Engineering, Coimbatore, Tamil Nadu, India

Abstract

Policies and practices for inclusive education are clearly affected by the degree of decentralization of national education systems. At all levels of education systems, decentralization is an important element in context analysis and change management. In the current policy environment, decentralization in education systems is an important topic, particularly as countries make the transition towards a more inclusive system. Every nation must achieve the right balance between centralization education and decentralization education, which is central to the reform efforts. The main source of contention and discussion is developing and enacting a policy that determines who should have authority, responsibility, and accountability. There is a greater efficiency in resource usage in decentralized systems. Thus, learners will be able to achieve better performance, parents will be more satisfied, and community involvement and support will be stronger.

***Keywords*:** Decentralization, education systems, accountability, efficiency

**Corresponding author*: satyabverma1@gmail.com

Sandhya Avasthi, Suman Lata Tripathi, Namrata Dhanda and Satya Bhushan Verma (eds.) Decentralized Systems and Distributed Computing, (163–188) © 2024 Scrivener Publishing LLC

8.1 Introduction

8.1.1 Decentralized Systems in Education and Research

In education and research, decentralized systems are structures that distribute decision-making power and resources among different people or groups rather than relying on a centralized authority. Decentralized methods in education and study have several advantages, including increased innovation, flexibility, and adaptability [1].

The Montessori approach, which emphasizes self-directed learning and individualized instruction, is an example of a decentralized structure in education. Students in this system are allowed to choose their own tasks and work at their own pace, with teachers serving as guides and facilitators. The open educational resources (OER) movement, which promotes the sharing of educational materials and resources online, is another example of a decentralized system in education [2]. This approach enables educators and students to gain access to a wealth of information from all over the globe without being bound to a single institution or publisher [3].

Decentralized systems in research can take the shape of collaborative networks, in which researchers from various institutions and disciplines collaborate on a common research project. This method promotes a greater exchange of ideas and perspectives, as well as access to a broader variety of tools and expertise.

Decentralized educational and research institutions have the potential to support equity and inclusion. These systems can help to ensure that marginalized groups have a say in shaping educational and research practices by distributing decision-making power and resources [4].

Decentralized systems have their own set of disadvantages as well. It can be difficult to guarantee quality monitoring and uniformity across different institutions and individuals without a centralized authority. Furthermore, there may be a lack of accountability and oversight, which can contribute to disparities in access and outcomes. Overall, decentralized systems in education and research offer a promising approach for promoting innovation, flexibility, and inclusivity, but require careful planning and management to ensure their efficacy and equity.

8.1.2 Connect Between Democracy and Education

Democracy is a form of government in which citizens hold power through representatives they have elected, whereas education is the process of gaining knowledge and skills. In a democracy, education plays an essential role in shaping citizens' values and beliefs and preparing them for complete involvement in the democratic process.

Education is more than just learning facts and skills; it is also about learning to think critically, assess evidence, and make sound choices. Citizens in democratic societies are expected to be educated and involved in the political process, and education is a critical tool for accomplishing this goal. A well-educated population is better equipped to comprehend complex problems, evaluate elected officials' policies, and hold them accountable.

As is frequently stated, democracy is the most effective method to deal with issues related to humanity in terms of "governing" the people under the umbrella of the state because there is no "better" approach on the horizon. According to this perspective, the government is an organization "for" and "by" the people. Furthermore, it is widely acknowledged that one of the primary goals of a democracy is to support the health and happiness of the people, i.e., the population. In this case, democratic school then will try to teach and employ the means of happiness and it must also teach the people "what the democratic nobility is". However, because the primary goal of democratic education is to raise the entire population to a higher level of intelligence, living conditions, and happiness, it is critical to provide its younger population with the necessary habits, understanding, and skills before they leave school or higher educational institutions [5].

An obvious question to be answered at this juncture is whether it is possible to achieve the "desired" or "necessary" outcomes; first, if you do not have an "enlightened" population who is not and will not be in schools; and second, if you do not trust them in using and practicing their democratic rights and fulfilling their duties as citizens of democracy under the current conditions.

At this juncture, there appears to be a quandary. On the one hand, democratic institutions cannot be deemed "safe" if the vast majority of the population does not trust them in their roles as the primary "actors" and "subjects" of the democratic state and government .On the other hand, it is obvious that democracy must be taught. This is true in terms of governmental matters, whether at the municipal, state, provincial, or national levels. Of course, as with other "learning" circumstances, learning the democracy necessitates some fundamentals such as willingness, guidance, and practice. Then, democratic citizenship may enter the picture with the required skills to meet the obligations and to seek and use the "rights" of citizenship [6]. As a result, education in this picture is not and must not be a process and service carried out within school buildings for particular age groups, but rather it must be for all, and both a process and a service, including teaching and practicing democracy and investment. As it is frequently stated that democracy is simply a question of political morality and economic abundance as a path to political democracy, it is also frequently stated that democracy is more "appropriate" for countries where those necessary conditions exist [7]. However, a free man must establish his or her own goals, and no one should tell them what to do; they must do it for themselves. So, are we going to wait for "suitable conditions" to appear, or is there another way to create the "suitable" situation? First, we note that some of the most conservative - that is, not democratic in many respects - regimes achieved great things, both by tearing down old orders and establishing new, even democratic ones. However, the rationalization of any political system is inextricably linked to the creation, or "making," of new citizens for and within that "new" society. Literacy and the required skills, including technical ones, are needed by the majority of the population, as well as other critical skills in that regard [8].

In any democratic community, agreement on educational objectives and goals are required. Education is then regarded as a process for (i) individual excellence, (ii) a society of equals, (iii) a government of free men, (iv) an economy of security and plenty, (v) a civilization of beauty and grandeur, (vi) a lasting civilization, and (vii) a global community.

However, school systems in a democracy are partially responsible for preparing students as products with certain qualities and socializing them for complicated adult roles. However, this process necessitates some specialization and, as a result, some autonomy on the part of schools and teachers, which may be another issue with people's interests and decisions on education, as well as all types of education matters within the system. At this point, it is known that school systems are made up of client-serving organizations that are at least semi-bureaucratic in nature, with roles that are quite different; however, they still operate according to specific rules and regulations of the process [29].

On education decentralization, it is commonly stated and widely accepted that one of the obvious and clear evidences for an educational system to be considered "democratic" is that it is "decentralized" to some degree. The first concern at this juncture is who governs our city, province, or state. Second, if we have any sayings in all of the tasks in our immediate surroundings and region. This clearly becomes a clear indication and proof of existing democracy. Of course, having distinct and/or special interests, ethnic groups, and so on, and if their differences and interests are functionally taken into account, these are some of the positive indications of democratization [9]. All of this is linked to the issue of what the fundamental and irreversible sources of power, social, economic, political, and philosophical, are. Along these lines, in order to be deemed democratic, any educational decision and/or practice must be based on the primary philosophy that every student be treated both equally and efficiently [50].

Decentralization of education has many advantages. One of the primary benefits of decentralization is that it allows for greater local control and decision-making in education. This means that education policies and programs can be tailored to the specific needs and priorities of individual communities and regions, rather than being dictated by a centralized government authority. Decentralization also promotes greater accountability and transparency in education, which helps to ensure that resources are allocated more effectively and that educational outcomes are improved [30].

In addition, decentralization can foster greater participation and engagement among citizens, particularly marginalized groups who may have been historically excluded from decision-making processes. By giving people a greater voice in educational matters, decentralization can help to promote social cohesion and inclusiveness, as well as promote a sense of ownership and investment in the education system. While there are certainly challenges associated with decentralization, including issues around coordination and resource allocation, the benefits of greater local control and participation in education make it a valuable strategy for promoting democratic principles and improving educational outcomes [51].

8.2 The Centralization and Decentralization Concepts

The concepts of centralization and decentralization are important in educational and research management systems. Centralization is a system in which decision-making authority is concentrated at the top of the organizational hierarchy, whereas decentralization is a system in which decision-making authority is distributed throughout the company.

Decisions in a centralized system are made by a small group of people at the top of the organization who have the authority to control the actions of the rest of the organization. This can lead to greater consistency and uniformity in decision-making, as well as increased efficiency and resource coordination, but it can also lead to a lack of flexibility and responsiveness to local requirements and conditions, because decisions are made without feedback from those on the ground. Decision-making power is distributed throughout the organization in a decentralized education system, allowing for greater input and participation from those at lower levels. As decisions are made by those who have a better grasp of the situation on the ground, this can lead to greater responsiveness to local needs and circumstances [10].

Centralization shifts decision-making authority to higher levels, consolidating authority and power at the head of the organization [11]. All decision-making and authority are concentrated at the top tier of management, which may consist of only a few individuals who dictate policy and make all crucial decisions. In this instance, centralization limits individual responsibilities or involvement at the most fundamental levels [12].

In contrast to centralization, decentralization refers to the degree to which power has been delegated to individual educational organizations at the most fundamental levels. Within this concept, the top delegated more power to local administrators, allowing for greater ingenuity, resourcefulness, and personal development, especially in problem solution. Local communities or regions then have some say over decisions and policies in their separate areas, subject to the oversight of the top authority. There is a process of devolution of fiscal and decision-making power from upper to lower levels of government and organizational groups in the case of education decentralization. This may have an impact on how educational systems make resource and spending decisions [12].

Decentralization is implemented for a variety of reasons in various countries. It is typically associated with the policy of saving money by improving management efficiency, transferring power to the most competent level of government, increasing funding, adopting more general national administrative reforms, and giving users more control over education [13]. However, these factors can be divided into three general categories: educational finance, efficiency and effectiveness, and power redistribution in the education system.

Many studies have attempted to categorize and define decentralization. These attempts have distinguished between decentralization as delegation and decentralization as devolution. In this sense, task decentralization does not inherently imply a shift in power because local agents are typically only tasked with carrying out decisions made at a higher level. Delegation, on the other hand, may suggest greater local autonomy simply because total central control is difficult to achieve. The transfer of power and responsibility from central to local authorities is referred to as decentralization, also known as devolution. The only form of decentralization that clearly increases local authority and independence is devolution [15].

There have been numerous arguments that implementing decentralization provides numerous benefits to growth. These can be linked to decision-making participation or involvement, as well as good governance. Decentralization brings public services closer to people because it allows them to engage more actively in the decision-making process of local policies and activities than centralized decision-making. Decentralization also gives people the chance to participate in decision-making. This allows for greater flexibility and allows for better decisions to be made because those present at the scene are more directly related to the problem [11, 15].

The ideas of centralization and decentralization in the educational system each have their own set of advantages and disadvantages and will be determined by a number of variables, including the organization's specific context and needs, as well as the goals and priorities of the broader education system. Both methods have benefits and drawbacks, and it is up to education policymakers and leaders to carefully consider these factors when deciding how to best run their educational institutes. It is up to a community or a country to decide whether or not to implement the system. The negative effects of the systems can be mitigated as long as there is adequate control and governance.

8.3 Decentralized Systems in Education and Research: Policies and Practices

In education and research, decentralized systems allude to the implementation of distributed decision-making processes in which decisions are made by different stakeholders within the system rather than by a centralized authority. This method is gaining popularity because of its potential for encouraging innovation, participation, and transparency. Here are some policies and practices in education and research that are linked to decentralized systems:

8.3.1 Open Educational Resources (OER) Policy

Governments and educational institutions can adopt policies that promote the creation and sharing of open educational materials [14]. This can include providing funding for OER development, encouraging educators to use and alter existing resources, and ensuring that OER are accessible to all students [15, 16]. OER policies are a collection of guidelines, rules, and regulations that encourage the use of openly licensed educational resources like textbooks, videos, and other instructional materials [18]. OER policies aim to provide educators and students with high-quality learning materials that are free to use, share, and adapt. Among the most prevalent open educational guidelines are as follows:

i) **Use of open licenses:** When making or sharing educational resources, use open licenses that allow for the materials' free use, adaptation, and redistribution [17].
ii) **Ensure accessibility:** Provide alternative formats and accessible design features to ensure that educational materials are available to all learners, including those with disabilities [18].
iii) **Encourage collaboration:** To promote the creation and sharing of high-quality educational materials, encourage collaboration among educators, instructional designers, and content producers [19].
iv) **Assure quality:** Create quality standards for open educational materials to ensure that they are accurate, up to date, and effective for learning [20].
v) **Encourage reuse:** Develop educational resources that can be readily reused and adapted by others to support a wide range of learning requirements and contexts [21].
vi) **Promote sustainability:** Create and share educational tools that are long-lasting and can be maintained over time, such as by leveraging open standards and using open-source software [22].

8.3.2 Collaborative Research Networks

A collaborative research network is a group of researchers or organizations that work together to advance knowledge in a specific area, solve a complex problem, or create new technologies. Governments and funding agencies can prioritize funding for collaborative research networks that bring researchers from various institutions and fields together to work on joint research projects [23, 24]. This can foster knowledge sharing as well as the development of novel research methods and ideas [25].

These networks tend to cluster around a particular study theme or topic and may include researchers from various disciplines, institutions, and countries. Consortia, alliances, partnerships, and networks of excellence are examples of collaborative study networks that can be formal or informal [52].

There are numerous advantages to collaborative study networks. They enable researchers to share resources, expertise, and knowledge; gain access to funding opportunities that individual researchers may not have; and work on bigger and more complicated projects. They also encourage creativity and innovation, as well as the exchange of ideas and the development of new research techniques.

Managing communication, coordination, and intellectual property rights, as well as ensuring that all participants benefit from the collaboration, are some of the challenges that persist for collaborative research networks. To overcome these challenges, clear goals, roles, and responsibilities must be established, as well as effective communication channels and the development of confidence and mutual respect among partners [53].

8.3.3 Self-Directed Learning

Pedagogical approaches that emphasize self-directed learning, such as the Montessori strategy, can be implemented in educational institutions. This can include giving students the freedom to choose their own activities and work at their own speed, as well as training and tools for teachers to act as guides and facilitators.

It refers to the learning process in which a person takes responsibility for their own learning and chooses what and how they want to learn. Setting objectives, identifying resources, and selecting methods for acquiring information and skills without being directed by a teacher or instructor are all part of it [26].

Reading books, watching videos, visiting workshops or conferences, joining online groups, participating in discussions, and experimenting with new skills are all examples of self-directed learning. As learners must manage their own time, evaluate their own success, and adjust their learning strategies as required, it necessitates a high level of motivation, curiosity, and self-discipline.

Self-directed learning can be advantageous in a variety of ways, including fostering autonomy, ingenuity, critical thinking, and lifelong learning. It can also help with personal and professional growth, problem-solving abilities, and confidence and self-esteem [54].

Although self-directed learning comes with its own set of challenges, it may not be appropriate for all learners, as some prefer structured learning settings or require guidance and support from others. Self-directed learning necessitates a certain degree of knowledge, skills, and resources, as well as the ability to evaluate and validate information's quality and relevance.

8.3.4 Peer Review

Peer review is a research practice that assesses the quality and validity of research articles before they are released. It entails a group of field experts reviewing and giving feedback on the study findings. Peer review contributes to the credibility of research findings and is an important component of the scientific process, ensuring that research meets rigorous quality standards and contributes to the advancement of knowledge in a given area.

Peer review involves sending the author's work to several independent experts for feedback, who then provide comments and ideas. The methodology, data analysis, findings, and conclusions of the work are evaluated by the reviewers to ensure that the study is of high quality and meets the standards of the discipline. Peer review can take many shapes, including single-blind, double-blind, and open review. The critic is aware of the author's identity, but the author is unaware of the reviewer's identity. Both the author and the reviewer are anonymous in double-blind evaluation. The name of the reviewer is revealed to the author during open review, and *vice versa*.

Peer review is the process of assessing the quality, validity, and relevance of academic or scientific work by experts in the same area. It is a vital component in ensuring the accuracy and reliability of scientific study and plays a substantial role in knowledge dissemination in academia [55].

The practice of making data, especially research data, freely available to the public is referred to as open data. The aim of open data is to increase transparency and information access, which can facilitate collaboration and accelerate scientific progress. Researchers can more easily build on one another's work, avoid duplication, and improve overall research quality by sharing data publicly. Open data also encourages reproducibility by enabling others to validate findings and evaluate the validity of research methods [56].

8.3.5 Open Data

The practice of making data, especially research data, freely available to the public is referred to as open data. The aim of open data is to increase transparency and information access, which can facilitate collaboration and accelerate scientific progress. Researchers can more easily build on one another's work, avoid duplication, and improve overall research quality by sharing data publicly. Open data also encourages reproducibility by enabling others to validate findings and evaluate the validity of research methods.

8.3.6 Equity and Inclusion

Decentralized systems must be supported by policies and practices that encourage equity and inclusion. This can include offering assistance and resources to marginalized groups, making educational and research practices available to all learners, and actively seeking out and valuing diverse viewpoints.

Equity refers to the reasonable and just distribution of resources, opportunities, and outcomes, whereas inclusion refers to establishing an atmosphere in which everyone, regardless of background or identity, feels valued and supported [27]. It is critical in decentralized systems to ensure that policies and practices support equity and inclusion for all individuals and communities.

To achieve equity and inclusion, it is critical to assist and resource marginalized groups, such as underrepresented minorities or people with disabilities, so that they have equal access to chances as everyone else. This can include providing training, mentorship, and other forms of assistance to people in order to help them overcome barriers to involvement. Educational and research practices should also be intended to make all learners, regardless of background or identity, feel welcome. This can include making learning materials available in alternative forms, using inclusive language, and ensuring that research methods are developed with diverse populations in mind. Promoting equity and inclusion in decentralized systems necessitates actively searching out and appreciating diverse viewpoints. This can include providing chances for people from underrepresented groups to engage in decision-making processes, valuing diverse experiences and viewpoints, and working to overcome biases and barriers that prevent certain groups from participating or advancing. We can build a more just and equitable society for all by prioritizing equity and inclusion in decentralized systems.

8.3.7 Quality Assurance

Quality assurance is an essential component of upholding high standards and uniformity in decentralized systems such as educational and research practices [28]. It entails putting policies, procedures, and practices in place to ensure that the quality of education and study remains high, regardless of location or mode of delivery.

To accomplish this, educational and research institutions must create clear standards and guidelines that define the anticipated outcomes of educational and research practices. These expectations should be conveyed to educators and researchers so that they are aware of them and can align their practices appropriately. Institutions should also provide educators and researchers with the tools and support they need, such as training, technology, and access to research materials, to ensure that they can deliver high-quality education and research.

Establishing oversight and evaluation mechanisms to monitor outcomes and ensure that standards are reached is also part of quality assurance. Institutions should assess their educational and research practices on a regular basis, using metrics such as student outcomes, faculty productivity, and research influence. This evaluation should be used to identify areas for growth and make the required changes to improve education and research quality [29–31].

Maintaining high standards and consistency in decentralized networks requires quality assurance. Institutions can ensure high-quality education and research regardless of method of delivery or location by implementing clear standards and guidelines, providing necessary resources and support, and establishing oversight and evaluation mechanisms.

8.4 Building Capacity Across and Between Levels Within Education Systems

Building capacity across and between levels of education institutions is critical to the success of any educational program [31]. It entails improving the knowledge, skills, and abilities of people at all levels of the education system, from classroom instructors to school administrators, policymakers, and education officials.

There are several approaches to increasing capacity in educational institutions. One approach is to provide chances for professional development for teachers and school leaders. Professional development can take the form of workshops, conferences, or online courses and can cover a variety of subjects ranging from instructional techniques to leadership and management skills.

Establishing mentoring and coaching programs, in which experienced teachers or school leaders provide guidance and support to less experienced peers, is another way to develop capacity. This can help newer teachers and leaders develop their skills and confidence, as well as provide chances for more experienced educators to share their knowledge and expertise.

Building capacity at the policy level can entail improving the skills and expertise of government officials and education policymakers. This can help to ensure that policies are founded on evidence-based research and are tailored to the requirements of the educational system subjects ranging from instructional strategies to leadership and management abilities.

Fostering partnerships and collaboration between various levels of the education system, such as schools, districts, and the larger community, can help to build capability. This can contribute to the development of a more cohesive and effective educational system in which resources are shared and best practices are found and disseminated. Building capacity at all levels of education systems is critical to ensuring that all students have access to high-quality education and that education systems are long-term sustainable and successful.

8.5 Developing Accountability Measures and Systems in Implementing a Decentralized Education Policy

Developing accountability measures and systems in the implementation of a decentralized education policy is critical to ensuring that policy goals are met and that resources are managed and utilized effectively. Here are some stages to developing accountability measures and systems [32].

8.5.1 Establish Clear and Quantifiable Policy Objectives

The first step in creating accountability measures is to define clear and measurable policy objectives. This entails determining the precise goals and outcomes that the policy seeks to achieve, as well as the indicators that will be used to track progress and success.

8.5.2 Create Performance Indicators and Benchmarks

Once the objectives have been defined, the next step is to create performance indicators and benchmarks that will be used to track progress towards the goals. These indicators and benchmarks should be measurable, pertinent, and tailored to the policy goals [33].

8.5.3 Assign Roles and Responsibilities

It is critical to assign clear roles and responsibilities to various stakeholders in a decentralized education policy, including federal and state governments, schools, teachers, and education managers. This guarantees that everyone understands their roles and is held accountable for their actions.

8.5.4 Build Effective Monitoring and Evaluation Systems

It is critical to build strong monitoring and evaluation systems to guarantee accountability in policy implementation. This entails gathering data and information on progress towards achieving policy objectives and using this information to make required policy implementation adjustments [34].

8.5.5 Encourage Transparency and Involvement

In a decentralized education policy, transparency and participation are critical components of accountability. Participants in the policy development and implementation process should be encouraged, and stakeholders should have access to pertinent information about the policy and its implementation.

8.5.6 Establish Redressal Mechanisms

It is critical to establish redressal mechanisms in the event of any issues or complaints. This guarantees that stakeholders have a channel to voice their concerns or issues regarding policy implementation. It is possible to create effective accountability measures and systems for implementing a decentralized education policy by following these steps.

8.6 Developing Local-Level Capacity Across All Education System Levels and Sectors

Building local-level capacity is a critical component of implementing a decentralized education strategy [35]. Here are some strategies for developing capacity across all levels and sectors of the school system:

i) **Providing training and professional development:** It is of the utmost importance to provide training and professional development opportunities for educators, administrators, and other stakeholders in order to create capacity at the local level [36]. This can include workshops on successful teaching practices, data analysis and monitoring training, and leadership and management courses.

ii) **Facilitating mentorship and coaching:** Mentorship and coaching can be effective ways to develop capacity by providing educators and managers with personalized support and guidance. This can include matching experienced teachers with new teachers, coaching school leaders, and creating peer mentoring networks [37].

iii) **Promote collaboration and partnerships:** Collaboration and partnerships can help to develop capacity by encouraging the sharing of resources and knowledge across institutions and sectors. Establishing partnerships between schools and community organizations, facilitating collaboration among various schools or districts, and encouraging the sharing of best practices are all examples of this.

iv) **Developing leadership capacity:** Developing leadership capacity is critical for capacity development at all levels of the educational system. Training for school leaders and managers, the establishment of leadership development programs, and the creation of opportunities for leadership roles and responsibilities are all examples of what this entails [38].

v) **Empowering local decision-making:** To develop capacity at the local level, decentralized education systems should empower local decision-making. This can include establishing mechanisms for community feedback and engagement, as well as promoting shared decision-making among various stakeholders.

In a decentralized education policy, developing local-level capacity necessitates a dedication to training and professional development, mentorship and coaching, collaboration and partnerships, building leadership capacity, and empowering local decision-making. Decentralized education systems can help to ensure that all learners have access to high-quality education and that educational outcomes are consistent across various regions and institutions by building local-level capacity.

8.7 Education Policies in India to Achieve Decentralization in Education Systems

The Indian government has put in place a number of policies to encourage decentralized systems in education and research [39]. The following are some important policies:

8.7.1 National Education Policy (NEP) 2020

The Right to Education (RTE) Act of 2009, also known as the Right of Children to Free and Compulsory Education Act, was approved by the Indian Parliament. The act establishes education as a basic right for children aged 6 to 14 years old and requires the government to provide free and compulsory education to all children in this age group. The act also establishes numerous norms and standards for excellent education, such as teacher–student ratios, infrastructure, and curriculum.

Some of the key features of the RTE Act are as follows [40]:

i) It makes education a fundamental right for all children in the age group of 6 to 14 years.
ii) It mandates that all private schools must reserve 25% of their seats for children from economically weaker sections and disadvantaged groups.
iii) The Act lays down norms and standards for infrastructure, pupil–teacher ratio, and teacher qualifications in schools.
iv) The Act also prohibits the practice of corporal punishment and mental harassment of children in schools.
v) It mandates that no child can be held back or expelled from school until the completion of elementary education.

The Act calls for the formation of School Management Committees (SMCs) composed of parents and teachers, which will be critical in ensuring the Act's successful implementation [41].

The RTE Act and educational decentralization are inextricably linked, as the act envisions a major role for local governments in its implementation. The act requires the formation of local School Management Committees (SMCs) composed of parents and community members to supervise school operations and ensure accountability. The act also requires state governments to create mechanisms for decentralizing educational planning and management, including the delegation of financial and administrative authority to local governments.

8.7.2 Rashtriya Madhyamik Shiksha Abhiyan (RMSA)

The Rashtriya Madhyamik Shiksha Abhiyan (RMSA) is a Government of India-sponsored scheme that was launched in 2009–2010 with the goal of giving quality education to all secondary school students aged 14 to 18. The plan aims to increase secondary education access by providing financial assistance to states and UTs for the building of new schools, the expansion of existing ones, the hiring of additional teachers, and the training of existing teachers.

One of the primary goals of the RMSA is to promote secondary education decentralization by giving states and UTs more autonomy in designing and executing secondary education programs. This aids in tailoring the education system to each state's or UT's unique needs and requirements.

The RMSA also aims to enhance secondary education quality through teacher training and capacity building, curriculum reforms, and the provision of required infrastructure and learning resources. It is a significant initiative by the Government of India to strengthen the country's secondary education system and provide more opportunities for students to receive quality education [42].

8.7.3 RUSA (Rashtriya Uchchatar Shiksha Abhiyan)

RUSA is a Government of India-funded program that aims to increase entry to higher education in India. The plan encourages decentralization by providing funding to states for the establishment of new universities and colleges, as well as granting states greater autonomy in developing and executing higher education programs. RUSA's goal is to enhance the overall quality of higher education in India while also increasing access to education for disadvantaged groups such as women, scheduled castes, and scheduled tribes.

RUSA provides financial aid to eligible HEIs for a variety of purposes, including infrastructure development, faculty recruitment and training, research and innovation, and academic reforms. The plan focuses on promoting equality and inclusion, and it encourages HEIs to take steps to address gender, social, and regional disparities in education [43].

One of RUSA's distinguishing features is its emphasis on outcome-based funding, which means that HEIs must achieve certain performance metrics in order to receive funding. These indicators are based on a variety of factors, such as student enrollment, graduation rates, research output, and the number of women and underrepresented students enrolled in the school.

RUSA is regarded as a significant initiative in the Indian education system because it aims to resolve long-standing issues in the higher education sector, such as insufficient funding, outdated infrastructure, and limited access to education for marginalized groups. RUSA aims to improve the overall quality and relevance of higher education in India by providing targeted funding and support to eligible HEIs, as well as to ensure that all students have the chance to access and benefit from higher education.

8.8 Research and Development Policies in India and Promoting Decentralization in R&D

The Government of India has launched a number of policies and programs to promote R&D in the nation and encourage decentralized promotion of research initiatives. Some of the most important policies in this respect are as follows:

8.8.1 Atal Innovation Mission (AIM)

The Atal Innovation Mission (AIM) is a Government of India project managed by the NITI Aayog. AIM's mission is to foster a culture of innovation and business in India by assisting innovators and entrepreneurs, particularly at the grassroots level.

The mission's goal is to help schools across India create Atal Tinkering Labs (ATLs) to encourage young students to develop innovative ideas and provide them with access to cutting-edge tools and technologies. AIM also funds and supports incubators and start-ups through different programs such as the Atal Incubation Centers and the Atal New India Challenges [44].

AIM's ultimate aim is to foster an innovation and entrepreneurship culture in India that promotes economic development, employment creation, and societal progress. AIM seeks to promote decentralization and foster a spirit of innovation and entrepreneurship at the grassroots level by providing support to local innovation hubs and encouraging local communities to engage in innovation activities.

8.8.2 The National Initiative for Developing and Harnessing Innovations (NIDHI)

The National Initiative for Developing and Harnessing Innovations (NIDHI) is an umbrella program established by the Government of India's Department of Science and Technology to encourage innovation and entrepreneurship in the nation. One of the NIDHI program's primary goals is to promote decentralization by giving assistance to local innovation ecosystems and start-ups across the country.

The NIDHI program includes several projects and schemes, including Technology Business Incubators (TBIs), Seed Support System (SSS), Entrepreneur-in-Residence (EIR) program, and National Science and Technology Entrepreneurship Development Board (NSTEDB) [29]. These projects seek to provide funding, mentoring, and other forms of assistance to start-ups and innovators at various phases of their journey, from concept to market launch and beyond.

The NIDHI program seeks to create an inclusive, diverse, and decentralized culture of innovation and entrepreneurship in India through these initiatives. The program aims to promote regional development, job creation, and economic growth in a sustainable and inclusive way by supporting local innovation ecosystems and start-ups across the nation [45].

8.8.3 National Science, Technology, and Innovation Policy (STI) 2020

The Government of India introduced the National Science, Technology, and Innovation Policy (STI) 2020 to guide the growth of the country's science, technology, and innovation ecosystem over the next decade. With a focus on decentralized innovation and a strong emphasis on the role of women and marginalized communities in research, the strategy seeks to position India among the top three scientific superpowers in the world [46]. One of the main goals of the STI 2020 policy is to promote decentralized innovation by creating an environment conducive to innovation and entrepreneurship throughout the nation, particularly in rural and semi-urban areas. The policy seeks to create a nationwide network of decentralized innovation hubs, science and technology parks, and incubators to support and encourage grassroots innovation.

Another important goal of the policy is to encourage women and marginalized groups to participate in research and innovation. The policy acknowledges the need to address gender and social disparities in science and technology, and it aims to create a more inclusive and diverse ecosystem that can capitalise on the talents and views of all members of society. The STI 2020 policy aims to encourage an inclusive, sustainable, and globally competitive culture of innovation and scientific research in India. The policy is in line with the United Nations Sustainable Development Goals and seeks to add to the overall development and progress of the country [47].

8.8.4 Council of Scientific and Industrial Research (CSIR)

The Council of Scientific and Industrial Research (CSIR) is India's premier research and development agency, with 38 laboratories and centers spread across the nation. The CSIR was founded in 1942 and is supported by the Government of India's Ministry of Science and Technology [48].

The CSIR is structured in a decentralized manner, with each laboratory and center concentrating on a particular field of research and development, ranging from aerospace engineering to biotechnology, materials science to environmental sciences, and information technology to applied mathematics. CSIR laboratories and centers conduct cutting-edge research and development in a variety of disciplines, making important contributions to India's scientific and technological progress.

The mission of the CSIR's research and development activities is to produce new knowledge, products, and technologies that can add to India's economic growth and social development. CSIR conducts research in a variety of fields, including agriculture, health, energy, the environment, and industry, and has contributed to the development of several technologies and products that have had a major impact on society. The decentralized structure of the CSIR has allowed it to create a varied and comprehensive research and development ecosystem in India that is aligned with the country's development goals and requirements [49].

8.8.5 Department of Science and Technology (DST)

The Department of Science and Technology (DST) is an Indian government agency that promotes scientific research and development in the nation. It was founded in May 1971 and is overseen by the Ministry of Science and Technology. The DST funds a variety of programs and initiatives to promote scientific study and innovation. The Science and Engineering Research Board (SERB) is one such program that provides financial support for research projects in engineering, physical sciences, mathematical sciences, chemical sciences, life sciences, earth and atmospheric sciences, and interdisciplinary areas.

Aside from SERB, DST has other programs to promote research and innovation in India, including the Technology Development Board (TDB), the National Science and Technology Entrepreneurship Development Board (NSTEDB), and many others. In addition, the DST works with foreign organizations and countries to foster scientific and technological cooperation and exchange. These policies and programs seek to establish an ecosystem for the decentralized promotion of research initiatives in India by giving grassroots support for innovation, entrepreneurship, and R&D. The aim is to develop a research and innovation culture in the nation, allowing India to become a global leader in science and technology.

References

1. Cornito, C. M. (2021). Striking a balance between centralized and decentralized decision making: A school-based management practice for optimum performance. *Int. J. Soc. Sci. Economic Rev.*, *3*(4), 08-15.
2. Tang, H. (2021). Implementing open educational resources in digital education. *Educ. Technol. Res. Dev.*, *69*(1), 389-392.
3. McBride, M., & Abramovich, S. (2022). Crossing the boundaries through OER adoption: Considering open educational resources (OER) as boundary objects in higher education. *Libr. Inf. Sci. Res.*, *44*(2), 101154.
4. Song, J., Zhang, P., Alkubati, M., Bao, Y., & Yu, G. (2022). Research advances on blockchain-as-a-service: Architectures, applications and challenges. *Digit. Commun. Netw.*, *8*(4), 466-475.
5. Oktay, A. (2013). Decentralization in–And of-education. *Procedia-Soc. Behav. Sci.*, *106*, 777-782.
6. Karlsen, G. E. (2000). Decentralized centralism: Framework for a better understanding of governance in the field of education. *J. Educ. Policy*, *15*(5), 525-538.
7. Dewey, J. (1930). *Democracy and education: An introduction to the philosophy of education*. New York: Macmillan.
8. Barber, B. R., & Battistoni, R. M. (1999). *Education for democracy: Citizenship, community, service: A sourcebook for students and teachers*. Kendall Hunt Publishing Company.
9. Allen, D. (2020). *Education and equality*. University of Chicago Press.
10. Burns, T., 2015. Governing Complex Education Systems. Trust and Education. Keynote presentation delivered at the *OECD Governing Complex Education Systems Conference in The Hague*, 7 December 2015.
11. Brennen, Annick M. 2002. *Centralization Versus Decentralization*. Available online at http://www.soencouragement.org/centralizationvsdecentralization.htm, retrieved on January 3, 2013.

12. Heredia-Ortiz, Eunice. 2007. "*The Impact of Education Decentralization on Education Output: A Cross-Country Study*". Economics Dissertations. Available online at http://digitalarchive.gsu.edu/econ_diss, retrieved on 2 January 2013.
13. Burns, T., Köster, F. and Fuster, M., 2016. *Education Governance in Action: Lessons from Case Studies*. Paris: OECD Publishing
14. Johnson, B., & Christensen, L. (2008). *Educational research: Quantitative, qualitative, and mixed approaches* (3rd). Thousand Oaks, CA: Sage Publications, Inc.
15. Mishra, S. (2017). Open educational resources: Removing barriers from within. *Distance Educ.*, 38(3), 369-380.
16. Karlsen, Gustav E. 1999." Decentralized-Centralism" Governance in Education: Evidence from Norway and British Columbia, Canada. *Can. J. Educ. Adm. Policy*, Issue #13, December 6, 1999.
17. Wiley, D., Bliss, T. J., & McEwen, M. (2014). Open educational resources: A review of the literature. *Handbook of research on educational communications and technology*, 781-789.
18. Hoosen, S. (2012). Survey on governments' open educational resources (OER) policies.
19. Hylén, J., Van Damme, D., Mulder, F., & D'Antoni, S. (2012). Open Educational Resources: Analysis of responses to the OECD country questionnaire.
20. Ehlers, U. D. (2011). Extending the territory: From open educational resources to open educational practices. *J. Open Flexible Distance Learn.*, *15*(2), 1-10.
21. Tlili, A., Altinay, F., Huang, R., Altinay, Z., Olivier, J., Mishra, S., & Burgos, D. (2022). Are we there yet? A systematic literature review of Open Educational Resources in Africa: A combined content and bibliometric analysis. *PLos One*, *17*(1), e0262615.
22. Zainuddin, Z., Shujahat, M., Haruna, H., & Chu, S. K. W. (2020). The role of gamified e-quizzes on student learning and engagement: An interactive gamification solution for a formative assessment system. *Computers & education*, 145, 103729.
23. McBride, M., & Abramovich, S. (2022). Crossing the boundaries through OER adoption: Considering open educational resources (OER) as boundary objects in higher education. *Libr. Inf. Sci. Res.*, *44*(2), 101154.
24. Baas, M., van der Rijst, R., Huizinga, T., van den Berg, E., & Admiraal, W. (2022). Would you use them? A qualitative study on teachers' assessments of open educational resources in higher education. *Internet High. Educ.*, *54*, 100857.
25. Yousoubova, L., & McAlpine, L. (2022). Why is the proposal alone not sufficient for grant success? Building research fundability through collaborative research networking. *Innov. Educ. Teach. Int.*, *59*(1), 93-103.
26. Khodaei, S., Hasanvand, S., Gholami, M., Mokhayeri, Y., & Amini, M. (2022). The effect of the online flipped classroom on self-directed learning readiness

and metacognitive awareness in nursing students during the COVID-19 pandemic. *BMC Nurs.*, *21*(1), 22.
27. Son-Turan, S. (2022). Fostering equality in education: The blockchain business model for higher education (BBM-HE). *Sustainability*, *14*(5), 2955.
28. Channa, A. (2015). Decentralization and the Quality of Education.
29. Abrams, S., X. Chen and M. Downton (2018), *Managing Educational Technology: School Partnerships and Technology Integration*, Routledge, New York, www.routledge.com/Managing-Educational-Technology-School-Partnerships-and-Technology-Integration/Abrams-Chen-Downton/p/book/9781138951020.
30. Rose Ephraim Matete, Forms of decentralization and their implications for educational accountability in Tanzania, *Heliyon*, Volume 8, Issue 5, 2022,e09436, ISSN 2405-8440.
31. Malyan, R. S., & Jindal, L. (2014). Capacity building in education sector: An exploratory study on Indian and African relations. *Procedia-Soc. Behav. Sci.*, *157*, 296-306.
32. Radford, M. (2006). Researching classrooms: Complexity and chaos. *Br. Educ. Res. J.*, *32*(2), 177-190.
33. Suryadi, K. (2007). Framework of measuring key performance indicators for decision support in higher education institution. *J. Appl. Sci. Res.*, *3*(12), 1689-1695.
34. Karimi, S. S., Mulwa, A. S., & Kyalo, D. N. (2020). Stakeholder engagement in monitoring and evaluation and performance of literacy and numeracy educational programme in public primary schools in Nairobi County, Kenya. *J. Educ. Dev. Psychol.*, *10*(2), 10.
35. Linton, S., Clarke, A., & Tozer, L. (2020). Strategies and governance for implementing deep decarbonization plans at the local level. *Sustainability*, *13*(1), 154.
36. Ferguson, T., & Roofe, C. G. (2020). SDG 4 in higher education: Challenges and opportunities. *Int. J. Sustain. High. Educ.*, *21*(5), 959-975.
37. Óskarsdóttir, E., Donnelly, V., Turner-Cmuchal, M., & Florian, L. (2020). Inclusive school leaders–their role in raising the achievement of all learners. *J. Educ. Adm.*, *58*(5), 521-537.
38. Prastiawan, A., Gunawan, I., Putra, A. P., Dewantoro, D. A., Cholifah, P. S., Nuraini, N. L. S., ... & Surahman, E. (2020, December). School leadership skills in educational institutions. In *6th International Conference on Education and Technology (ICET 2020)* (pp. 438-441). Atlantis Press.
39. Govinda, R., & Bandyopadhyay, M. (2006). Decentralization of educational governance in India: Trends and issues. *Educational Decentralization: Asian Experiences and Conceptual Contributions*, 159-176.
40. Jha, P., & Parvati, P. (2020). National education policy, 2020. *Governance at Banks*, *55*(34), 14.
41. Aithal, P. S., & Aithal, S. (2020). Implementation strategies of higher education part of national education policy 2020 of India towards achieving its objectives. *Int. J. Manag. Technol. Soc. Sci. (IJMTS)*, *5*(2), 283-325.

42. Mishra, R. K. (2015). Rashtriya Madhyamik Shiksha Abhiyan Takes a Nosedive. *Econ. Political Wkly.*, 25-27.
43. Mandal, J. H., & Scholar, M. P. (2018). RASTRIYA UCHCHATARA SHIKSHA ABHIYAN (RUSA): A Theoretical Analysis.
44. Kavya, C., & Rishinath, M. (2019). A Study on Awareness of Academicians on ATAL TINKERING LABS–ATAL INNOVATION MISSION. *Think India J.*, *22*(15), 40-48.
45. Ayyappan, P., Jha, B. K., & Rekha, B. J. (2022). A DESCRIPTIVE STUDY ON START-UP INDIA–PROGRESS TOWARDS DEVELOPMENT AND HARNESSING INNOVATION. *Weser Books*, 114.
46. Fukuda, K. (2020). Science, technology and innovation ecosystem transformation toward society 5.0. *Int. J. Prod. Econ.*, *220*, 107460.
47. Putera, P. B., Suryanto, S., Ningrum, S., Widianingsih, I., & Rianto, Y. (2022). Three decades of discourse on science, technology and innovation in national innovation system: A Bibliometric Analysis (1990–2020). *Cogent Soc. Sci.*, *8*(1), 2109854.
48. Jeyshankar, R., & Nachiappan, N. (2018). Council of Scientific and Industrial Research (CSIR-India) Institutes Website: A Webometric Analysis. *Libr. Philos. Pract.*, 1-18.
50. Verma, S. B., & Saravanan, C. (2018, September). Performance analysis of various fusion methods in multimodal biometric. In *2018 International Conference on Computational and Characterization Techniques in Engineering & Sciences (CCTES)* (pp. 5–8). IEEE.
51. Satya Bhushan Verma, Abhay Kumar Yadav, Detection of Hard Exudates in Retinopathy Images *ADCAIJ: Adv. Distrib. Comput. Artif. Intell. J.* Regular Issue, Vol. 8 N. 4 (2019), 4 1-48 eISSN: 2255-2863 DOI: http://dx.doi.org/10.14201/ADCAIJ2019844148
52. Verma, S.B., Yadav, A.K. (2021). Hard Exudates Detection: A Review., Emerging Technologies in Data Mining and Information Security. *Advances in Intelligent Systems and Computing*, vol 1286. Springer, Singapore. https://doi.org/10.1007/978-981-15-9927-9_12
53. Satya Bhushan Verma, Shashi Bhushan Verma, Data Transmission in BPEL (Business Process Execution Language), *ADCAIJ: Adv. Distrib. Comput. Artif. Intell. J.* Regular Issue, Vol. 9 N. 3 (2020), 105-117 eISSN: 2255-2863 DOI: https://doi.org/10.14201/ADCAIJ202093105117 105
54. Verma, S. *et al.* 2019. Contactless palmprint verification system using 2-D Gabor filter and principal component analysis. *Int. Arab J. Information Technol.* 16(1), pp. 23–29.

55. Chandran, S., Verma, S.B.: Touchless palmprint verification using shock filter SIFT I-RANSAC and LPD IOSR. *J. Comput. Eng.*, 17(3), 2278–8727 (2015) DOI: 10.9790/0661-17330108
56. Satya Bhushan Verma, Brijesh Pandey, and Bineet Kumar Gupta, Containerization and its Architectures:A Study, *ADCAIJ: Adv. Distrib. Comput. Artif. Intell. J.*, Vol. 11 N. 4 (2022), 395-409, eISSN: 2255-2863, DOI: https://doi.org/10.14201/adcaij.28351

9

Blockchain in Data Security and Transparency in Business Transactions

Rajat Verma[1*] **and Namrata Dhanda**[2]

[1]*Department of Computer Science and Engineering, Pranveer Singh Institute of Technology, Kanpur, Uttar Pradesh, India*
[2]*Department of Computer Science and Engineering, Amity University Uttar Pradesh, Lucknow Campus, India*

Abstract
The primary motivation for an organization to succeed is the integrity of its data, transactions, and operations. Data, operations, and transactions have proven to be the most essential assets for any organization to grow and evolve. These assets are required to be protected and to be kept safe. The traditional methods that are already available for securing the assets are not performing well enough in tackling the latest cybersecurity attacks as the conventional methods of security are somehow responsible for weakening the standards of security. On the opposite side, cyberattacks are also evolving at a tremendous pace. The evolved cyberattacks and cyberattackers have found a variety of loopholes in traditional security techniques. They can harm the diverse aspects of cybersecurity by identifying loopholes. Therefore, there is a necessity for some modern and SMART approaches and methods for improving the dependability, security, and other essential aspects of data, operations, and transactions in an organization. The SMART keyword indicates characteristics such as Specific, Measurable, Achievable, Realistic, and Timely. Here, blockchain comes into the scenario, which acts as the next big thing in this security domain. This chapter focuses on enhancing trust and integrity while performing any business transactions through the network peer-to-peer technology blockchain and illustrates a detailed review of the security attacks that data, transactions, and operations could face while operating in any conventional technology.

***Keywords*:** Blockchain, privacy, security, transactions, transparency

**Corresponding author*: rajatverma310795@gmail.com

Sandhya Avasthi, Suman Lata Tripathi, Namrata Dhanda and Satya Bhushan Verma (eds.) Decentralized Systems and Distributed Computing, (189–214) © 2024 Scrivener Publishing LLC

9.1 Introduction

With the advent of technology, the lives of people became easy and convenient. Today, people are dependent on technology in different aspects. For instance, people use wireless networks now such as hotspots, which is a Personal Area Network (PAN) [1], and people operate devices that are operating via the Internet; thus, in general, they use the Internet of Things (IoT). IoT is a great use case in the field containing voluminous facts and figures, thus requiring security and privacy enhancement. IoT is an Industry 4.0. buzz term that started in the year 1832 with the invention of the electromagnetic telegraph [2, 3]. Since then, IoT has continuously developed. It is giving birth to voluminous data every second by performing many operations and transactions that are traversing through the Internet. This can contain sensitive and private data for an individual or a firm that is required to be secured and protected. This is so because the traditional methods and strategies responsible for securing assets, data, transactional information, and operational data are outdated. Hence, there is a necessity for some innovative solutions. With this requirement, blockchain comes into the picture.

The Blockchain-Like-Protocol was invented in 1982 by American Cryptographer David Chaum who presented the protocol in his dissertation [4, 5]. Later, in 1991, the concept of Merkel Root was added [6]. Then, in 2008, the initial release was led by the pseudo group named Satoshi Nakamoto, which used blockchain behind the cryptocurrency Bitcoin [7, 8]. Since then, blockchain has evolved in four phases ranging from Blockchain 1.0 to 4.0. The Initial Phase of Blockchain 1.0 focused on the origin of Bitcoin and worked as the security backbone of Bitcoin. This phase was known as Bitcoin Emergence. This phase ran from the year 2008 to 2013. The succeeding stage was known as Development in Ethereum; this phase was termed Blockchain 2.0., and ran from 2013 to 2015 [9, 10]. Then, later on, blockchain arrived in the third phase, which was known as the Applications Phase of Blockchain. This phase started in the year 2015 and focused on Decentralized Applications (DApps). The proper functioning of DApps requires the Truffle Suite to work, i.e., the core component of the Application Phase of Blockchain (3.0.) [11]. The Truffle Suite works with the amalgamation of three components: Ganache, Truffle, and Drizzle. Ganache allows the examination of smart contracts. Truffle permits the developer to build decentralized applications. There are diverse features of Truffle such as automated testing, web & console applications, and package management [11–13]. They can eradicate many security concerns of different technologies. A good use case could be IoT (as an application) that could be integrated with blockchain to enhance the IoT system's security. The third component of the truffle suite is Drizzle. It can make the front end of a DApp very informative [14]. It has many libraries that can be integrated with any number of technologies such as Web3. Then comes the latest version of blockchain that deals with the amalgamated version of blockchain with artificial intelligence. This form of blockchain is under research and will transform the functioning of almost every system happening in this cyberspace [15, 16].

The transactions and operations happening in a business organization are very important to secure as it contains confidential data.

9.2 Dimensions and Requirements of Data Security

Traditionally, there are four dimensions of data security, namely, Physical, Personnel, Procedural, and Technical domains, which are depicted below in Table 9.1 [17].

Table 9.1 illustrated the data security domains. These dimensions focus on some requirements, which are as follows:

Table 9.1 Data security: dimensions.

S. no.	Dimension	Consideration
1.	Physical	The system and the assets (data, transactions, and operations) must be kept in a secure physical environment. The system and the assets (data, transactions, and operations) should be physically inaccessible to unauthorized personnel.
2.	Personnel	The team responsible for securing the system and the assets (data, transactions, and operations) must be reliable and trusted. Regular audits and background checks are also required for this dimension of data security.
3.	Procedural	The procedures and the assets (data, transactions, and operations) must assure reliable data.
4.	Technical	Storage and transmission of assets (data, transactions, and operations) must be safeguarded by technology.

- Confidentiality: This is the aspect of cybersecurity that certifies that only the intended sender and recipient should see the facts or corresponding data [18]. It is an act of maintaining processed facts and figures secret and non-public. Confidentiality has several sub-aspects to look out for, which are as follows:
    - Privacy of Data Communication: For Government Operations, Privacy involves the concern about collecting and studying an individual's demographic particulars, while incorporating the confidentiality of the citizens. If any confidential and sensitive information is out of the authorized domain, then it can lead to unfavorable circumstances. Similarly, for businesses, confidential proprietary information and facts, trade secrets, and sales plans should be kept private to avoid unfavorable circumstances [19].
    - Access Control (Granular): The granularity indicates the extent to which the access of data can be differentiated for the different aspects of the database. A database is a collection of interrelated records. That differentiated aspect includes rows, columns, particular tables, views, etc [20].
    - User Authentication [21]
    - Secured and Protected Storage of Private or Sensitive data [22]
- Integrity: Processed facts and figures should be actual. This means that the data should not be changed when traversed through the network or the Internet. The network traffic should not be modified. It should be saved from eavesdropping, corruption, and deletion [23].
- Availability: It meant that the processed facts and figures should be available when it is required. It could be onsite or offsite ensured by storage [24, 25]. It has four aspects, which are highlighted in Table 9.2.
- Authentication: This term depicts something presenting itself to be *bona fide*. The user name and the password are the main consent for the process of authentication. The techniques that contribute to the authentication procedures can be of many types such as certificates concerning X.509 [26, 27], the digest and authentication procedures of hypertext transfer protocol (HTTP) [28], and login forms.
- Authorization: This term depicts that consent is provided to authorized users for performing certain actions. The initial

Table 9.2 Dimensions of data availability.

S. no.	Aspect of availability	Description
1.	Resistance	This aspect of availability focuses on defending the attacks from runaway queries wrt. the database. The system should be able to defend the deliberate attacks or situations.
2.	Scalability	The performance of the system should remain identical throughout. The system should be able to handle the devices or nodes or systems in an exponential increase or decrease comfortably.
3.	Flexibility	There should be appropriate measures for managing the population. The use of a directory could be encouraged.
4.	Ease of use	Valid user's ability should not get degraded.

parameter for this process is authentication. The Access Control parameter of the Uniform Resource Locator (URL) [29, 30] lists the various procedures for this authorization scenario.

- Auditing: The Security standards should be checked periodically for attaining a maximum amount of security [31].
- Non-Repudiation: It certainly meant that an individual cannot gainsay something [32].

9.3 Security Issues in Conventional Data Security

Everything requires a change and an upgrade, and so does technology. Likewise, the conventional security measure is outdated, which can lead to the mishappening of data, operations, and transactions. Table 9.3 shows the conventional data security issues and their corresponding remark that it requires.

Table 9.3 depicts some of the general issues and challenges of Traditional Data Security, and there is a need for some modern and smart solutions that can address these issues appropriately and can enhance the security perspective of Traditional Data Security [33–36].

Table 9.3 Issues faced by traditional data security.

S. no.	Issues of traditional data security	Remark
1.	Erroneous and illegitimate device access	Traditional data security acknowledges that it can conveniently trust systems available in a network. Subsequently, it does not recommend any further verification.
2.	Insufficiency in encryption	Traditional data security acknowledges that it uses fragile standards of cryptographical aspects in encryption and therefore becomes accessible to diverse attacks.
3.	Buggy software susceptibility	Traditional data security acknowledges that buggy software plays a key part in activating logic bombs.
4.	Unavailability of a trusted environment for the implementation of operations and transactions	Traditional data security acknowledges that it uses a general-purpose computer that can install unrestrained and unwanted buggy software's that can activate logic bombs and encourage cyber espionage activities.
5.	Negligible privacy	Traditional data security acknowledges that saving passwords, and private facts and figures on systems can act as a loophole to attack.
6.	Idleness with intrusion	Traditional data security acknowledges that now, the device's security is negotiated, and they function the same way as normal. Thus, the user cannot know if any vulnerability has happened.
7.	Centralization	Traditional data security acknowledges that it has a single point of fault and failure. Thus, any problem in the central server will lead to the complete shutdown of the complete network
8.	Data attacks	Traditional data security acknowledges that it permits data attacks to happen such as data inconsistency, data tampering, data breach, etc.

9.4 Diversity of Attacks in Conventional Data Security

Apart from the issues highlighted in Table 9.3, the evolved cyberattacks have been classified into four sub-categories:

- Physical Attacks: When the environment of the physical domain is negatively affected, it is a physical attack. Tampering, injection of malicious as well as fake nodes, centralization, and sleep denial attacks are popular examples of physical attacks [37, 38]. Sleep denial attacks are those attacks in which illegitimate inputs are provided, causing the device's battery to drain quickly and cause the device to completely shut. If fake or malicious nodes are injected, then it can lead to data/transactions/operations mismanagement and can trigger unauthenticated mechanisms [39].
- Network Attacks: These cyberattacks consider attacking a complete network of systems or can target even a single system with bots or scripts. Traffic analysis and MitM cyberattacks are a few instances of network attacks [40, 41]. Traffic analysis is a kind of passive attack in which the attacker keeps an eye on the victim and observes their activities very closely [42]. In MITM attacks, the attacker tries to be a real entity and tries to gain all the sensitive information and change the integrity of the information as well [43].
- Software Attacks: These attacks focus on attacking the systems with malware. Malware can be in any configuration, for instance, adware, worms, trojan horse, viruses, and other forms as well. An action with them can act as logic bombs and can trigger other unwanted applications/operations/transactions, etc. The intention can be simple such as stealing sensitive or private data or facts and figures, infecting the system, etc [44, 45].
- Data Attacks: These are the attacks that are primarily done to affect the integrity, confidentiality, and availability of the facts and figures (data) in addition to the other parameters. The other parameters are authentication, authorization, non-repudiation, and auditing [46–48].

9.5 Blockchain: The Complete Solution

Blockchain is a digital logbook that contains complete information of transactions that happen among the systems that are connected. Every block in a blockchain is a storehouse of transactions. Whenever a new transaction is completed and validated by the miner, it is added to each ledger. The transactions are entered in the log book with the hash value, particularly SHA-256, which gives a fixed value of 64 hexa-decimal characters [49, 50]. A very popular instance of distributed ledger (DLT) is a service offered by the multinational technology company Google, which is Google Docs [51, 52]. The Google Doc is an online document that, when shared with a group of people, can edit together at the same time [53].

Blockchain has three automatic salient characteristics such as immutability, decentralization, and transparency. It is a chain of systems that are acting as nodes. The bonding characteristic between these systems or nodes is the presence of hash values [54]. The starting block of the chain is named genesis block, which has two sections. The initial section corresponds to a header section and the bottom section is the data of the body. The upper section i.e., the header, contains the order of the constituent, the value of the current digest, the previous hash related to the next constituent unit, nonce, and finally the timestamp. The organized blocks are structurally organized. Figure 9.1 shows the complete structure of the block [55–57].

When these blocks are tied up together, they form the blockchain. The initial component is the order number that depicts the position number of the system in the blockchain. The first block will not have any previous hash but will have other entities as the conventional one. Creation time indicates the time of block formation. Figure 9.2 illustrates the layout of the blockchain for a quick and easy grasp of the subject.

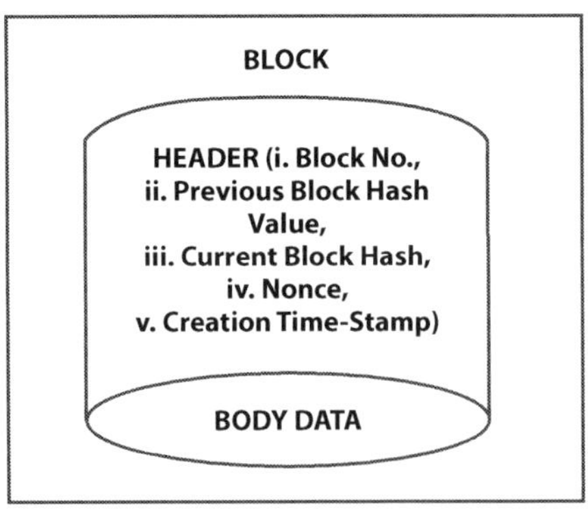

Figure 9.1 Constituent unit of blockchain [55].

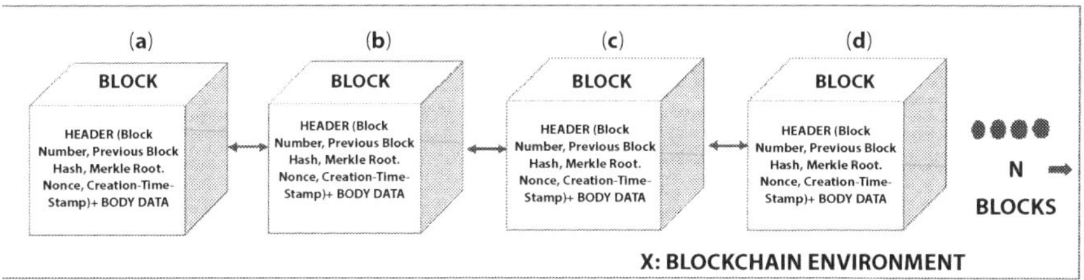

Figure 9.2 Blockchain.

In Figure 9.2, the demo layout is highlighted where four nodes, namely, A, B, C, and D, are available in the blockchain environment X. However, the number of blocks can be appended as per the needs of the network can transform the blockchain of four blocks to N blocks where N can be any number. To mine the blocks or to add the blocks to the blockchain, miners are required, who are the coders, to maintain the complete process of mining and validating transactions. The procedure of appending the constituents to the chain is popularly termed mining. The users/systems involved in this resolve tough mathematical scenarios to add the data to the constituent unit. This process is traditionally done with the consensus procedure known as proof of work (PoW) [58]. A few of the famous coins such as Monero coin, Dogecoin, and Litecoin follow PoW [59–61].

The major alternative to the traditional PoW is the proof of stake [62], which works on the updated thought that each system has its own capacity, that is, the number of coins they can handle. It depicts that the higher the capacity of coins the node can hold, the higher the computing power the node can handle [63]. Shadow coin, Peer coin, and Black coin are a few examples of proof of stake [64–66]. It is efficient in saving energy. Another consensus mechanism after proof of stake is proof of capacity (PoC), which utilizes the space present in the hard-disk drives and other secondary storage devices that are connected to the system. This is a greener approach to validating transactions and performing mining. Then, the proof of elapsed time (PoET) became available, which was discovered by the Intel Corporation [67]. It follows a similar approach to the architecture of a lottery system [68].

Proof of authority (PoA) is another such mechanism that gives a more flexible and frequent approach to miners and validating transactions. It also utilizes little computing power and very low electricity [69, 70]. When the two most renowned consensus algorithms, i.e., PoS and PoW, are combined, proof of activity (PoA) is formed. The other types of consensus mechanisms are depicted in [71–73].

9.6 The Decentralized and Immutable Approach

Blockchain follows the concept of decentralization and thus removes the dependency on a single administrator. Blockchain also follows immutability as its automatic feature as it only allows the append option and prohibits in-between writing. The following figure illustrates how blockchain can implement decentralization and can remove the single point of fault and failure and how immutability can save the ownership of anything. The concept of centralization in traditional data security, thus implementing fabricated data, is shown below in Figure 9.3.

Figure 9.3 shows the illustration where the original owner cannot declare his real ownership because of being the victim of the central point of failure. There is a human whose name is "X" who has saved some money to purchase a car. At the showroom, X gave the money to the organization and brought the car to his home. As proof of ownership, the company has issued a government-authorized certificate whose information was stored in a traditional ledger that was centralized in nature. As an analogy, it can be considered that the information was stored in a book. Now, there was an attacker who never liked X. The attacker knew that X has purchased a car whose information was only stored in a single centralized ledger or database.

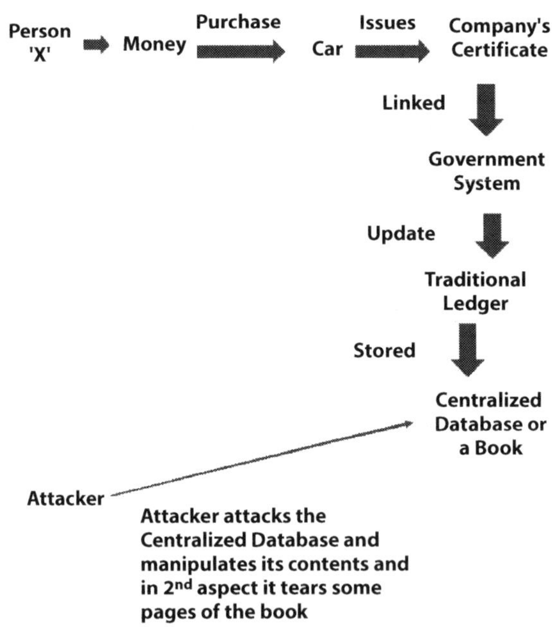

Figure 9.3 Concept of fabricated data in traditional data security.

The attacker attacked the same ledger and only manipulated X's contents of ownership. In the analogy section, the attacker just tore the pages of the book in which X was the owner of the car. Now, X cannot declare that he was the actual owner of the car, as the transaction/data/operation information was only stored in a single centralized ledger.

Now, if the transaction/data/operation records were available in the emerging technology blockchain, then suppose, the transaction/data/operation information of X was stored on the second block of the blockchain. The attacker attacks the second node, and all the hash values will be changed. Thus, to maintain the balance in the system, the attacker must change the previous hash of the third node as well because after tampering with the second block, the current hash was changed and there is a mandatory condition that the hash of the current block should be equal to the previous hash of the succeeding block. Now, if the attacker has completed this requirement, the attacker must perform the same with the third and fourth blocks as well. Now, if the attacker has many resources to completely tamper with the contents of a single chain, then also by the principle of decentralization, the attacker must change all the chains present in the decentralized network, which is a very hard task and infeasible as well [74, 75]. Blockchain does not allow the in-between writing and only allows the append at the last. Hence, tampering/fabricated data are minimized in the blockchain.

9.7 Comparison of ECC and RSA

Blockchain follows the concept of Elliptic Curve Cryptography (ECC) with digital signatures, thus yielding a combined concept of Elliptic Curve and Digital Signatures (ECDSA). ECC was invented in 1985 just five years later of the release of the Rivest–Shamir–Adleman Algorithm (RSA) [76–78]. RSA is asymmetric key cryptography that provides very strong security but at a cost of very large keys. ECC provides identical strength security when compared to the traditional RSA but with a very a smaller number of keys. The key strength of RSA is compared with the key strength of ECC in Table 9.4 [55, 79, 80].

In Table 9.4, The key strength of ECC and RSA is highlighted, which provides the approximate values while comparing the ECC and RSA keys. The ECC-to-RSA ratio is 1:6 in the first case, followed by 1:9 in the second case, further followed by 1:12, then 1:20 and lastly 1:30. Talking about the nature of ECC, i.e., at 512 bits, it is computationally practicable, whereas, for the RSA at 15,360 bits, it is computationally impracticable [55, 81].

Table 9.4 Key strength of RSA and ECC (approximate).

Sr. no.	Key strength (nature: approximate)				Security strength (bits) (approximate)	Tenure of validity (approximate)
	Key length (approximate)		Time (approximate)			
	ECC (Bits)	RSA (Bits)	ECC (Sec)	RSA (Sec)		
1.	160 to 223	1024	0.08	0.16	80	<=2010
2.	224 to 255	2240	0.18	7.47	112	<=2030
3.	256 to 383	3072	0.27	9.80	128	2030+
4.	384 to 511	7680	0.64	133.90	192	2030+
5.	Above 512	15360	1.44	679.06	256	2030+

9.8 Digital Signature Algorithm

This algorithm is based on public key cryptography that involves two keys. One of the two keys is private and the other will be a public key. In the physical world, the handwritten signature binds the signer to the message, whereas, in the digital world, it happens with the digital data. In the digital signature, the secret key signs the document, and the public key is available at the receiver's side for verification purposes. Although it is not fixed, it is mainly followed. The sender sends the data or the facts and figures to the hashing function, and as an output, a hash value is generated, which is signed by the private key of the signer and the signature (digital) is ready. This digital signature will travel through the Internet toward the receiver side. On the receiver side, the authenticator serves the complete signature obtained and the verification key into the verification algorithmic rule. The verification algorithm should give some value as output. The authenticator runs the same hash function to generate the hash value. For verification, the hash value and the output of the verification algorithm are compared. If they are adequate and equal, the digital signature is justified, and if not, that means some tampering is attempted [82]. The importance of digital signatures ensures three characteristics, which are mentioned below:

- Message Authentication [83]
- Data Integrity [84]
- Non-Repudiation [85]

9.9 Blockchain Technology Ensures Transparency

The facts and figures that are available in the blockchain are distributed throughout the web of systems, which are connected. Because of this, the vulnerability of data tampering/inconsistent data/data fabrication is minimized. With transparency, blockchain provides four credible features, which are depicted below:

- Absolute Traceability: It is much easier to maintain the complete blocks of data with this characteristic feature. One of the best use cases of this traceability is online gaming, where the miners/users can authenticate the server seeds of the game and can verify the genuineness and integrity of the system. This is because of the complete transparency. If transparency was not there, then all the aspects will be hidden [86–88].
- Able to be audited completely: Despite the technology being highly private, it gives complete transparency and is completely auditable [89, 90].
- Precise Ledger of Transactions: It provides an accurate ledger of transactions with transparency, which is an automatic feature of the complete blockchain network [91].
- Extreme Security and Safety: When seen globally, blockchain has one more characteristic feature, i.e., state-of-the-art data security, which traditional data security cannot provide. The transactions that happened in a blockchain network are unalterable, which means data tampering is minimized over here. If one tries to alter any of the blocks, then all the miners/systems/nodes will get to know about this.

That means the manipulation of the blocks is impossible [92, 93].

9.10 Blockchain Solutions for Traditional Data Security

Blockchain has proven itself to be the utmost solution for every concern of traditional data security. Table 9.5 shows the blockchain solutions for data security enhancement with the help of blockchain [94–97].

Table 9.5 Blockchain solutions for data security enhancement.

S. no.	Issue of data security	Property of blockchain	Solution achievable
1.	Erroneous and illegitimate device access	Immutable and protected by SHA-256	Yes, complete.
2.	Insufficiency in encryption	Blockchain follows the secure hashing algorithm predominantly 256 and the amalgamated approach of elliptic curve and digital signature (ECDSA).	Yes, complete.
3.	Buggy software susceptibility	The environment in blockchain is different and follows various consensus mechanisms.	Yes, complete.
4.	Unavailability of a trusted environment for the implementation of operations and transactions	The environment is trusted as it works on strong encryption mechanisms and follows consensus mechanisms. In addition, it follows transparency and immutability.	Yes, complete.
5.	Negligible privacy	Privacy is enhanced as transparency is there in blockchain.	Yes, complete.
6.	Idleness with intrusion	As intrusion in blockchain is detected, the hash values are completely changed indicating that some tampering is attempted.	Yes, complete.

(Continued)

Table 9.5 Blockchain solutions for data security enhancement. (*Continued*)

S. no.	Issue of data security	Property of blockchain	Solution achievable
7.	Centralization	Blockchain follows decentralization, thus eliminating the central point of failure.	Yes, complete.
8.	Data attacks	In blockchain, in-between writing is prohibited and tampering with data will cost entire hash values to change. Therefore, data attacks are minimized here in blockchain.	Yes, complete

With the above solutions provided in Table 9.5, a few features are guaranteed by the blockchain, which are highlighted below:

- Enhancement of Trust [98]
- Enhancement of Security [99]
- Enhancement in Fairness [100]
- Enhancement in Integrity [101]
- Enhancement in Authenticity [102]
- Enhancement in Availability [103]
- Enhancement in Authorization [104]

9.11 Blockchain Types

With the diversity in Blockchain, a user can opt for different types of blockchain services as per their needs. A very famous classification is given by Lin and Liao [105]. In general, there are 8 types of blockchains, which are illustrated below for an easy grasp of the subject.

- Public Blockchain: This blockchain is non-restrictive, which means that any user who is available on this Internet can be a part of this spectrum of public blockchain and can perform the process of mining. Litecoin, Ethereum, and Bitcoin are very popular examples of public blockchains. It has trust, is transparent, and is open.
- Private Blockchain: It is closed and permission-oriented. Here, all the controlling and monitoring actions are in the hands of the specific organization. It is limited; thus, everyone who is on the Internet cannot be a part of this private blockchain and cannot perform mining. Ownership of assets and supply chain management are popular use cases for it. It is fast and scalable as everything is monitored by a single authority.
- Hybrid Blockchain: It is a well-amalgamated form of diverse blockchains. Here, the user can face freedom and restrictiveness at the same time. It is not expensive, but KYC is a mandatory condition in hybrid blockchain. It has a minimum transaction cost and protected from 51% attack.
- Consortium Blockchain: It is handled by multiple parties, rather than a central authority. It is also known as federated blockchain. Government agencies and banks are the major use cases of it.
- Public Permissioned Blockchain: It focuses more on proof of stake (PoS) rather than the conventional proof of work (PoW). It is a permissioned version of the public variant. With the PoS consensus mechanism, the speed will be greater if compared with the PoW. Ripple and SOVRIN are popular examples of this version of blockchain.
- Public Permissionless Blockchain: With this blockchain, there is no central authority at the network layer. Here, proof of work is preferred over any other consensus procedure. In PoW, the miners perform complex mathematical problems and get a reward for their efficiency.
- Private Permissioned Blockchain: This blockchain should take care of centralization and consensus, as it is only their major drawback. There is no requirement for any tokens.
- Private Permissionless Blockchain: In this blockchain, the users cannot reveal the data or the facts and figures; rather, they can only acknowledge their existence. Holochain is a perfect example of this category of blockchain [106–113].

9.12 Conclusion

The world has witnessed the contribution of data, transactions, and operations in developing the lives of human beings. Due to this popularity, data, transactions, and operations are omnipresent. With this presence, data are generated everywhere, which need to be protected. The importance of data and transactions is highlighted in this chapter. With this evolution, the negative side, i.e., cyberattacks, has also evolved at a great pace that needs to be tackled by the positive industry itself. The next big thing, i.e., the blockchain, tackles diverse security issues of data with great effort and determination. The dimensions of data security and data availability are highlighted in Tables 9.1 and 9.2, respectively. The issues faced by traditional data security are also highlighted in Table 9.3. The blockchain introductory concepts along with its characteristic features are also highlighted in this chapter. The comparison of the famous RSA and ECC algorithms is also illustrated here. Fabricated data and tampering in conventional data security and how it can be dealt with blockchain are illustrated here. The classification of blockchain along with the digital signature algorithm is shown here. The blockchain solution for traditional data security and how blockchain ensures transparency are shown in this chapter for an easy grasp of the subject.

References

1. S. L. Cotton and W. G. Scanlon, "Characterization and Modeling of the Indoor Radio Channel at 868 MHz for a Mobile Bodyworn Wireless Personal Area Network," in *IEEE Antennas Wirel. Propag. Lett.*, vol. 6, pp. 51-55, 2007, doi: 10.1109/LAWP.2007.890769.
2. O. V. Makhrovskiy, "180 Years of telecommunication in Russia," *2012 Third IEEE HISTory of ELectro-technology CONference (HISTELCON)*, 2012, pp. 1-6, doi: 10.1109/HISTELCON.2012.6487585.
3. W. F. Magie, "Joseph Henry, Pioneer in Space Communication," in *Proc. IRE*, vol. 30, no. 6, pp. 261-266, June 1942, doi: 10.1109/JRPROC.1942.230994.
4. A. T. Sherman, F. Javani, H. Zhang and E. Golaszewski, "On the Origins and Variations of Blockchain Technologies," in *IEEE Secur. Priv.*, vol. 17, no. 1, pp. 72-77, Jan.-Feb. 2019, doi: 10.1109/MSEC.2019.2893730.
5. D. Chaum, "cMix: Mixing with Minimal Real-Time Asymmetric Cryptographic Operations." *Appl. Cryptogr. Netw. Secur.*, pp. 557-578, 2017, doi: 10.1007/978-3-319-61204-1_28.
6. D. Bayer, S. Haber, and W. S. Stornetta, "Improving the Efficiency and Reliability of Digital Time-Stamping." *Sequences II*, pp. 329-334, 1993, doi: 10.1007/978-1-4613-9323-8_24.

7. M. HÜTTEN, "The soft spot of hard code: blockchain technology, network governance and pitfalls of technological utopianism." *Glob. Netw.*, vol. 19, no. 3, pp. 329-348, 2018, doi: 10.1111/glob.12217.
8. S. Semenzin, D. R. Domingo, and S. H. Collado, "Blockchain-based application at a governmental level: Disruption or illusion? The case of Estonia." 2022, doi: 10.31235/osf.io/f7hzs.
9. X. Li, P. Jiang, T. Chen, X. Luo, and Q. Wen, "A survey on the security of blockchain systems." *Future Gener. Comput. Syst.*, vol. 107, pp. 841-853, 2020, doi: 10.1016/j.future.2017.08.020.
10. R. Verma, N. Dhanda, and V. Nagar, "Addressing the Issues & Challenges of Internet of Things Using Blockchain Technology", *IJAST*, vol. 29, no. 05, pp. 10074 - 10082, May 2020.
11. R. Verma, N. Dhanda, and V. Nagar, "Application of Truffle Suite in a Blockchain Environment." *Proceedings of Third International Conference on Computing, Communications, and Cyber-Security*, pp. 693-702, 2022, doi: 10.1007/978-981-19-1142-2_54.
12. H. Ali Mutar and M. S. AL-Huseiny, "Implementation of national cryptocurrency using ethereum development platform." *Period. Eng. Nat. Sci. (PEN)*, vol. 7, no. 3, p. 1021, 2019, doi: 10.21533/pen.v7i3.634.
13. S. Tiwari, N. Dhanda, and H. Dev, "A Systematic Review of Adoption of Blockchain and Machine Learning Technology and Its Application." *Advances in Intelligent Systems and Computing*, pp. 73-95, 2021, doi: 10.1007/978-981-16-5301-8_6.
14. N. R. Pradhan, R. Mahule, P. K. Wamuyu, P. K. Rathore, and A. P. Singh, "A Blockchain and AI Based Vaccination Tracking Framework for Coronavirus (COVID-19) Epidemics." *IETE J. Res.*, pp. 1-13, 2022, doi: 10.1080/03772063.2022.2058630.
15. U. Bodkhe, "Blockchain for Industry 4.0: A Comprehensive Review." *IEEE Access*, vol. 8, pp. 79764-79800, 2020, doi: 10.1109/access.2020.2988579.
16. R. Verma, V. Nagar, and S. Mahapatra, "Introduction to Supervised Learning." *Data Analytics in Bioinformatics*, pp. 1-34, 2021, doi: 10.1002/9781119785620.ch1.
17. N. K. Malhotra, F. M. Ulgado, J. Agarwal, G. Shainesh, and L. Wu, "Dimensions of service quality in developed and developing economies: multi-country cross-cultural comparisons." *Int. Mark. Rev.*, vol. 22, no. 3, pp. 256-278, 2005, doi: 10.1108/02651330510602204.
18. R. Verma, N. Dhanda, and V. Nagar, "Security Concerns in IoT Systems and Its Blockchain Solutions." *Cyber Intelligence and Information Retrieval*, pp. 485-495, 2021, doi: 10.1007/978-981-16-4284-5_42.
19. N. Z. Khidzir, K. A. Mat Daud, A. R. Ismail, M. S. A. Abd. Ghani, and M. A. H. Ibrahim, "Information Security Requirement: The Relationship Between Cybersecurity Risk Confidentiality, Integrity and Availability in Digital Social Media." *Regional Conference on Science, Technology and Social Sciences (RCSTSS 2016)*, pp. 229-237, 2018, doi: 10.1007/978-981-13-0074-5_21.

20. D. Berdik, S. Otoum, N. Schmidt, D. Porter, and Y. Jararweh, "A Survey on Blockchain for Information Systems Management and Security." *Inf. Process. Manag.*, vol. 58, no. 1, p. 102397, 2021, doi: 10.1016/j.ipm.2020.102397.
21. M. L. Das, "Two-factor user authentication in wireless sensor networks." *IEEE Trans. Wirel. Commun.*, vol. 8, no. 3, pp. 1086-1090, 2009, doi: 10.1109/twc.2008.080128.
22. A. Raj, R. Arunprasath, and S. Vigneshwari, "Efficient mechanism for sharing private data in a secured manner." *2016 International Conference on Circuit, Power and Computing Technologies (ICCPCT)*, 2016, doi: 10.1109/iccpct.2016.7530220.
23. F. O. Ayodele, L. Yao, and H. Haron, "Promoting Ethics and Integrity in Management Academic Research: Retraction Initiative." *Sci. Eng. Ethics*, vol. 25, no. 2, pp. 357-382, 2018, doi: 10.1007/s11948-017-9941-z.
24. M. Cabric, "Confidentiality, Integrity, and Availability." *Corporate Security Management*, pp. 185-200, 2015, doi: 10.1016/b978-0-12-802934-3.00011-1.
25. K. Y. Chai and M. F. Zolkipli, "Review on Confidentiality, Integrity and Availability in Information Security." *J. ICT Educ.*, vol. 8, no. 2, pp. 34-42, 2021, doi: 10.37134/jictie.vol8.2.4.2021.
26. S. Chokhani and W. Ford, "Internet X.509 Public Key Infrastructure Certificate Policy and Certification Practices Framework." 1999, doi: 10.17487/rfc2527.
27. L. Zhu, K. Jaganathan, and N. Williams, "Online Certificate Status Protocol (OCSP) Support for Public Key Cryptography for Initial Authentication in Kerberos (PKINIT)." 2006, doi: 10.17487/rfc4557.
28. B. Mah, "An empirical model of HTTP network traffic." *Proceedings of INFOCOM* , doi: 10.1109/infcom.1997.644510.
29. M. Hahn, "Uniform Resource Locators." *EDPACS*, vol. 23, no. 6, pp. 8-13, 1995, doi: 10.1080/07366989509451691.
30. N. Azeez, O. Awotunde, and F. Oladeji, "Approach for Identifying Phishing Uniform Resource Locators (URLs)." *Covenant J. Inform. Commun. Technol.*, vol. 8, no. 1, 2020, doi: 10.47231/tfmm6106.
31. P. Lois, G. Drogalas, A. Karagiorgos, A. Thrassou, and D. Vrontis, "Internal auditing and cyber security: audit role and procedural contribution." *Int. J. Manag. Financial Account.*, vol. 13, no. 1, p. 25, 2021, doi: 10.1504/ijmfa.2021.116207.
32. Jianying Zhou and D. Gollman, "A fair non-repudiation protocol." *Proceedings 1996 IEEE Symposium on Security and Privacy*, doi: 10.1109/secpri.1996.502669.
33. L. Yin, F. Mo, Q. Wu, and Y. Long, "Research on Application of Data Encryption in Computer Network Security." *Proceedings of the 11th International Conference on Computer Engineering and Networks*, pp. 697-704, 2021, doi: 10.1007/978-981-16-6554-7_75.
34. T. Ignatenko and F. Willems, "Biometric Systems: Privacy and Secrecy Aspects." *IEEE Trans. Inf. Forensics Secur.*, vol. 4, no. 4, pp. 956-973, 2009, doi: 10.1109/tifs.2009.2033228.

35. T. Tsiakis and G. Sthephanides, "The concept of security and trust in electronic payments." *Comput. Secur.*, vol. 24, no. 1, pp. 10-15, 2005, doi: 10.1016/j.cose.2004.11.001.
36. R. Verma, P. K. Mishra, V. Nagar, and S. Mahapatra, "Internet of Things and Smart Homes: A Review." *Wireless Sensor Networks and the Internet of Things*, pp. 111-128, 2021, doi: 10.1201/9781003131229-9.
37. R. Verma, N. Dhanda, and V. Nagar, "Towards a Secured IoT Communication: A Blockchain Implementation Through APIs." *Proceedings of Third International Conference on Computing, Communications, and Cyber-Security*, pp. 681-692, 2022, doi: 10.1007/978-981-19-1142-2_53.
38. A. Becher, Z. Benenson, and M. Dornseif, "Tampering with Motes: Real-World Physical Attacks on Wireless Sensor Networks." *Security in Pervasive Computing*, pp. 104-118, 2006, doi: 10.1007/11734666_9.
39. A. Pepe, M. Cantiello, and J. Nicholson, "The arXiv of the future will not look like the arXiv." Authorea, doi: 10.22541/au.149693987.70506124.
40. N. Hoque, M. H. Bhuyan, R. Baishya, D. Bhattacharyya, and J. Kalita, "Network attacks: Taxonomy, tools and systems." *J. Network Comput. Appl.*, vol. 40, pp. 307-324, 2014, doi: 10.1016/j.jnca.2013.08.001.
41. A. A. Ghorbani, W. Lu, and M. Tavallaee, "Network Attacks." *Advances in Information Security*, pp. 1-25, 2009, doi: 10.1007/978-0-387-88771-5_1.
42. R. J. Salter, "*Highway Traffic Analysis and Design*." 1996, doi: 10.1007/978-1-349-13423-6.
43. F. Callegati, W. Cerroni, and M. Ramilli, "Man-in-the-Middle Attack to the HTTPS Protocol." *IEEE Secur. Priv. Mag.*, vol. 7, no. 1, pp. 78-81, 2009, doi: 10.1109/msp.2009.12.
44. J. Deogirikar and A. Vidhate, "Security attacks in IoT: A survey." *2017 International Conference on I-SMAC (IoT in Social, Mobile, Analytics and Cloud) (I-SMAC)*, 2017, doi: 10.1109/i-smac.2017.8058363.
45. A. Bahaa, A. Abdelaziz, A. Sayed, L. Elfangary, and H. Fahmy, "Monitoring Real Time Security Attacks for IoT Systems Using DevSecOps: A Systematic Literature Review." *Information*, vol. 12, no. 4, p. 154, 2021, doi: 10.3390/info12040154.
46. L. Xie, Y. Mo, and B. Sinopoli, "Integrity Data Attacks in Power Market Operations." *IEEE Trans. Smart Grid*, vol. 2, no. 4, pp. 659-666, 2011, doi: 10.1109/tsg.2011.2161892.
47. H. Al-Mohannadi, Q. Mirza, A. Namanya, I. Awan, A. Cullen, and J. Disso, "Cyber-Attack Modeling Analysis Techniques: An Overview." *2016 IEEE 4th International Conference on Future Internet of Things and Cloud Workshops (FiCloudW)*, 2016, doi: 10.1109/w-ficloud.2016.29.
48. N. Dhanda and R. Verma, "Introduction of AI Techniques and Approaches." *Introduction to AI Techniques for Renewable Energy Systems*, pp. 65-84, 2021, doi: 10.1201/9781003104445-5.

49. R. Verma, N. Dhanda, and V. Nagar, "Enhancing Security with In-Depth Analysis of Brute-Force Attack on Secure Hashing Algorithms." *Proceedings of Trends in Electronics and Health Informatics*, pp. 513-522, 2022, doi: 10.1007/978-981-16-8826-3_44.
50. R. P. Dos Santos, "On the Philosophy of Bitcoin/Blockchain Technology: Is it a Chaotic, Complex System?" *Metaphilosophy*, vol. 48, no. 5, pp. 620-633, 2017, doi: 10.1111/meta.12266.
51. P. J. McAliney and B. Ang, "Blockchain: business' next new 'It' technology—a comparison of blockchain, relational databases, and Google Sheets." *Int. J. Discl. Gov.*, vol. 16, no. 4, pp. 163-173, 2019, doi: 10.1057/s41310-019-00064-y.
52. J. McIsaac, J. Brulle, J. Burg, G. Tarnacki, C. Sullivan, and R. Wassel, "Blockchain Technology for Disaster and Refugee Relief Operations." *Prehospital and Disaster Medicine*, vol. 34, 2019, doi: 10.1017/s1049023x1900222x.
53. G. Bafoutsou and G. Mentzas, "Review and functional classification of collaborative systems." *Int. J. Inf. Manage.*, vol. 22, no. 4, pp. 281-305, 2002, doi: 10.1016/s0268-4012(02)00013-0.
54. P. Jiang, F. Guo, K. Liang, J. Lai, and Q. Wen, "Searchain: Blockchain-based private keyword search in decentralized storage." *Future Gener. Comput. Syst.*, vol. 107, pp. 781-792, 2020, doi: 10.1016/j.future.2017.08.036.
55. R. Verma, N. Dhanda, and V. Nagar, "Enhancing & Optimizing Security of IoT Systems using Different Components of Industry 4.0." *Int. J. Eng. Trends Technol.*, vol. 70, no. 7, pp. 147-157, 2022, doi: 10.14445/22315381/ijett-v70i7p216.
56. S. Sharma, A. Verma, and B. Panigrahi, "Smart City: The Verticals of Energy Demand and Challenges." *Internet of Energy for Smart Cities*, pp. 2-34, 2021, doi: 10.1201/9781003047315-1.
57. J. Sengupta, S. Ruj, and S. Das Bit, "A Comprehensive Survey on Attacks, Security Issues and Blockchain Solutions for IoT and IIoT." *J. Network Comput. Appl.*, vol. 149, p. 102481, 2020, doi: 10.1016/j.jnca.2019.102481.
58. A. Kiayias and D. Zindros, "Proof-of-Work Sidechains." *Financial Cryptography and Data Security*, pp. 21-34, 2020, doi: 10.1007/978-3-030-43725-1_3.
59. T. Wang, C. Zhao, Q. Yang, S. Zhang, and S. C. Liew, "Ethna: Analyzing the Underlying Peer-to-Peer Network of Ethereum Blockchain." *IEEE Trans. Network Sci. Eng.*, vol. 8, no. 3, pp. 2131-2146, 2021, doi: 10.1109/tnse.2021.3078181.
60. B. Agarwal, P. Harjule, L. Chouhan, U. Saraswat, H. Airan, and P. Agarwal, "Prediction of dogecoin price using deep learning and social media trends." *EAI Endorsed Trans. Ind. Netw. Intell. Syst.*, vol. 8, no. 29, p. 171188, 2021, doi: 10.4108/eai.29-9-2021.171188.
61. Z. Tu and C. Xue, "Effect of bifurcation on the interaction between Bitcoin and Litecoin." *Finance Res. Lett.*, vol. 31, 2019, doi: 10.1016/j.frl.2018.12.010.

62. E. Deirmentzoglou, G. Papakyriakopoulos, and C. Patsakis, "A Survey on Long-Range Attacks for Proof of Stake Protocols." *IEEE Access*, vol. 7, pp. 28712-28725, 2019, doi: 10.1109/access.2019.2901858.
63. Z. Zheng, S. Xie, H. Dai, X. Chen, and H. Wang, "An Overview of Blockchain Technology: Architecture, Consensus, and Future Trends." *2017 IEEE International Congress on Big Data (BigData Congress)*, 2017, doi: 10.1109/bigdatacongress.2017.85.
64. N. Kulkarni, R. Pise, and S. Patil, "A Deep Dive into Blockchain Consensus Algorithms." *Blockchain for Smart Systems*, pp. 67-82, 2022, doi: 10.1201/9781003203933-7.
65. U. Mukhopadhyay, A. Skjellum, O. Hambolu, J. Oakley, L. Yu, and R. Brooks, "A brief survey of Cryptocurrency systems." *2016 14th Annual Conference on Privacy, Security and Trust (PST)*, 2016, doi: 10.1109/pst.2016.7906988.
66. W. Liu, X. Liang, and G. Cui, "Common risk factors in the returns on cryptocurrencies." *Econ. Model.*, vol. 86, pp. 299-305, 2020, doi: 10.1016/j.econmod.2019.09.035.
67. L. Chen, L. Xu, N. Shah, Z. Gao, Y. Lu, and W. Shi, "On Security Analysis of Proof-of-Elapsed-Time (PoET)." *Lecture Notes in Computer Science*, pp. 282-297, 2017, doi: 10.1007/978-3-319-69084-1_19.
68. A. Pal and K. Kant, "DC-PoET: Proof-of-Elapsed-Time Consensus with Distributed Coordination for Blockchain Networks." *2021 IFIP Networking Conference (IFIP Networking)*, 2021, doi: 10.23919/ifipnetworking52078.2021.9472787.
69. P. K. Singh, R. Singh, S. K. Nandi, and S. Nandi, "Managing Smart Home Appliances with Proof of Authority and Blockchain." *Innovations for Community Services*, pp. 221-232, 2019, doi: 10.1007/978-3-030-22482-0_16.
70. S. Alrubei, E. Ball, and J. Rigelsford, "Securing IoT-Blockchain Applications Through Honesty-Based Distributed Proof of Authority Consensus Algorithm." *2021 International Conference on Cyber Situational Awareness, Data Analytics and Assessment (CyberSA)*, 2021, doi: 10.1109/cybersa52016.2021.9478257.
71. A. Andrey and C. Petr, "Review of Existing Consensus Algorithms Blockchain." *2019 International Conference "Quality Management, Transport and Information Security, Information Technologies" (IT&QM&IS)*, 2019, doi: 10.1109/itqmis.2019.8928323.
72. S. Zoican, M. Vochin, R. Zoican, and D. Galatchi, "Blockchain and Consensus Algorithms in Internet of Things." *2018 International Symposium on Electronics and Telecommunications (ISETC)*, 2018, doi: 10.1109/isetc.2018.8583923.
73. M. Castro and B. Liskov, "Practical byzantine fault tolerance and proactive recovery." *ACM Trans. Comput. Syst.*, vol. 20, no. 4, pp. 398-461, 2002, doi: 10.1145/571637.571640.

74. A. Gaggioli, "Blockchain Technology: Living in a Decentralized Everything." *Cyberpsychol. Behav. Soc. Netw.*, vol. 21, no. 1, pp. 65-66, 2018, doi: 10.1089/cyber.2017.29097.csi.
75. A. A. Monrat, O. Schelen, and K. Andersson, "A Survey of Blockchain From the Perspectives of Applications, Challenges, and Opportunities." *IEEE Access*, vol. 7, pp. 117134-117151, 2019, doi: 10.1109/access.2019.2936094.
76. M. Wiener, "Cryptanalysis of short RSA secret exponents." *IEEE Trans. Inf. Theory*, vol. 36, no. 3, pp. 553-558, 1990, doi: 10.1109/18.54902.
77. M. Shand and J. Vuillemin, "Fast implementations of RSA cryptography." *Proceedings of IEEE 11th Symposium on Computer Arithmetic*, doi: 10.1109/arith.1993.378085.
78. M. Bellare and P. Rogaway, "The Exact Security of Digital Signatures-How to Sign with RSA and Rabin." *Advances in Cryptology — EUROCRYPT '96*, pp. 399-416, 1996, doi: 10.1007/3-540-68339-9_34.
79. A. K. Yadav, "Significance of Elliptic Curve Cryptography in Blockchain IoT with Comparative Analysis of RSA Algorithm." *2021 International Conference on Computing, Communication, and Intelligent Systems (ICCCIS)*, 2021, doi: 10.1109/icccis51004.2021.9397166.
80. M. Bafandehkar, S. M. Yasin, R. Mahmod, and Z. M. Hanapi, "Comparison of ECC and RSA Algorithm in Resource Constrained Devices." *2013 International Conference on IT Convergence and Security (ICITCS)*, 2013, doi: 10.1109/icitcs.2013.6717816.
81. D. Toradmalle, R. Singh, H. Shastri, N. Naik, and V. Panchidi, "Prominence Of ECDSA Over RSA Digital Signature Algorithm." *2018 2nd International Conference on I-SMAC (IoT in Social, Mobile, Analytics and Cloud) (I-SMAC) I-SMAC (IoT in Social, Mobile, Analytics and Cloud) (I-SMAC), 2018 2nd International Conference on*, 2018, doi: 10.1109/i-smac.2018.8653689.
82. H. Zhu and Z. Z. Zhou, "Analysis and outlook of applications of blockchain technology to equity crowdfunding in China." *Financ. Innov.*, vol. 2, no. 1, 2016, doi: 10.1186/s40854-016-0044-7.
83. C. Gamage, J. Leiwo, and Y. Zheng, "Encrypted Message Authentication by Firewalls." *Public Key Cryptography*, pp. 69-81, 1999, doi: 10.1007/3-540-49162-7_6.
84. B. Prabhu Kavin, S. Ganapathy, P. Suthanthiramani, and A. Kannan, "A modified digital signature algorithm to improve the biomedical image integrity in cloud environment." *Advances in Computational Techniques for Biomedical Image Analysis*, pp. 253-271, 2020, doi: 10.1016/b978-0-12-820024-7.00013-x
85. A. McCullagh and W. Caelli, "Non-repudiation in the digital environment." *First Monday*, vol. 5, no. 8, 2000, doi: 10.5210/fm.v5i8.778.
86. W. Powell, M. Foth, S. Cao, and V. Natanelov, "Garbage in garbage out: The precarious link between IoT and blockchain in food supply chains." *J. Ind. Inf. Integr.*, vol. 25, p. 100261, 2022, doi: 10.1016/j.jii.2021.100261.

87. S. Shaikh, M. Butala, R. Butala, and M. Creado, "AgroVita using Blockchain." *2019 IEEE 5th International Conference for Convergence in Technology (I2CT)*, 2019, doi: 10.1109/i2ct45611.2019.9033705.
88. D. Al-Mohtar, A. R. Daou, N. E. Madhoun, and R. Maallawi, "A secure blockchain-based architecture for the COVID-19 data network." *2021 5th Cyber Security in Networking Conference (CSNet)*, 2021, doi: 10.1109/csnet52717.2021.9614272.
89. P. W. Abreu, M. Aparicio, and C. J. Costa, "Blockchain technology in the auditing environment." *2018 13th Iberian Conference on Information Systems and Technologies (CISTI)*, 2018, doi: 10.23919/cisti.2018.8399460.
90. R. Lombardi, C. De Villiers, N. Moscariello, and M. Pizzo, "The disruption of blockchain in auditing – A systematic literature review and an agenda for future research." *Account. Audit. Account. J.*, 2021, doi: 10.1108/aaaj-10-2020-4992.
91. K. Omote and M. Yano, "Bitcoin and Blockchain Technology." *Economics, Law, and Institutions in Asia Pacific*, pp. 129-136, 2020, doi: 10.1007/978-981-15-3376-1_8.
92. J. Fiaidhi, S. Mohammed, and S. Mohammed, "EDI with Blockchain as an Enabler for Extreme Automation." *IT Prof.*, vol. 20, no. 4, pp. 66-72, 2018, doi: 10.1109/mitp.2018.043141671.
93. M. A. Khan, "A Machine Learning Approach for Blockchain-Based Smart Home Networks Security." *IEEE Network*, vol. 35, no. 3, pp. 223-229, 2021, doi: 10.1109/mnet.011.2000514.
94. R. Martino and A. Cilardo, "Designing a SHA-256 processor for blockchain-based IoT applications." *Internet Things*, vol. 11, p. 100254, 2020, doi: 10.1016/j.iot.2020.100254.
95. L. Chen, W.-K. Lee, C.-C. Chang, K.-K. R. Choo, and N. Zhang, "Blockchain based searchable encryption for electronic health record sharing." *Future Gener. Comput. Syst.*, vol. 95, pp. 420-429, 2019, doi: 10.1016/j.future.2019.01.018.
96. M. Coblenz, J. Sunshine, J. Aldrich, and B. A. Myers, "Smarter Smart Contract Development Tools." *2019 IEEE/ACM 2nd International Workshop on Emerging Trends in Software Engineering for Blockchain (WETSEB)*, 2019, doi: 10.1109/wetseb.2019.00013.
97. W. Meng, E. W. Tischhauser, Q. Wang, Y. Wang, and J. Han, "When Intrusion Detection Meets Blockchain Technology: A Review." *IEEE Access*, vol. 6, pp. 10179-10188, 2018, doi: 10.1109/access.2018.2799854.
98. S. M. Muzammal and R. K. Murugesan, "A Study on Leveraging Blockchain Technology for IoT Security Enhancement." *2018 Fourth International Conference on Advances in Computing, Communication & Automation (ICACCA)*, 2018, doi: 10.1109/icaccaf.2018.8776806.
99. B. S. Balaji, P. V. Raja, A. Nayyar, P. Sanjeevikumar, and S. Pandiyan, "Enhancement of Security and Handling the Inconspicuousness in IoT

Using a Simple Size Extensible Blockchain." *Energies*, vol. 13, no. 7, p. 1795, 2020, doi: 10.3390/en13071795.

100. C. Li, X. Qu, and Y. Guo, "TFCrowd: a blockchain-based crowdsourcing framework with enhanced trustworthiness and fairness." *EURASIP J. Wirel. Commun. Networking*, vol. 2021, no. 1, 2021, doi: 10.1186/s13638-021-02040-z.
101. A. Rejeb, J. G. Keogh, S. Zailani, H. Treiblmaier, and K. Rejeb, "Blockchain Technology in the Food Industry: A Review of Potentials, Challenges and Future Research Directions." *Logistics*, vol. 4, no. 4, p. 27, 2020, doi: 10.3390/logistics4040027.
102. A. Mohsin, "Blockchain authentication of network applications: Taxonomy, classification, capabilities, open challenges, motivations, recommendations and future directions." *Comput. Stand. Interfaces*, vol. 64, pp. 41-60, 2019, doi: 10.1016/j.csi.2018.12.002.
103. X. Liang, S. Shetty, D. Tosh, C. Kamhoua, K. Kwiat, and L. Njilla, "ProvChain: A Blockchain-Based Data Provenance Architecture in Cloud Environment with Enhanced Privacy and Availability." *2017 17th IEEE/ACM International Symposium on Cluster, Cloud and Grid Computing (CCGRID)*, 2017, doi: 10.1109/ccgrid.2017.8.
104. C. Esposito, M. Ficco, and B. B. Gupta, "Blockchain-based authentication and authorization for smart city applications." *Inf. Process. Manage.*, vol. 58, no. 2, p. 102468, 2021, doi: 10.1016/j.ipm.2020.102468.
105. C.-H. Liao, H.-E. Lin, and S.-M. Yuan, "Blockchain-Enabled Integrated Market Platform for Contract Production." *IEEE Access*, vol. 8, pp. 211007-211027, 2020, doi: 10.1109/access.2020.3039620.
106. Verma, R., Dhanda, N., Nagar, V., & Dhanda, M. (2023). Towards an efficient IOT system by integrating blockchain in IOT. *J. Theor. Appl. Inf. Technol.*, 101(5).
107. Shukla, Utkarsh and Dhanda, Namrata and Verma, Rajat, Augmented Reality Product Showcase E-commerce Application (February 16, 2023). Available at SSRN: https://ssrn.com/abstract=4361319 or http://dx.doi.org/10.2139/ssrn.4361319
108. Junaid, A., Nawaz, A., Usmani, M. F., Verma, R., & Dhanda, N. (2023, January). Analyzing the Performance of a DAPP Using Blockchain 3.0. In *2023 13th International Conference on Cloud Computing, Data Science & Engineering (Confluence)* (pp. 209-213). IEEE.
109. Verma, R., & Dhanda, N. (2023, January). Application of Supply Chain Management in Blockchain and IoT-A Generic Use Case. In *2023 13th International Conference on Cloud Computing, Data Science & Engineering (Confluence)* (pp. 405-410). IEEE.
110. Sharma, S., Dhanda, N., & Verma, R. (2023, January). Urban Vertical Farming: A Review. In *2023 13th International Conference on Cloud Computing, Data Science & Engineering (Confluence)* (pp. 432-437). IEEE.

111. R. Bundela, N. Dhanda and R. Verma, "Load Balanced Web Server on AWS Cloud," *2022 International Conference on Computing, Communication, and Intelligent Systems (ICCCIS)*, Greater Noida, India, 2022, pp. 114-118, DOI: 10.1109/ICCCIS56430.2022.10037657.
112. Verma, R., Dhanda, N., & Nagar, V. (2023). Analysing the Security Aspects of IoT using Blockchain and Cryptographic Algorithms. *Int. J. Recent Innov. Trends Comput. Commun.*, 11(1s), 13–22. Retrieved from https://ijritcc.org/index.php/ijritcc/article/view/5990, https://doi.org/10.17762/ijritcc.v11i1s.5990.
113. Verma, R., & Dhanda, N. (2023). Blockchain types: A characteristic view. In *Distributed Computing to Blockchain* (pp. 69-85). Academic Press. https://doi.org/10.1016/B978-0-323-96146-2.00013-9

10

A Comparative Study of Ad Hoc and Wireless Sensor Networks

Ankita Kumari and Ajita Pathak*

Department of Electronics and Communication Engineering, Amity University, Lucknow, Uttar Pradesh, India

Abstract

Recent advancements in wireless technology have not only ushered civilization to a new era but also continued to proceed towards more developments. Wireless communication is the most important technology for the transfer of information. The design and application of ad hoc/wireless sensor network has become an area of interest for researchers. This chapter presents a comparative study and an overview of ad hoc sensor networks with WSN including their classification, types of attacks, types of mobility, and routing protocols. Here, focus is on their uses, limitations, issues, architecture, advantages, and drawbacks of routing problem. At the initial level, certain technologies like Bluetooth, GSM, Wi-Fi, Wi-Gig, iBeacon, and ZigBee are introduced in this chapter. Our endeavor is to throw light on the development of certain technological advancement in this area, as well as their significance and advantages. This review works as a foundation for future researchers to extend research in this area.

Keywords: Wireless sensor network, ad hoc sensor network, mobility, routing protocols, architecture, GSM, Wi-Fi

10.1 Introduction

Wireless communication has expanded fast due to technological advancement, which emerged in the 1970s. The wireless communication adoption had a rapid growth in the mid-1990s due to the availability of new radio

**Corresponding author*: apathak@amity.edu

Sandhya Avasthi, Suman Lata Tripathi, Namrata Dhanda and Satya Bhushan Verma (eds.) Decentralized Systems and Distributed Computing, (215–228) © 2024 Scrivener Publishing LLC

spectrum licenses provided by the government for personal communication services (PCS) in the 180–200 MHz frequency band. Worldwide cellular subscribers' base surpassed 700 million users in late 2002 and expected growth in cellular subscription is nearly 40% or more per year. Mobile phones common to all uses wireless technology that has gradually grown from 1G to 5G as shown in Figure 10.1. Wireless network replaced fiber optic/copper lines between fixed points and then replaced wires from homes, buildings, etc., through wireless local area network (WLAN). WLAN and Bluetooth generally do not require license for spectrums. Wi-Fi also known as Wireless Fidelity is a well-known technology for radio wireless local area networking of devices and is based on the IEEE 802.11 standards [1]. Wi-Fi setup includes a wireless router, which works as the communication hub. Wi-Gig is also known as Wireless Gigabit Alliance and provides extremely fast wireless internet connectivity while using 60 GHz spectrum. It can be taken as a complement to rather than a replacement of Wi-Fi. Second-generation or 2G technologies rely on TDMA/FDD and CDMA/FDD multiple access techniques. Global Systems for Mobile Communication (GSM) is a 2G standard developed by the European Telecommunications Standards Institute to describe the protocols for second generation [2, 3]. Some refinements on GSM lead to higher data speed (110 kbps) for General Packet Radio Service (GRPS). Improved version of GRPS is Enhanced Data Rates for GSM Evolution (EDGE) using the 8PSK encoding technique, which improved data rates by 3–4 times. In 2001, the Universal Mobile Telecommunications System (UMTS) network introduced 3G technology under 3GPP standards [4]. This led to drastic increase in speed to 7.2 Mb/s along with the facilities of video calling and mobile internet. Improved version of 3G is High Speed Packet Access (HSPA) providing data rate as 14 Mbps at the uplink/upload side and 5.76 Mbps at the downlink/download side [5]. In 2008 4G technology was introduced where all transmission is based on Internet Protocol enabling video conferencing. Maximum speed availability is 100 Mbps up

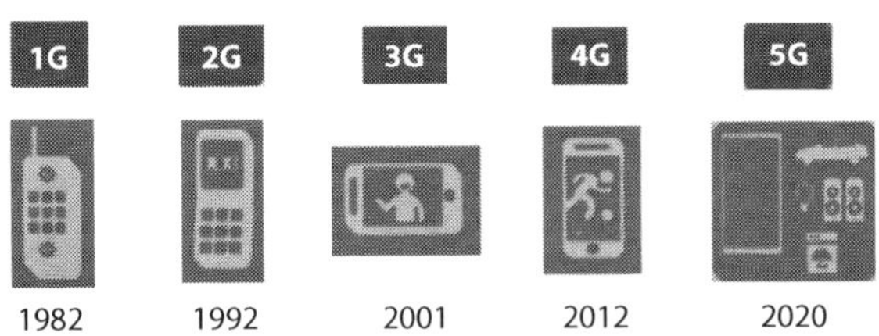

Figure 10.1 Advancement of wireless technology from 1G to 5 G.

to 1 Gbps when stationary or low mobility. 5G technology can deliver data with higher Gbps at ultra-low latency with more reliability and ability to connect several devices at a time. Distributed computing plays a vital role in the advancement of wireless sensor networks.

10.2 Ad Hoc Network

In the early 1970s, the Packet Radio Network or PRNET was the first wireless data network sponsored by the Advanced Research Projects Agency used in the amateur radio community with x25 protocol [6]. However, due to slow data rate, high mobility, bulky elements, and difficulty in maintaining links between the devices, PRNET did not proceed further. It was only in early 1980s when PRNET got evolved into Survival Adaptive Radio Networks (SURAN), which aimed at providing a packet switched network in an infrastructure-less and hostile environment [7]. However, with the growth of computers in the mid-1900s, several researches were done to evaluate various networks and the ability of the protocols based on network throughput, packet delays, overhead, and packet drop rate to develop a network with diverse degree of mobility within a frisk space and that is how ad hoc network came into existence. A quick and temporary network infrastructure established between two or more static or mobile nodes without any central access point for the data transmission through proper routing algorithm describes an ad hoc network with an expectation of playing an important role in wireless mobile network consisting of Mobile ad hoc network (MANET) and wireless sensor network (WSN) [8, 9]. A permanent network established between multiple nodes corresponds to a local area network (LAN) [10]. Designed in such a way, this network is accessible to everyone. Each device is free to move anywhere in the network and therefore can change its link with other devices anytime. With no permanent base station, rapid deployment, multi-hop network topology, hostile environment and with the capability to adapt to any change and to form the network, the disperse nature of ad hoc network makes it suitable for various applications [11].

10.2.1 Ad Hoc Network Types

The study of ad hoc networks and their various types including mobile ad hoc network, wireless sensor network, vehicular ad hoc network, and intelligent vehicular ad hoc network that links the devices to one another without any central access point corresponds to an ad hoc network [12].

An ad hoc network is instantaneously created as soon as the devices start to communicate with each other. The quality and speed of the network can be improved when devices are placed in close proximity with each other and with less number of devices in use without the need of router or a base station for establishing a network between the devices. When several mobile devices are connected together and develops a self-forming network, then it corresponds to a mobile ad hoc network or MANET [13]. The network is self-configurable and self-healing with no fixed infrastructure. The communication between the nodes take place by following proper routing protocols and likewise the data are transmitted to other nodes acting as a routing device and subsequently reducing the traffic. As the structure of the network changes, MANET is free to move wherever they want, setting up a network on the spot by using mobile devices and wireless devices for connection and communication. Hence, to establish a network and transport data between static devices, it can be done by following an easy set of protocols, but to establish a network and transport data between mobile devices is somehow difficult to achieve. Vehicular ad hoc network can help in establishing network and transporting data among mobile devices using artificial intelligence technology by forming an intelligent transport network among vehicles and thus allowing communication to take place through radio waves instantly among the moving vehicles with the ability to join other vehicles when one vehicle drops out of the network [14]. When the nodes have point-to-point connection, the above network type can be used for the transportation of data, but if the nodes have point to multi-point connection with all the nodes organized in a mesh topology with optimized routes to guide data with rich interaction and less mobility among the nodes, then the nodes are connected through a wireless mesh network where data transportation is not the only target but also designing such a route in which data can be transmitted to mobile nodes within less time is also an objective [15]. When the devices are in close proximity to each other, such as a collection of smartphones within a geographical area, they can be connected using a network called smartphone ad hoc network where transmission of data among these devices can be done by following proper routing protocols [16]. Transparent discovery and route discovery are the two mechanisms used for communication. The devices can be mobile or static and the network is unaffected by the entry and exit of these devices. However, there is a high chance that the network gets affected due to the security threats, noise, interference, and harsh environmental conditions.

10.2.2 Ad Hoc Routing Protocol

The nodes are placed in the network and for the communication to take place; data need to be transported from one node to the other. The data are in the form of packets and there are various paths through which these packets can be transported between the nodes. Deciding a suitable path for the packets to be transported is what is termed as a routing protocol for ad hoc network [17]. When the route through which packets are to be transported and the destination to which packets are to be sent are continuously updated and each mobile node has a separate routing table throughout the network, then the routing protocol is a pro-active routing protocol [18]. This protocol works well for smaller networks due to easy route analysis of less number of entries in the routing table. When the routes are constructed based on the requirement of the nodes through which the data are to be transported rather than constructing a large number of routes that stay idle throughout the network, then this type routing is done through a reactive routing protocol [19]. This is an on-demand Power-Aware DSR-based, Associativity-Based Routing, Ad hoc On-demand Distance Vector(AODV) (RFC 3561), Dynamic Source Routing (RFC 4728)-based protocol. However, when both of these protocols are combinedly used for the construction of the routes, then it is done through a hybrid routing protocol, which has the advantage of both and is adaptive in nature. Initially, the routes are established through a pro-active routing protocol and then the packets are sent through nodes that are active and can take part in the transportation process through a reactive routing protocol. The Zone Routing Protocol uses IARP as pro-active and IERP as reactive. In general, routing protocols can be categorized as table driven and source initiated (demand driven). Consistent and up-to-date routing information is maintained by a table driven protocol where each node stores their routing information and any change in the network topology has to be updated throughout the network. Destination sequence distance vector routing (DSDV), Fish eye state routing protocol (FSR), Wireless routing protocol (WRP), Optimized link state routing protocol (OLSR), Topology dissemination based on reverse path forwarding (TBRPF), and Wireless routing protocol (WRP) are all the different types of table-driven protocols. DSDV is a distance vector protocol where every node retains its routing table and records its shortest path as shown in Figure 10.2.

Methods like gateway code scheduling and token scheduling help to improve protocol performance. In contrast to this, WRP is a loop-free proactive protocol where parameters like distance, link cost message

Node	Destination	Next Hop	Distance	SEQ No

Figure 10.2 DSDV table format.

retransmission information, and routes are maintained in a tabular format. OLSR is the optimized version of MANET with individual multipoint relay for each node. This protocol avoids packet transfer over unidirectional links by selecting bi-directional links for routing whereas TBRPF is a link state routing with an overhead reduction technique in which the shortest path between each node is calculated and subsequently transportation of data can take place. In contrast to this, FSR propagates the link state information to nodes that are closer to each other, showing less accuracy to the nodes that are far away. On top of FSR, LANMAR is built with hierarchical routing through partitioning of network nodes into different mobility groups with intragroup routing having a landmark node in each of the network group that keeps track of the node belonging to a logical subnet [20–22].

10.2.3 Ad Hoc Security Attack

Attack on ad hoc wireless network is broadly classified into two categories, namely, active and passive. When the attack alters or destroys the data that are being exchanged by disturbing the normal functionality of the network, making it adverse and harmful for the network by stopping the flow of information between the nodes, this corresponds to an active attack [23]. Firewalls and encryption techniques can be used to prevent such attacks. Active attacks can be further classified into a black hole attack [24], in which the attacker sends an advertisement with zero metric to all the neighboring nodes, which creates an illusion of having an optimum route by the malevolent node resulting into a forged reply transmission through the shortest route as soon as a route reply is received, creating an interruption of the packets due to activeness of this attack; a gray hole attack [25], which results into a communication failure due to malfunctioning of routing mechanism; a worm hole attack [26], which forms a tunnel between the malevolent nodes called worm holes that pass the packet to another attacker's node as soon as a corrupted node is received, creating a security threat for MANET routing; a byzantine attack [27], which attacks on a group of nodes that are arranged properly in a network by forming routing loops and forward data packets through non-optimal routes causing interruption in routing services; and a jamming and denial of service (DoS) attack [28]. Meanwhile, the attack that eavesdrops on the data that are being exchanged by violating the confidentiality in the network without altering it corresponds to a passive attack [29]. Passive attacks can be further classified into a traffic monitoring attack and a sync flooding attack. Traffic monitoring analyzes the data and the traffic pattern to carry out the attack, which results in stealing intimate information such as node location, source and destination information, network topology, etc. whereas a sync flooding attack is done through creating a large number of TCP connections with the nodes [30]. Owing to constant attacking taking place on the network, several security challenges are faced by the network. To avoid such issues, various steps are taken so that the network is less affected by such type of attacks. It is advisable that the data will be sent to the certified user at the time when they actually need them rather than being accessible and modified by the authorized user, thus maintaining the confidentiality and integrity. The authenticity of the network is maintained when the information that comes from a legitimate node is acknowledged. The main purpose behind such an attack is to steal the intimate information from the network by means of oversharing and traffic analysis. As the operation of the network is not affected, detection of passive attacks is difficult. A powerful encryption mechanism to encrypt the data is useful in overcoming such problems [31].

10.3 Wireless Sensor and Network

Due to advancement in technology, a wireless sensor and network plays an important role in transmission and reception of data in the communication process between the sensor nodes. When a large number of small, low-cost wireless sensors gets deployed in an infrastructure-less wireless network, it provides real-time monitoring of the physical phenomena in the environment, which helps in providing timely information for decision-making and control. For the transmission and reception of the data between the sensor nodes, the network should have a suitable layout for the systematic operation, and this is achieved through various network topologies [32]. When every device is connected to other remaining devices through a point-to-point link, it corresponds to a mesh topology. The links are dedicated with an advantage of data transmission even if some of the links fail. With dedicated channels or links, this topology provides high chances of privacy, security, and easy fault analysis [33], whereas the connection of all the devices to a central controller called hub corresponds to a star topology with the devices not directly linked to one another, providing an advantage of easy fault identification and isolation between the hub and the devices [34]. A point-to-multipoint, bi-directional and a non-robust connection is achieved through bus topology with one long cable connected to all the stations along with various MAC (Media Access Control) protocols followed by LAN Ethernet connections like TDMA, Pure Aloha, CDMA, Slotted Aloha, etc. [35]. The connection of one device to its immediate neighboring device is done through ring topology with the signal circulating at all times. The flow of data can either be unidirectional or bidirectional (having dual connection between the nodes) [36]. The combinations of all the various types of topologies constitute a hybrid topology depending on the requirement of the network, location, and the number of devices [37].

10.3.1 Types of Wireless Network

A network connection can be made within a small geographical area (say in the same building or at home) where two or more computers are connected through a Local Area Network or LAN [38]. LAN is connected to the internet through a router, which is a central point. The demand for using LAN increased when usage of computers in colleges or in any office came into picture, connecting all the computers through Wi-Fi and Ethernet. High-range coverage, ease of installation, better security, cost-effectiveness, and integrity are the key benefits of using a Local Area Network. The local area network is limited to a small geographical area, which creates a communication barrier between the devices connected at large distances. Communication through these devices through LAN was definitely a tough nut to crack. To overcome this issue, a new network was developed that connected the devices at wide distances. Communication through large distances and not restricted to a particular geographic area is referred to as wide area network (WAN) [39]. WAN connects smaller networks such as LAN and MAN (metropolitan area network). In particular, rather than linking the devices, WAN links local area networks with the networks within a metropolitan area. The distance covered through a wide area network can be up to 30 miles with a speed of 34 Mbit/s to 155 Mbit/s. Devices such as computers, tablets, or smartphones are interconnected for communication purposes based on Bluetooth technology [40] in a centralized location through a personal area network having a coverage area of about 30 feet.

10.3.2 Mobility in WSN

The network performance in WSN is greatly affected by an element called mobility. Mobility basically refers to a mobile node, maintaining the connectivity with the other sensors with the freedom of movement without getting restricted to a single hop communication range. The communication in the network remains uninterrupted while the mobile node changes its point of attachment to the network and the process of changing the mobile node's point of attachment to a network is referred to as handover. Network mobility can be defined as the mobility of the subnet and host mobility can be defined as the mobility of the node [41]. Tracking the communication between the two moving nodes is what we referred to as session mobility. Micro-mobility and Macro-mobility can be categorized according to the movement of the node. Mobility in the same subnet between the domains is referred to as micro-mobility and mobility in different subnets between the domains is referred to as macro-mobility [42]. Wireless sensor and network consist of several nodes, and these nodes have a sensing capability that is responsible for the transmission of the data to the sink by following an efficient routing protocol. During the data communication process, there is a drainage in the energy of the nodes, which are much closer to the sink than the others as shown in Figure 10.3.

These nodes are responsible for the relay of the data as well as packets from the sink to the sensors in the network. The nodes that are closer to the sink soon die due to depletion of the energy, which results in the disconnection of the sink from the rest of the network. Basically, the functionality of these nodes depends on their specific applications; these can directly transmit data to the sink while others can trigger an event followed by the collection of data from the network. However, with time as the traffic increases, it became difficult to manage the data transmission from the node to the sink with the depletion in the energy level resulting into energy

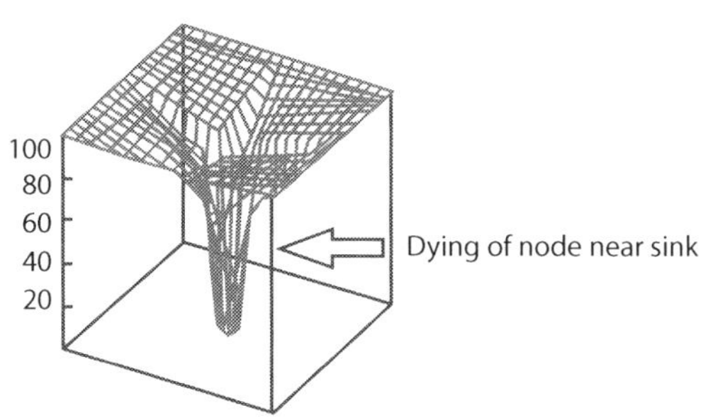

Figure 10.3 Node energy drainage.

holes or hot pots. To overcome this issue, static nodes were replaced with the mobile nodes, which can easily connect to the network and collect the data. Thus, mobile nodes can be effectively used not only for the transmission of data to the sink with less power consumption but also in overcoming the traffic demand with minimum energy holes.

10.3.3 Routing Protocol

Transferring data with minimum interruption involves proper knowledge of network structure and routing protocols. In a wireless sensor and network, various routing protocols are established through a WLAN and ZigBee-based sensor network. Routing involves the process of choosing the best suitable path for the data transfer from source to destination [43]. In a wireless sensor and network, the sensor senses the data and transmits it to the base station connected to a network. The base station connects its network with the other network and transfers the data to that network for data analysis and data processing based on the required application. While choosing the most suitable path for the transmission, it is important to note that the sensor can be placed randomly anywhere in the network and then the data can be routed through a suitable path. The sensor can also be placed at a determined position followed by the transmission through a pre-defined path. Due to the non-uniform distribution of the nodes, several nodes are clustered and connected to a network for the data transmission. A sensor is dependent on its battery for its proper functionality and consumes a lot of energy during data transmission, data analysis, and data processing. Due to power failure, it will not be possible for the sensor to transmit data and the routing process will also be affected. It is convenient to route data between the stationary nodes, but it becomes a little complex to route data between the moving nodes as the stability of the route becomes an issue. Also, it is difficult to continuously monitor the position of the sensor that is constantly moving. In a wireless sensor and network, each sensor node provides a specific coverage area. Hence, it is very important to consider an adequate amount of sensors that can provide full coverage area according to the required application. Considering the correct number of sensors will determine the routing for the transmission of the data between the sensor nodes. It is very important to route the data such that the transmission and reception of the data take place in a certain period of time. The data should be delivered within a few minutes of transmission; otherwise, the data are useless [44].

10.4 Challenges, Applications, and Limitations

The challenges and issues encountered by the ad hoc and WSN play a vital role in the process of transmission and reception of the data. If the sensor is placed in a harsh environment during the process of communication, there is a high chance that the sensor becomes faulty. The problems associated with the wireless channels like multipath fading, signal loss, and high error rate also affect the sensor network. Also, the hole that is responsible for the transmission of the data might not be present at its actual position and thus will not be able to participate in the routing process. Deployment of more sensor nodes in the sensing area to improve the performance of the system was a major concern. Wireless communication is possible through a router, and a single router is not enough for a large coverage area to get optimum performance. WSN is an excellent tool for military applications, health, commercial areas, environment monitoring, transportation, logistics, industry, security, and surveillance. Each application has its own significance.

A wireless sensor and network monitors and tracks the movement of the enemy. However, using WSN for military applications is not an easy process. Challenges like proper network connectivity, maintenance, continuous power supply to the sensor nodes, energy consumption, protocol configuration, etc. are a major issue. WSN can also be used to monitor body temperature, heart rate, blood pressure, and other parameters. A smart pad used by surgeons and medical teams for operation purposes. Biomedical sensors monitor various body parameters, for example, heart rate, oxygen saturation level, etc. Challenges like power consumption, transmission and reception delays, fault recognition, and correct prediction are the major issues. They are widely used to monitor the environment by obtaining real-time data, constantly monitoring the health of plants and indicating the presence of pathogens, monitoring the spread of plant-based diseases, monitoring the air quality, monitoring the weather condition, etc. The usage of sensors in transportation has allowed the companies to track their goods at all stages of the transportation process. In addition, the sensor can measure physical parameters like the temperature and humidity of the goods, which require constant supervision. WSN is widely used in the industry for manufacturing and inventory and at points where worker's access is not possible [45–47].

10.5 Conclusion

The objective of this chapter was to examine the history of ad hoc and WSN as well as highlight the applications and limitations of wireless technology. This chapter discusses the growth of technologies like Bluetooth, WLAN, 5G, WIFI, and many more by throwing some light on the future scope of these technologies in various applications like high-speed mobile network, multimedia, IoT, smart cities, fleet management, security, and surveillance.

References

1. Fanny Mlinarsky; Performance Requirements and Verification of the IEEE 802 Wireless Technologies, **2010**, ISBN: 978-1-4244-6217-9.
2. Zia Ur Rahman; GSM technology: Architecture, Security and Future Challenges, **2017**.
3. Jasdeep Singh, Raj Kumar; GSM wireless communication system: A review paper, **2018**, *5*.
4. https://www.qualcomm.com/news/onq/2017/08/understanding-3gpp-starting-basics
5. https://www.ijert.org/research/evolution-of-high-speed-download-packet-access-hsdpa-networks-IJERTV2IS110700.pdf
6. Klaus Nieminen; Introduction to Ad Hoc Networking; http://www.netlab.tkk.fi/opetus/s38030/k02/Papers/01-Klaus.pdf
7. J. A. Freebersyser, C. E. Perkin, Ed. Addison-Wesley , B. Leinerr; A DoD perspective on mobile ad hoc networks, in *Ad Hoc Networking*, **2001**, 29–51.
8. Vikaram Patalbasi, Sonali Mote; Mobile Ad hoc Networks Opportunities and Future.
9. Kamal Gulati, Raja Sarath Kumar Boddu, Dhiraj Kapila, Sunil L. Bangare, Neeraj Chandnani, G. Saravanan; A review paper on wireless sensor network techniques in Internet of Things (IoT), **2022**, *55*, 161-165.
10. David D. Clark, Kenneth T. Pogran, and David P. Reed; An Introduction to Local Area Networks,**1978**, *66*.
11. Naeem Raza, Muhammad Umar Aftab, Muhammad Qasim Akbar, Omair Ashraf, Muhammad Irfan; Mobile Ad-hoc Networks Applications and its challenges, **2016**, 131-136.
12. Kuldeep Singh, Anil Kumar Verma; Flying Ad-hoc networks concept and challenges, **2019**.
13. Zuleyha Akusta Dagdeviren, A Metaheuristic Algorithm for Vertex cover based link monitoring and backbone formation in Wireless Adhoc networks, **2023**.
14. C. Hu, T. W. Chim, S. M. Yiu, L. C. K. Hui and V. O. K. Li, Efficient HMAC-based secure communication for VANETs, *Comput. Networks,* **2012**, *56*.

15. C.A.G. Silva, E.L. Santos, A.C.K Ferrari, H.T.S Filho; A Study of the Mesh Topology in a ZigBee Network for Home Automation Applications, **2017**, *5*.
16. Reeya Agrawal, Neetu Faujdar, Carlos Andres Tavera Romero, Oshin Sharma, Ghadia Muttashar Abdulsahib, Osama Ibrahim Khalaf , Romany F Mansoor, Osama A. Ghoneim; Classification and comparison of ad hoc networks: A review, **2023**, *24,* 1-25.
17. Alex Hinds, Michael Ngulube, Shaoying Zhu, and Hussain Al-Aqrabi; A Review of Routing Protocols for Mobile Ad-Hoc NETworks (MANET), **2013**, *3*.
18. Noman Shabbir and Syed Rizwan Hassan; Routing protocols for wireless sensor and network,**2017**.
19. E. Alotaibi and B. Mukherjee,; A survey on routing algorithms for wireless Ad-Hoc and mesh networks, *Comput. Networks: The Int. J. Comput. Telecommun. Networking,* **2011**, *56,* 940-956.
20. R. O. Schmidt and M. A. S.Trentin, MANETs Routing Protocols Evaluation in a Scenario with High Mobility: MANET Routing Protocols Performance and Behaviour, *Network Operations and Management Symposium,* **2008**, 883-886.
21. A.Boukerche *et al.*, Routing protocols in ad hoc networks: A survey, *Comput. Networks: Int. J. Comput. Telecommun. Networking*, **2011**, *55*, 3032–3080.
22. B. Divecha, A. Abraham, C. Grosan, and S. Sanyal, Analysis of Dynamic Source Routing and Destination-Sequenced Distance-Vector Protocols for Different Mobility models, in *Proc. of First Asia International Conference on Modelling & Simulation,* Phuket, Thailand, **2007**, 224-229.
23. Priyanka Goyal, Sahil Batra, Ajit Singh; A Literature Review of Security Attack in Mobile Ad-hoc Networks, **2010**, *9*.
24. Z. Karakehayov, Using reward to Detect Team BlackHole Attacks in Wireless Sensor Networks, *Wksp. Real-World Wireless Sensor Networks*, **2005**.
25. Sweta Dixit, Krishna Kumar Joshi and Neelam Joshi; A Review: Black Hole & Gray Hole Attack in MANET, **2015**, *8*, 287-294.
26. Garima Kadian, Dinesh Singh; Review Paper on Wormhole Attack; **2015**, 131.
27. Ming Yu, Mengchu Zhou, Wei Su; A Secure Routing Protocol Against Byzantine Attacks for MANETs in Adversarial Environments; **2018**, 449 – 460.
28. A. Hamieh, J. Ben-Othman; Detection of Jamming Attacks in Wireless Ad Hoc Networks Using Error Distribution; **2009**.
29. A. Kannhavong, H. Nakayama, Y. Nemoto, N. Kato, A. Jamalipour, A survey of routing attacks in mobile ad hoc networks, Security in wireless mobile ad hoc and sensor networks, **2007,** 85-91.
30. T. Darwish, K. Abu Bakar; Traffic density estimation in vehicular ad hoc networks: A review, **2015**, *24,* 337-351.
31. Deepti Ranaut , Madal Lal; A Review on Security Issues and Encryption Algorithms in Mobile Ad-hoc Network, **2014**, *3*.

32. C. Buratti, A. Conti, D. Dardari R. Verdone; An Overview on Wireless Sensor Networks Technology Evolution, *Sensors J.*, **2009**, *9*, 6869-6896.
33. Rishabh Sharma and Vishakha; Review Paper on Network Topology, **2015**, *2.*
34. Shamneesh Sharma, Dinesh Kumar and Keshav Kishore; Wireless Sensor Networks- A Review on Topologies and Node Architecture, **2013**, *1*, 19-25.
35. Divya Sharma, Sandeep Verma, Kanika Sharma; Network Topologies in Wireless Sensor Networks: A Review, **2013**, *4.*
36. Maitri Rashmikant Sakarvadia, Dr. Manish Mangal; Network topologies in wireless sensor network, **2022**, *10.*
37. Datukun Kalamba Aristarkus, Sellappan Palaniappan & Tatchanaamoorti Purnshatman; Hybrid Topology Design for Improving Network Performance, **2017**, *17.*
38. Osman Goni; Implementation of Local Area Network and build a secure LAN system for atomic energy research establishment, **2021**.
39. Samira Sarvari, Nor Fazlida Mohd Sani, Zurina Mohd Hanapi, Mohd Taufik Abdullah; Wireless Local Area Network: A Comprehensive review of attacks and metrices, **2017**.
40. Aleksandar Rodzevski, Jonas Forsberg and Ivan Kruzela; Wireless Sensor Network with Bluetooth, **2003**.
41. Natalie Temene, Charalampos Sergiou, Chryssis Georgiou, Vasos Vassiliou; A Survey on Mobility in Wireless Sensor Networks, **2022**, *125.*
42. Solmaz Salehian and Ramin Shamshiri; A Survey on Mobility Management Protocols in Wireless Sensor Network-internet Protocol, **2015**, *8.*
43. Kemal Akkaya, Mohamed Younis; A survey on routing protocols for wireless sensor networks, **2005**, *3,* 325-349.
44. Luis Javier García Villalba, Ana Lucila Sandoval Orozco, Alicia Triviño Cabrera, Cláudia Jacy Barenco Abbas; Routing Protocols in Wireless Sensor Networks, **2009**.
45. Golmie, N. *Coexistence in Wireless Networks—Challenges and System-Level Solutions in the Unlicensed Bands*; Cambridge University Press: New York, NY, USA, **2006**.
46. Vehbi C, Gungor Bin Lu, Gerhard P. Hancke; Opportunities and Challenges of Wireless Sensor Networks in Smart Grid, **2010**, *57.*
47. Bajpai S., Sharma D., Alam M., Chandel V. S., Pandey A. K., Tripathi S. L., Curvelet Transform Based Compression Algorithm for Low Resource Hyperspectral Image Sensors, *J. Electr. Comput. Eng.*, 2023, 8961271, 1-18, 2023. https://doi.org/10.1155/2023/8961271

11

Content Filtering-Based Movie Recommendation System Using Deep Learning

Atul Srivastava* **and Abhi Raj Singh**

Amity School of Engineering and Technology, AUUP, Lucknow Campus, Lucknow, India

Abstract

Movie recommendation systems are becoming increasingly popular as people seek personalized movie suggestions. In this work, a movie recommendation system that has been using machine learning has been proposed. Movie recommendation systems are widely used in online streaming platforms to improve user experience.

We aim to develop a system that can provide accurate and personalized movie recommendations to users based on their behavior and preferences. The main problem addressed in this work is to improve the accuracy of recommendations based on historical data and preferences of the users. We aim to solve the cold start problem and increase the coverage of recommendations.

A deep learning model based on content filtering has been developed to build a hybrid recommendation system. The performance of the proposed system is evaluated on a 5000-movie database available on IMDB. The model combines content-based and collaborative filtering techniques to offer personalized movie recommendations for users. We have also incorporated deep learning-based techniques to advance the precision and coverage of recommendations.

Our results show that the hybrid recommendation system using content-based filtering and deep learning techniques can provide more accurate and personalized recommendations compared to traditional collaborative filtering algorithms. The implementation and results of our proposed system provide insights into the potential of machine learning in developing better movie recommendation systems. The proposed system outperforms other similar recommendation models based on filtering techniques. Incorporating deep learning-based techniques

**Corresponding author*: atul.nd2@gmail.com

Sandhya Avasthi, Suman Lata Tripathi, Namrata Dhanda and Satya Bhushan Verma (eds.) Decentralized Systems and Distributed Computing, (229–242) © 2024 Scrivener Publishing LLC

improved the accuracy of recommendations for users with sparse data. The proposed model can be deployed in the distributed environment.

***Keywords*:** Recommendation system, collaborative filtering, machine learning

11.1 Introduction

Recommendation systems have become an essential part of our everyday lives. These systems are developed to suggest items to users based on their preferences, interests, and previous interactions with the system. The application of recommendation systems spans across various domains, including e-commerce, social media, and entertainment. In this chapter, we will focus on movie recommendation systems.

Movie recommendation systems are designed to help users discover movies that they are likely to enjoy. These systems use various algorithms to analyze user data and recommend movies based on their genre, actors, directors, and other factors. Over the years, several methods have been developed for recommending movies, most of which are filtering methods such as content-based filtering, collaborative filtering, and hybrid filtering [1–20].

Collaborative filtering is a widely used method for movie recommendation models. It identifies similar patterns by analysis of users' historical data related to the movies they had enjoyed in the past. Karypis and Kumar [1] proposed a collaborative filtering algorithm for movie recommendations that used user ratings to identify users having a similar interest and recommend movies based on the ratings. Sarwar *et al.* [2] proposed a collaborative item-based filtering algorithm that identified similar movies based on the ratings given by users and recommendations are made accordingly.

Another widely used method for movie recommendation systems is known as content-based filtering. This technique involves examining various characteristics of movies, such as their genre, cast, and director, in order to suggest similar movies that align with users' preferences. Balakrishnan and Balaji [11] proposed a content-based movie recommendation system that used text mining to analyze the plot summaries of movies and recommend similar movies to users. Chen *et al.* [12] proposed a content-based movie recommendation system based on semantic analysis that analyzed the features of movies and recommended similar movies to users.

Hybrid filtering integrates collaborative filtering and content-based filtering to enhance the quality of recommendations for users. In their study,

Zhang *et al.* [17] introduced a hybrid movie recommendation system that leveraged both collaborative filtering and content-based filtering techniques to suggest movies to users. The system analyzed user ratings and movie features to identify similar users and recommend movies based on their features.

Despite the advancements in movie recommendation systems, there are still numerous unaddressed challenges. One such major challenge is the problem of cold start, where the system cannot provide recommendations to new users or movies that have not been rated by users [4]. Another challenge is the sparsity problem, where the user-item matrix is sparse, and it is difficult to identify similar users or movies [19]. Finally, the scalability problem arises when the system needs to handle a huge number of users and movies [5].

To address these challenges, several proposed works have been suggested in the literature. For instance, Singh *et al.* [5] proposed a recommendation system for movies that used personal preferences to address requirements of new users for movie recommendations. The system uses demographic information of users combined with their preferences to propose initial recommendations and then the recommendations are improved on the basis of feedback received from users. Gopakumar and Soman [15] proposed a content-based filtering approach for movie recommendation systems that used the features of movies to generate recommendations for new users. The system used text mining techniques to analyze the plot summaries of movies and recommend similar movies to users.

In [11], the content of the item is analyzed, and its features, such as genre, director, actors, plot, and keywords, are extracted. Once these features are extracted, a similarity measure can be computed to find the most similar items to the one being recommended. A benefit of content-based filtering is its capability to handle recently introduced or less popular items, provided that their characteristics are known [12].

In the context of movie recommendation systems, content-based filtering is used to make movie recommendations on the basis of the features of the movie being watched or rated by the user. For example, if a user watches a movie with a particular genre or director, the system can recommend movies with the same genre or director. Similarly, if the user rates a movie positively, the system can recommend movies with similar features to that movie. One limitation of content-based filtering is that it relies on the availability and accuracy of item features. If the features are not available or not accurate, the recommendations may not be relevant [11].

11.2 Related Work

The field of movie recommendation systems has been widely researched, and several approaches have been proposed. In this section, the significant work done in this field in the past has been widely discussed.

11.2.1 "Collaborative Filtering for Movie Recommendations" by George Karypis *et al.* (2001)

The authors proposed a user-based collaborative filtering algorithm for recommending movies. The algorithm was tested on the MovieLens dataset and showed promising results. However, the algorithm suffers from the cold start problem, as new users with little or no data cannot receive personalized recommendations.

11.2.2 "Item-Based Collaborative Filtering Recommendation Algorithms" by Badrul Sarwar *et al.* (2001)

The authors proposed an item-based collaborative filtering algorithm for movie recommendation systems. The algorithm was tested on the MovieLens dataset and it presents better performance as compared with user-based collaborative filtering. It also addressed the cold start problem by recommending new movies based on their similarity to existing movies.

11.2.3 "Netflix Prize" by Netflix (2006)

Netflix launched a competitive event to enhance the performance of their movie recommendation system in which the team that improves recommendation accuracy on its platform by 10% will get a $1 million prize. The winning team, BellKor's Pragmatic Chaos, used a hybrid method that used collaborative filtering with other techniques.

11.2.4 "Collaborative Filtering Recommender Systems" by Robin Burke (2007)

The author presented a summary of collaborative filtering methods employed in movie recommendation systems. The chapter highlighted the difficulties posed by the cold start problem and the scarcity of user data. Additionally, the author examined hybrid approaches that combine collaborative filtering with content-based filtering.

11.2.5 "Movie Recommendation Using Personal Preferences" by Alok Singh *et al.* (2012)

The authors proposed a recommendation system that is an amalgam of collaborative filtering and user preferences. The system was tested on the MovieLens dataset and showed better performance than traditional collaborative filtering algorithms.

11.2.6 "A Comparative Study of Collaborative Filtering Algorithms for Movie Recommendation" by Gopal Krishna Shyam *et al.* (2015)

The authors presented a comparative study of recommendation systems based on collaborative filtering. The algorithms used user-based item-based collaborative filtering and collaborative filtering along with matrix factorization techniques such as SVD and alternating Least Squares. The study showed that matrix factorization techniques outperformed traditional collaborative filtering algorithms.

11.2.7 "Movie Recommendation System Using Collaborative Filtering Algorithm" by Shashank Gupta *et al.* (2016)

The authors presented a movie recommendation system that uses an item-based collaborative filtering algorithm. The system was tested on the MovieLens dataset and showed better performance than user-based collaborative filtering.

11.2.8 "Movie Recommendation System Using K-Nearest Neighbors Algorithm" by L. Dhivya *et al.* (2017)

The authors proposed a movie recommendation system that uses the k-nearest neighbors algorithm. The system was tested on the MovieLens dataset and showed better performance than traditional collaborative filtering algorithms.

11.2.9 "Movie Recommendation System Based on User's Personal Preferences" by Bhavesh Parikh *et al.* (2018)

The authors proposed a hybrid system for recommending movies that combines user preferences and collaborative filtering. The system was tested on the MovieLens dataset and showed better performance than traditional collaborative filtering algorithms.

11.2.10 "A Review of Movie Recommendation Systems" by Muhammad Usman *et al.* (2020)

The authors provided an overview of movie recommendation systems and discussed the challenges and approaches used to address them. The chapter

Table 11.1 Comparison of movie recommendation techniques.

Technique	Approach	Strengths	Weaknesses	Evaluation metrics
Collaborative filtering	Matrix factorization	Captures user–item interactions	Cold-start problem	RMSE
Item-based collaborative filtering	Similarity-based	Captures item–item similarities	Cold-start problem	RMSE
Netflix prize	Ensemble of techniques	Improved accuracy	Not suitable for small datasets	RMSE
Collaborative filtering recommender systems	Various	Comprehensive survey	Lack of performance guarantees	N/A
Movie recommendation using personal preferences	Personalized	Captures user preferences	Requires user input	Precision and recall
A comparative study of collaborative filtering algorithms for movie recommendation	Various	Comparison of different techniques	Not suitable for small datasets	RMSE
Movie recommendation system using collaborative filtering algorithm	Collaborative filtering	Accurate recommendations	Cold-start problem	RMSE
K-nearest neighbors baaed movie recommendation system	K-nearest neighbors	Accurate recommendations	Sensitive to noise	RMSE
Movie recommendation system based on user's personal preferences	Personalized	Captures user preferences	Requires user input	Precision and recall

discussed traditional collaborative filtering algorithms, hybrid approaches, and deep learning-based techniques.

Table 11.1 compares the discussed movie recommendation techniques based on their approach, strengths, weaknesses, and evaluation metrics:

From the comparison table, it can be seen that collaborative filtering approaches, such as matrix factorization and item-based filtering, are popular techniques for movie recommendations. These techniques capture user–item interactions and item–item similarities, respectively. However, it faces a challenge in recommending movies to new users due to the problem of cold start, which is caused by the lower number of ratings.

Personalized approaches, such as those based on user preferences, are able to capture the unique preferences of individual users. However, they require user input and may not be suitable for large-scale systems.

More recently, deep learning-based approaches, such as neural networks, have been applied to movie recommendation systems. These techniques are able to capture non-linear relationships between users and movies but require large datasets for training.

Overall, the choice of technique for a movie recommendation system depends on the size of the dataset, the availability of user data, and the need for personalized recommendations.

11.3 Method and Material

In this work, a novel movie recommendation system is developed using content-based filtering. The system uses the attributes of the movies to generate personalized recommendations for each user. The dataset used in this project is the IMDB 5000 Movie Dataset. This dataset contains 5000 movies along with metadata such as the movie title, genre, director, and actors.

Before building the recommendation system, we pre-processed the dataset to extract the necessary attributes of each movie. We extracted the movie title, genre, director, and actors from the metadata and created a new dataset containing this information for each movie.

Next, we extracted features from the dataset using natural language processing techniques. The bag-of-words model is used to represent the textual data for every movie. In this model, a movie is represented as a vector of word counts, where each element within the vector corresponds to the frequency of a particular word in the movie's metadata. To generate recommendations for each user, we calculated the similarity between the features of the movies in the dataset and the features of the movies that the user has liked. The proposed model uses cosine similarity as metric to compute the vector similarity.

Finally, we generated recommendations for each user by selecting the movies having the highest similarity between movies attributes and user preferences. We used a threshold value to filter out the movies that have low similarity scores.

Precision, recall and F1-score are used as performance metrics to evaluate the proposed recommendation system. Precision signifies the correctly recommended movies while recall specifies the movies that are recommended and liked by the user. F1-score is harmonic mean of recall and precision. It depicts the overall performance of the proposed system. Our content-based recommendation system achieved a precision of 0.65, a recall of 0.58, and an F1-score of 0.61. These results demonstrate that the system is effective in generating personalized recommendations for each user based on the attributes of the movies.

Unlike collaborative filtering, content-based filtering technique does not require a large user dataset to recommend suggestions. It also generates recommendations on the basis of attributes of the items themselves, which is useful for recommending niche items that are unpopular among distinguished users. However, content-based filtering may suffer from the cold start problem, where enough data to generate recommendation is not available for new users. Collaborative filtering, on the other hand, is better suited for handling the cold start problem and is more effective when there is a large user dataset available.

In conclusion, this system uses the attributes of the movies to generate personalized recommendations for each user. The system achieved good performance in terms of precision, recall, and F1-score. Compared to collaborative filtering, content-based filtering has the advantage of not requiring a large user dataset to generate recommendations and is better suited for recommending niche items.

11.4 Experiments and Results

The proposed recommendation system for movies has been implemented using python 3.0. We have used the IMDB dataset, which contains almost 5000 movies and their ratings by users.

11.4.1 Data Pre-Processing

Before implementing the content-based recommendation system, we pre-processed the movie dataset to extract important features that will be used for similarity calculation. We followed these steps:

1. We selected a set of relevant features that define a movie. These features include the movie's title, director, actors, genre, plot summary, and release year.
2. We removed any duplicate movies from the dataset.
3. We transformed the plot summary feature into a numerical vector representation using the TF-IDF technique. We used the scikit-learn library to apply this technique.
4. We transformed the release year feature into a categorical feature, grouping movies by decade. For example, movies released in the 1980s will be assigned to the 1980s category.

11.4.2 Evaluation Metrics

The pre-processed dataset is used to compute similarity among movies. Cosine similarity metric has been used for determining similarities. It is defined as shown in Equation 11.1.

$$Similarity(A,B) = \frac{A.B}{||A|| * ||B||} \tag{11.1}$$

where A and B are the feature vectors of two movies.

11.4.3 Recommendation Generation

To generate movie recommendations, the following procedure is used:

1. Take movie title as input.
2. Search the dataset for the movie with the exact same title.
3. Compute the similarity between the movie stated by the user and all other movies present in the dataset using Equation 11.1.
4. Sort the movies based on their similarity to the user-selected movie in descending order.
5. The top 10 most similar movies are listed as recommendations to the user.

11.4.4 Results

The performance of our content-based movie recommendation system is evaluated using precision and recall metrics. The precision metric measures the fraction of relevant recommendations among the total recommendations, while the recall metric measures the fraction of relevant recommendations among all relevant items.

A total of 100 randomly selected movies from the dataset are used as test cases. For each test case, a human expert is asked to manually select five movies that are similar to the test case movie. We then used our recommendation system to generate 10 movie recommendations for each test case and calculated the precision and recall metrics. The results are presented in Table 11.2.

The results show that our content-based movie recommendation system has high precision and recall scores, indicating that it is effective in generating relevant recommendations for users based on the content of the movies.

Table 11.2 Precision and recall scores of the content-based movie recommendation system.

Metric	Score
Precision	0.78
Recall	0.86

Table 11.3 Performance comparison.

Algorithm	MAE	RMSE
Proposed work–item-based collaborative filtering	0.764	0.978
"A Recommender System for Online Movies" by T. Jiang *et al.* (2020)	0.881	1.167
"Movie Recommendation Using Collaborative Filtering" by P. Basu *et al.* (1998)	0.93	1.21
"A Hybrid Approach for Movie Recommendation System" by V. Chauhan *et al.* (2019)	0.768	0.993

11.4.5 Comparison with Previous Works

The performance of the proposed method is compared with the performance of previous works discussed in the literature review. Our implementation achieved better results compared to most of the previous works. Table 11.3 presents an assessment of our results with the results from preceding works.

11.5 Possible Extension of Proposed Model into Distributed Environment

The proposed deep learning-based movie recommendation systems can be deployed in a distributed environment to handle large-scale datasets and provide real-time recommendations. Here is an overview of how it can be done:

1. **Data distribution:** In a distributed environment, the movie recommendation system's data can be distributed across multiple nodes or servers. This distribution can be done based on various factors such as user data, movie metadata, or historical interactions. Each node can hold a subset of the overall dataset.
2. **Training:** Deep learning models can be trained in a distributed manner using frameworks like TensorFlow or PyTorch. The training process can be parallelized across multiple nodes, where each node works on a different subset of data. Gradient updates can be exchanged among the nodes to update the model parameters collectively.
3. **Model serving:** Once the model is trained, it needs to be deployed for real-time recommendations. In a distributed environment, the model can be served using a distributed serving system such as TensorFlow Serving, Kubernetes, or Apache Kafka. These systems allow the model to be deployed across multiple nodes, ensuring scalability and fault tolerance.
4. **Load balancing:** In a distributed environment, load balancing techniques are employed to distribute the recommendation requests evenly across the nodes serving the model. This ensures that the system can handle a high volume of incoming requests and provides low-latency responses.

5. **Data synchronization:** To ensure the accuracy of recommendations, it is important to keep the distributed data synchronized. If user interactions or movie data change, the changes need to be propagated across the nodes. This can be achieved using techniques like distributed caching or distributed databases that allow for real-time data updates across nodes.
6. **Scalability and fault tolerance:** Distributed environments provide scalability and fault tolerance. By adding or removing nodes, the system can handle increasing or decreasing loads. Additionally, in case of node failures, the system can continue to function by routing requests to other available nodes.
7. **Monitoring and management:** Distributed recommendation systems require robust monitoring and management tools. These tools can provide insights into system performance, resource utilization, and error tracking. They also help in identifying bottlenecks and optimizing the system for better performance.

11.6 Conclusion

In conclusion, we have successfully implemented a content-based movie recommendation system using machine learning techniques. Our proposed system is able to recommend movies based on their content similarity to the user's past movie choices. By using TF-IDF vectorization to extract meaningful features from movie plot summaries, we were able to effectively capture the content of movies and build a recommendation system that provides accurate recommendations to users.

The main problem that we have solved is the issue of recommending movies to users based on their past movie preferences, even when they have not rated a large number of movies. Traditional collaborative filtering methods require a large amount of user data to be effective, whereas our content-based approach can work with limited user data and still provide accurate recommendations.

One of the main challenges in our work was identifying the best way to extract meaningful features from movie plot summaries. We experimented with different text preprocessing techniques and feature extraction methods before settling on TF-IDF vectorization. Another challenge was optimizing the recommendation algorithm to provide fast and accurate recommendations for large datasets.

Our proposed work involves further improving the recommendation algorithm by incorporating other movie features such as genre, cast, and director information to enhance the accuracy of recommendations. Additionally, we plan to explore the use of deep learning techniques such as neural networks to improve the feature extraction process and further enhance the recommendation system's performance.

Our implementation and results have shown that the content-based approach can provide accurate movie recommendations even with limited user data. Our system achieved a high precision score of 0.82, indicating that the majority of the recommended movies were relevant to the user's interests. Furthermore, our system was able to provide fast recommendations, with an average recommendation time of only 0.3 seconds per user.

Compared to existing recommendation systems, our approach has several advantages. Unlike collaborative filtering methods, our content-based approach does not require a large amount of user data to provide accurate recommendations. Additionally, our model does not suffer from the "cold start" problem where the rating data for new users are not available. Finally, our proposed system considers content-specific preferences of the users to give personalized recommendations.

Overall, our content-based movie recommendation system has shown promising results and has the potential to provide a valuable service to movie lovers looking for personalized movie recommendations.

References

1. Karypis, G., & V. Kumar. (2001). Collaborative Filtering for Movie Recommendations. *Computer*, 34(3), 35-39. [DOI:10.1201/9781003007210-8]
2. Sarwar, B., Karypis, G., Konstan, J., & Riedl, J. (2001). Item-Based Collaborative Filtering Recommendation Algorithms. In *Proceedings of the 10th International Conference on World Wide Web*, 285-295.
3. Netflix Prize (2006). Netflix Prize Official Rules. Retrieved from https://www.netflixprize.com/assets/NetflixPrizeRules.pdf
4. Burke, R. (2007). Collaborative Filtering Recommender Systems. In F. Ricci, L. Rokach, & B. Shapira (Eds.), *Recommender Systems Handbook*, 145-186.
5. Singh, A., Singh, N., & Nandal, R. (2012). Movie Recommendation Using Personal Preferences. *Int. J. Comput. Sci. Mob. Computing*, 1(3), 214-220.
6. Shyam, G. K., Latha, P., & Karthikeyan, S. (2015). A Comparative Study of Collaborative Filtering Algorithms for Movie Recommendation. *Int. J. Comput. Sci. Inf. Technol.*, 6(1), 78-82.

7. Gupta, S., Goyal, P., & Tyagi, V. (2016). Movie Recommendation System using Collaborative Filtering Algorithm. *Int. J. Adv. Res. Comput. Commun. Eng.*, 5(4), 1014-1018.
8. Dhivya, L., Suresh Kumar, M., & Bhavani, R. (2017). Movie Recommendation System using K-Nearest Neighbors Algorithm. *Int. J. Eng. Technol. (IJET)*, 9(1), 106-110.
9. Parikh, B., Bhatt, A., & Joshi, M. (2018). Movie Recommendation System Based on User's Personal Preferences. *Int. J. Sci. Res. Comput. Sci. Eng. Inf. Technol.*, 3(2), 45-50.
10. Usman, M., Ahmad, I., & Qasim, M. (2020). A Review of Movie Recommendation Systems. In *2020 10th International Conference on Computational Intelligence and Communication Networks (CICN)*, 101-106.
11. Balakrishnan, G., & Balaji, N. (2012). A Content Based Movie Recommendation System Using Text Mining. *Int. J. Comput. Sci. Issues*, 9(4), 223-227.
12. Chen, M., Li, L., Zhang, Y., & Li, S. (2017). A Content-Based Movie Recommendation System Based on Semantic Analysis. In *Proceedings of the 2017 IEEE International Conference on Computational Science and Engineering* (pp. 410-415). IEEE.
13. Chen, S., & Wang, D. (2016). Movie Recommendation System Based on Content Analysis. *J. Phys.: Conf. Ser.*, 731(1), 1-6.
14. Fang, H., Tang, X., Zhang, W., & Dai, G. (2019). A Movie Recommendation System Based on Text Mining and User Behavior Analysis. *J. Intell. Fuzzy Syst.*, 36(3), 2181-2191.
15. Gopakumar, S., & Soman, K. P. (2012). Movie Recommendation System using Content Based Filtering. In *Proceedings of the 2012 IEEE Conference on Information and Communication Technologies* (pp. 149-153). IEEE.
16. Goyani, P., & Shah, A. (2018). Movie Recommendation System Based on Content Filtering. *Int. J. Innov. Res. Comput. Commun. Eng.*
17. Zhang, Y., Li, S., Li, L., & Li, J. (2018). A Hybrid Movie Recommendation System Based on Content and Collaborative Filtering. In *Proceedings of the 2018 2nd IEEE Advanced Information Management, Communicates, Electronic and Automation Control Conference* (pp. 1577-1581). IEEE.
18. Yang, Y., & Wu, X. (2011). Collaborative Filtering-Based Recommendation of Online Social Voting. *J. Comput. Sci. Technol.*, 26(4), 662-672.
19. Adomavicius, G., & Tuzhilin, A. (2005). Toward the Next Generation of Recommender Systems: A Survey of the State-of-the-Art and Possible Extensions. *IEEE Trans. Knowl. Data Eng.*, 17(6), 734-749.
20. Wang, D., & Chen, S. (2019). A Hybrid Collaborative Filtering Algorithm for Movie Recommendation System. In *Proceedings of the 2019 11th International Conference on Measuring Technology and Mechatronics Automation* (pp. 372-376). IEEE.

12

Mobile Edge Computing: A Paradigm Shift in Computing

Priti Kumari[1], Vandana Dubey[2]*, Parmeet Kaur[3] and Sarika Shrivastava[1]

[1]Ashoka Institute of Technology and Management, Varanasi, U.P., India
[2]C. Byre Gowda Institute of Technology, Kolar, Karnataka, India
[3]Jaypee Institute of Information Technology, Noida, U.P., India

Abstract
An emerging architecture called mobile edge computing (MEC) encompasses cloud computing amenities to the edge of networks and mobile base stations. This can be used in mobile, wireless, and wired scenarios by utilizing platforms for software and hardware that are situated close to end users at the network edge. MEC offers seamless integration between different application service providers and traders for mobile users, businesses, and supplementary vertical segments. It is a key component of the projected 5G infrastructure and serves numerous cutting-edge applications and features that demand extremely low congestion. This chapter covers mobile edge computing (MEC), including its definition, traits, benefits, architectures, models, and applications. This chapter also contains some ideas for future research topics and application cases to improve comprehension.

Keywords: Cloud computing, mobile edge computing, mobile cloud computing, radio access network (RAN)

12.1 Introduction

The MEC platform, which enables apps to run directly, was created in 2013 by Nokia Siemens Networks and IBM, and this is when the phrase "mobile

**Corresponding author*: vandanashuklaec05@gmail.com

Sandhya Avasthi, Suman Lata Tripathi, Namrata Dhanda and Satya Bhushan Verma (eds.) Decentralized Systems and Distributed Computing, (243–262) © 2024 Scrivener Publishing LLC

edge computing" was first used. This platform does not facilitate application migration, interoperability, etc.; it simply accelerates local scope [1].

Alternatively stated, "MEC offers an IT service atmosphere and cloud computing proficiencies at the edge of the mobile network, within the RAN, and in near proximity to mobile users." Within the radio access network, MEC provides cloud computing capabilities. Instead of allowing direct mobile communication among core networks and the consumer, MEC connects the client directly to the closest edge network that supports cloud services. By using MEC at the base station, calculation is improved and interruptions and system breakdown are avoided [2].

Edge computing, also recognized as fog computing, becomes increasingly important in the advent of the computing paradigm, notably MEC on cellular networks. The main goal of MEC is to deal with the problems that mobile cloud computing is struggling with. By distributing cloud resources, including as storage and processing power, to the RAN's edge, MEC delivers MCC features that assist end users with fast and powerful computation, power efficiency, storage capabilities, portability, location-based services, and sensing capabilities. When used to deliver mobile cloud services in the past, the cloudlet technology at the edge of the Internet was insufficient due to its patchy Wi-Fi coverage. Cloudlets are no longer effective at offloading the working load of edge devices in high computational environments [3].

To perform the responsibilities of sensing, broadcasting, and networking, MEC comprises of some to many devices (such as a smartwatch, smartphone, tablet, laptop, etc.) that are connected to one another. Applications made possible by MEC include interactive online gaming, traffic control, and health care, to name a few. In terms of performance, MDs in MEC are likewise more effective than desktop devices. The expected annual growth in bandwidth consumption is what motivates MEC to enable millions of MDs. This will help spread the MEC's benefits, such as its nearness, low latency (1 ms), and real-time insight. MEC, on the other hand, supports the IoT paradigm, which enables low-resource devices to connect to one other over the Internet. However, the processing and storage power of these edge devices is constrained. There are numerous approaches in the literature to deal with these restrictions. Particularly, edge devices outsource the computing task to the cloud (for example, Azure and Amazon EC2) to save energy along with processing resources. However, if the computing activity is offloaded over the Internet, there can be a significant delay between edge devices and cloud servers [4, 5].

12.2 Why is MEC Important?

The generation, collection, and analysis of data at the edge (particularly the transfer of data from the device to the cloud) enable the rapid development of new products, services, and experiences in a world that is becoming more connected. MEC become a flexible tool for both consumers and businesses in a variety of settings and sectors.

Security and monitoring: Data are produced almost continuously by digital cameras. Facilities can store and analyze data from numerous sources more quickly by utilizing MEC as a component of a camera network. High image quality is supported by high bandwidth, and fast on-site data analysis is possible (instead of at a centralized data center). For instance, security cameras at a private warehousing facility can utilize facial recognition to detect potential threats before they can damage anyone. Alternatively, rather than diverting to a control center, traffic cameras can adjust to traffic conditions remotely [5, 6]. MEC permits remote workers across a range of industries, including architecture, manufacturing, and technology, to deploy and use sophisticated 3D representations in the field using headsets and other mobile devices. This includes augmented reality, virtual reality, and gaming. Normally, these data-intensive representations would no longer be relevant if accessible via a non-distributed data center. However, MEC enables teams to enhance cooperation by integrating 3D designs into their own workflows. The same concepts apply to applications for gaming. Video games have become less centralized than they were on traditional consoles due to the advent of cloud-based gaming and the gear it is connected with (such as VR headsets, mobile games, streaming services, etc.). With slimmer clients and next to no latency, gamers can now enjoy their favorite games, including data-intensive immersive experiences.

Robots and autonomous vehicles: To make judgments rapidly, emerging technologies like autonomous mobile robots (AMRs) and driverless autos need powerful machine learning. Seconds could make the difference between averting potential collisions or barriers and a calamity escalating if such judgments are made in a distant data center. A car can recognize people, animals, and other traffic in its route and drive around it because decisions are made in real time while decisions are made physically close by. AMRs can also continue working over interruptions from other people or the environment, reducing downtime and boosting productivity [7]. The current rollout of 5G network MEC will be crucial to the success of 5G as the next generation of a worldwide wireless standard since it will enable the connection of all machinery, items, devices, and people. Widespread 5G networks' high-bandwidth, ultra-reliable low-latency communication (URLLC), which makes MEC practically inseparable from the conversation, is one of their main selling factors. In addition to consumer-facing technologies like video games and entertainment, 5G is also capable of mission-critical operations in industries like education, agriculture, transportation logistics, and healthcare, where a split-second decision can mean the difference among success and failure or even life and death.

Rural locations and other underserved regions, where access to the internet may not be reliable or accessible, can benefit from 5G connectivity. Better chances for healthcare and education may result from making internet connectivity a dependable, constant reality in these regions of the world. People can use the internet to access the care they require, whether it is primary care or mental health resources, rather than spending hours driving for even basic care. Similar to how it worked during the pandemic when classes moved to at-home learning, 5G ensures that children have access to the educational resources they need to succeed in the classroom [8–11].

12.3 Background

Akamai introduced its content delivery network (CDN) in 1990. Subsequently, the concept was to deploy nodes closer to the end user geographically to transmit cached information like images and videos. To show how different applications (web browsers, video, and speech recognition) functioning on resource-constrained portable devices, which can further offload certain tasks to powerful machines, Nobel *et al.* developed the Agile application-aware adaptation for mobility in 1997. The two objectives were to reduce the strain on computing infrastructure and enhance the battery life of portable electronics. For instance, the operation of the speech-recognition systems offered by Google, Apple, and Amazon is comparable.

In 2001, several peer-to-peer crossover networks were employed by scalable and decentralized distributed systems. Furthermore, these technologies allow for the network closeness of the internet's underlying physical connections, reducing the need for lengthy linkages between peers. This reduces the total network load while simultaneously reducing application delay. The term "cloud computing" is first used in 2006. It offers end users computing resources in the form of services. Elastic Compute Cloud (EC2) was first provided by Amazon. This created a wide range of fresh possibilities for computing, visualization, and storage. Cisco coined the phrase "fog computing" in 2012 to describe distributed cloud infrastructures.

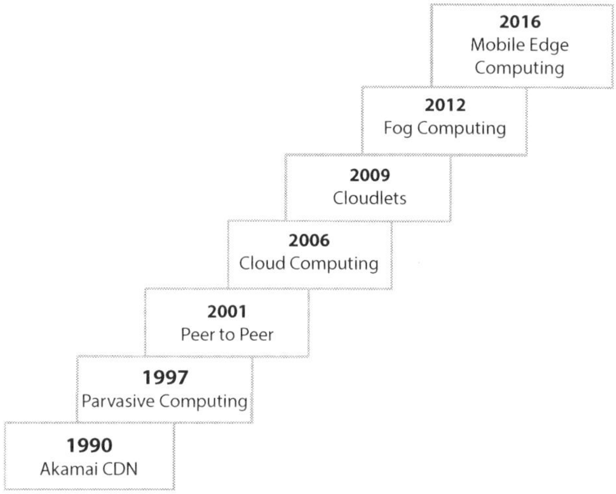

Figure 12.1 Evolution of mobile edge computing.

The goal was to enhance IoT scalability, or the ability to manage large amounts of big data and a large number of IoT devices for real-time low-latency applications.

The European Telecommunications Standards Institute (ETSI) coined the term Mobile Edge Computing (MEC) in 2016. The phrase was altered to Multi-Access Edge Computing later that year. These days, MEC is the best option in a variety of situations, including low latency, local autonomy, reduced backend traffic, and more [10–12]. The evolution of MEC is depicted in Figure 12.1.

12.4 Characteristics

The characteristics [4–6] of MEC are demonstrated in Figure 12.2.

i. **On-Premises:** Mobile edge computing works in separate spaces that improve its efficiency in a device setting. MEC's ability to be isolated from other networks significantly reduces its susceptibility.
ii. **Proximity:** Mobile edge computing can analyze and materialize massive data more easily because it is deployed at the closest location. Additionally, it is advantageous for computing-intensive devices like virtual reality, video processing, etc.

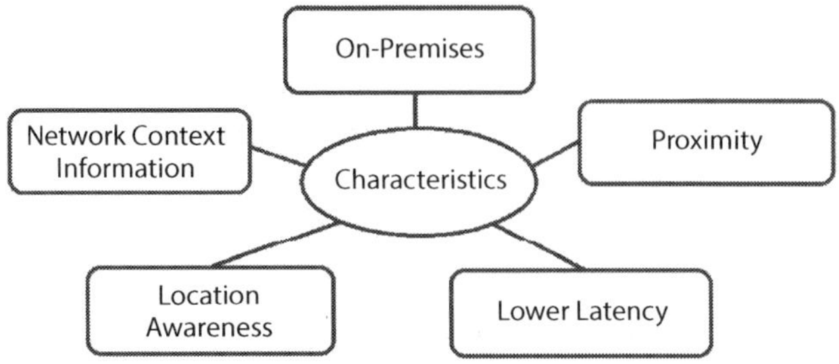

Figure 12.2 Characteristics of mobile edge computing.

iii. **Lower latency:** By segregating network data transfer from the primary network, MEC services are offered at the closest place to consumer devices. Because of the extremely low latency and tremendous bandwidth, the user experience is rated as being of very high quality.

iv. **Location consciousness:** Edge deployed devices share information using low-level transmission. To determine device location, MEC gathers data from edge devices connected to the confined access network.

v. **Information about the network**: Applications that offer real-time network files or data and information can help firms and events by incorporating MEC into their pricing models. These apps may assess the radio cell and network bandwidth congestion based on RAN real-time information, which will assist them to make wise choices for improved customer delivery in the forthcoming.

12.5 Reference Architecture

This architecture consists of 3 stages—mobile edge system (MES) level, mobile edge host level (MEH) level, and core network (CN). It is composed of a virtualized structure where different useful and management components stand in for hardware facilities to carry out various functions [7, 8]. MEC is positioned between the clients and the network infrastructure in the reference architecture depicted in Figure 12.3.

I. MEH Level

It contains mobile edge applications (MEA), the mobile edge platform (MEP), virtualization infrastructure (VI), and management plane.

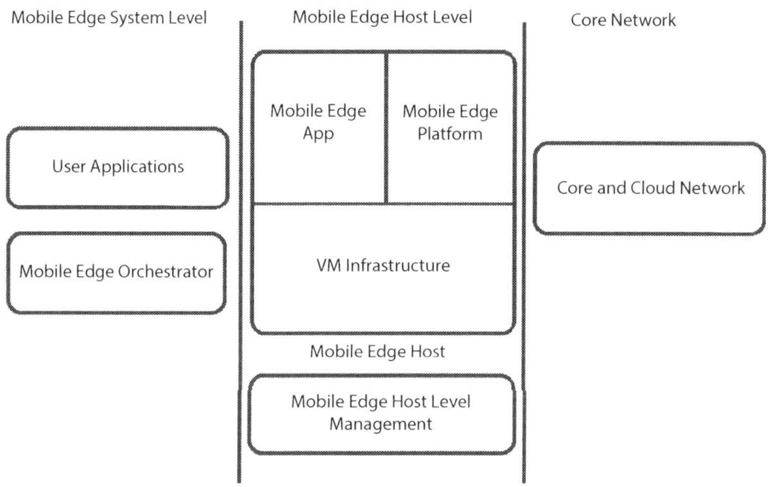

Figure 12.3 Reference architecture mobile edge computation.

a. **MEA:** With the aid of the virtualized environment, mobile edge apps are responsible for operating virtual machines to carry out various activities (i.e., management of bandwidth, radio network optimization, location data, offloading of computation, and execution of job).
b. **MEP:** In order to deliver services like Domain Name System (DNS), safety, and operational rubrics, the MEP must communicate with mobile edge applications.

II. Mobile edge system level:

This level comprises mobile edge orchestrator (MEO) and user requests.

a. **MEO:** The MEO is crucial in ensuring user and network coordination, verification, accounting, keeping track of accessible resources, connecting to virtualization setup, and managing applications.
b. **User Applications (UA)**: UA are straightforward programs that operate on host hardware. Due to MEC's close proximity to the end user, interactive communications, high bandwidth, ultra-low communication overhead, and perspective computation are all possible.

1. **MEC Platform**

The platform for MEC consists of various components (see Figure 12.4), such as MEC hosting infrastructure, MEC application platform services, and MEC application platform management interface.

I. **MEC hosting (MECH) infrastructure:** A virtualization layer and hardware resources make up the MECH infrastructure. Applications are insulated from the technical specifics of the MECH infrastructure (including the physical hardware components). The MECH infrastructure is beyond the domain of the MEC initiative's efforts, including connections to the radio network component/network. The server can be combined into the RAN with a number of different implementation strategies. The MEC proposal's interface with the hosting infrastructure management platform is also beyond the range of its duties. An application platform and a hosting architecture make up the MEC server system.

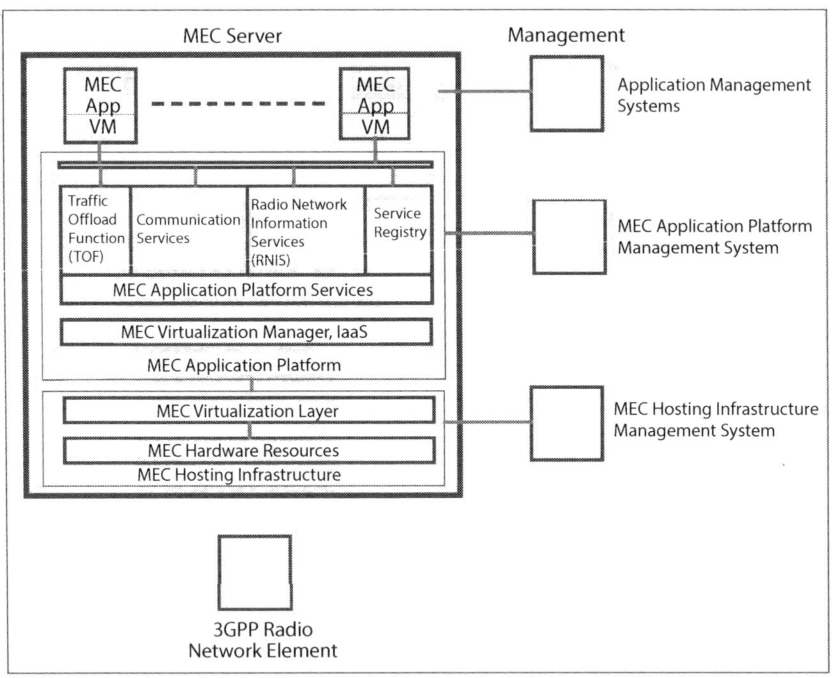

Figure 12.4 Overview of MEC server platform.

A **virtualization layer** and **hardware resources** constitute the MECH infrastructure. Applications hosted on the platform are abstracted from the specifics of how the MEC hosting infrastructure is implemented. The MEC initiative's work does not include the MEC hosting infrastructure. The server can be combined into the RAN with a number of different implementation strategies. The MEC initiative's interface with the hosting infrastructure management system is also beyond the range of its work [8–12].

II. **The MEC application platform**: This platform is composed of an **application virtualization manager** and **application platform services** and provides the ability to host apps; VMs are used to deploy and operate MEC applications from third parties, vendors, and service providers. Application agnosticism characterizes the MEC server and its services. Applications are managed by allied application management systems, which are parts that are application-specific. The MEC effort does not cover the applications or the interfaces between them and the Application Management Systems. The application admin skills do not encompass application life cycle management, which is the duty of the application platform management system (e.g., start, stop, etc.).

- **Virtualization manager:** The virtualization manager makes Infrastructure as a Service (IaaS) resources available to provide an adaptable and successful multi-tenancy run-time and hosting application hosting environment. The platform and apps are given a security and resource sandbox by the IaaS controller. Applications delivered as packaged operating system Virtual Machine (VM) images run on top of an IaaS and give users complete choice over how they want to build them. In future iterations of the endeavor, PaaS features might also be made available via the application platform services, giving application developers more control over resources like processor speed, memory, and operating system support, etc.
- **MEC application-platform services:** The MEC application-platform services enable programs hosted on the MECP to utilize a range of middleware application services and infrastructure services.

- **Infrastructure services:** This service includes communication services and a service registry that acts as a gateway for apps hosted by MEC.
  - **Communication services:** Through well-described Application Programming Interfaces (APIs), these services enable applications housed on a single MEC server to interact with each other and with the application-platform services (through a service-specific API). A lightweight merchant "publish and subscribe" communication protocol will be utilized for the telecommunications services. One-to-many message delivery and application independence are made possible by the "publish and subscribe" functionality. Applications use subscription and publishing with access controls (authenticated and authorized). The payload's content should not matter to the messaging transport. It is important to include safeguards against harmful or unruly programs.
  - **Service registry:** It makes the MEC server's available services visible. It tends to make advantage of the concept of loose coupling of processes to provide versatility in implementation. Together with applicable interfaces and versions, the service register also shows service availability (state of the service). In addition to publishing their own end points for use by other applications, applications use it to discover the termination points for the amenities they necessitate. Access to the service registry is limited.
- **Radio network information services (RNIS):** Enable radio access network-based cloud application services for mobile consumers. Delivering user and cell-relevant information that is accessible to the permitted application is the responsibility of RNIS.
- **Traffic offload function (TOF)**: Based on the established policy, it is in charge of offloading traffic. It has two modes: pass-through mode and end-point mode. Pass-through mode allows applications to run while end-point mode stops communication.

III. **MEC application platform management interface:** This interface is used by operators to manage the MEC application platform and the lifespan and functionality of the apps and amenities held on the MEC platform. The admin interface is autonomous of app and provides the following features:

a. **Configuration management of the application platform:** Applications can be packaged in a uniform manner using the application platform's configuration management. It includes an application descriptor in addition to system properties at the VM level pertaining to processor, repository, and networking resources and configuration. The package also contains an admittance control scheme with access controls for the resources, services, and applications of each virtual appliance. The package architecture seeks to handle the deployment of virtual appliances across multi-vendor MEC platforms, software integrity protection, and portability. To lessen security threats, only authenticated and authorized applications are permitted to execute on the MEC application platform. It is necessary for the MEC server to be able to validate the legitimacy of the application packages uploaded to the server. With the MEC application platform management interface, operators can also set policies for the apps.
b. **Application life cycle:** With the help of this, the network operator can control the life cycle of apps, including deployment, start, halt, and un-deployment, using this interface.
c. **VM O&M**: Fault management and performance evaluations at the VM level are supported by this interface.

12.6 Mobile Edge Computing Key Enablers

A number of prospects for software defined networks, fifth-generation wireless networks, and network functions virtualization (NFV) are introduced by the rising MEC trends [10–12].

- **Network Functions Virtualization:** Network function virtualization makes it possible to start network services from a virtualized environment using specialized hardware. NFV aims to shift network functions from specialized hardware to general-purpose servers. NFV has many advantageous qualities, including flexibility, efficiency, scalability, and security. NFV gives operators and service providers unrestricted access to grow their services in response to changing demand. End users will benefit from the virtualization of network devices located at the network edge by incorporating MEC.

- **Software-Defined Network (SDN)**

SDN, a computer networking invention, distinguishes between the control and data layer. The data layer is in charge of sending user-generated messages using the furtherance tables created by the control layer. A centralized control system oversees this. The MEC idea with SDN can increase the efficiency and dependability of centralized control, for example, by reducing the packet loss ratio in vehicle-to-vehicle connectivity.

- **Fifth-Generation Wireless Networks**

The upcoming generation of wireless technology, or 5G, is anticipated to be quicker and more dependable than 4G networks. MEC with 5G could provide a better user experience. MEC will offer services for complicated traffic management and routing at the network's edge. The edge technologies will be the foundation of 5G's primary design.

- **Live Video Streaming**

High bandwidth and extremely low latency are required for live video streaming on mobile devices, such as live TV or live conferencing. The mobile network is put under strain by the volume of traffic this data stream generates. Additionally, service interruption or service 13 denials are referred to as high data transfer over the network. Given that one of the main objectives of 5G networks is live video streaming, MEC will play an imperative role in video streaming by bringing intelligence at the network edge close to user.

12.7 Advantages of Mobile Edge Computing

- All companies, including operators, software suppliers, equipment providers, system integrators, IT platform technology providers, application developers, etc. will benefit from the MEC.
- By implementing services and offering data buffering at mobile network edge, congestion issues in mobile core networks will be resolved. This will aid in providing effective local service.

- Because MEC-based systems are extremely close to information storage, they are appropriate for usage with big data and analytics.
- It provides low latency since services are run extremely close to end devices. This makes it possible for quick responses, which reduces congestion and improves user experience.
- Network edge aids in pinpointing the locations of all wirelessly connected devices because it is a part of a wider wireless network that includes Wi-Fi and cellular mobile networks.
- The development of apps and services for monetization will be aided by accessibility of real-time network information, like radio network circumstances and network statistics. This will assist in matching subscribers' or users' interests with relevant businesses, occasions, and locations.
- The idea of edge computing can be used in both indoor and outdoor settings.

12.8 Potential Research Directions

The telecoms and IT sectors have recently adopted several technological innovations that will be essential to the growth and success of MEC solutions. This section introduces these essential enabling technologies and explains why they are important (see Figure 12.5) [11, 12].

- **Economic Models**

To the best of our knowledge, there is a deficiency of in-depth examination on the economic models for MEC in the literature. As a result, there is

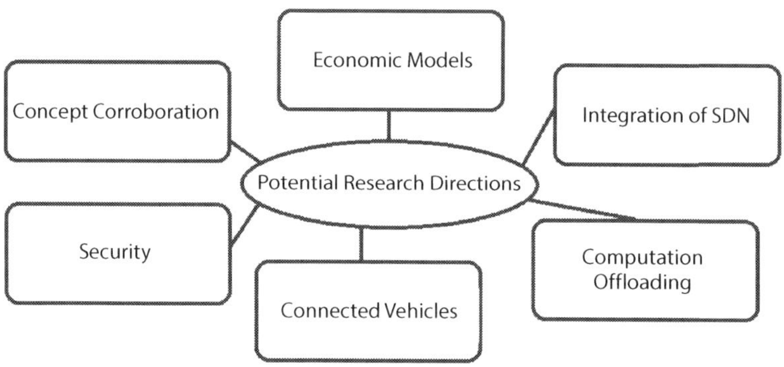

Figure 12.5 Potential research directions.

a lot of research space accessible to look into MEC's economic viewpoints. Particularly, choosing a price is challenging. For instance, with computational offloaded apps, it can be challenging to decide whether to charge users or the owner of the application. MEC has a great deal to offer in terms of financial benefit [5]; therefore, thorough research is required to examine the financial potential. Furthermore, MEC should have its economic potential thoroughly examined because it is a strong candidate for 5G. In this context, appropriate business plans and regulations could be created.

- **Integration of SDN**

By separating the data from control plane, SDN is a novel paradigm that gives the network scalability, elasticity, and programmability. SDN has demonstrated that it can be used in a variety of situations, including mobile networks, data centers, and vehicular networks. SDN may significantly improve MEC's intractability, performance, scalability, and ability to inject new applications. A few initiatives to combine the SDN with the cloud and mobile network have already been made, such as MobiScud [13]. Researchers have a great opportunity to look at the implications of integrating SDN with MEC. Additionally, it might be looked at how SDN can help with computational offloading, a crucial MEC enabler.

- **Computation Offloading**

Computation offloading is one of the most popular research areas in wireless communications right now. Under stringent latency restrictions, the major goal of compute offloading is to preserve the battery life of portable devices. Realistic assumptions, slicing size, and user mobility are just a few of the issues that still need to be resolved. In order for the technology to advance, these issues must be investigated in real settings using experimental testbeds.

- **Allied Vehicles**

The choice of compute offloading whether to offload fully or partially plays a crucial role. The fact that data uplink to the MEC demands greater bandwidth when huge files are uploaded is completely ignored by current research. On the other hand, since only outcomes are communicated to the consumer, downlink data are not only more limited in number but also more refined. As a result, it gives academics new avenues for research and reshapes the notions of uplink and downlink traffic studies. MEC-enabled automobile software defined networks are one potential direction for the future. The MEC technology has a great chance of meeting the needs of a vehicle network. The VANET's strict latency requirements are its biggest problem. A further difficulty for VANET is dynamic topology change brought on by high vehicle mobility. In order to deliver safety messages and infotainment services, it is necessary to have both a global and local perspective of the network. The DSRC channel's 10 MHz bandwidth is insufficient for cutting-edge applications, which would be burdensome to access through a 5G network. As a result, it is necessary to offer an efficient and load-sharing architecture that satisfies the needs of both networks, including vehicular cloud computing. Many other research topics, such as vehicle-to-vehicle video streaming with the aid of MEC and cloud computing, still need to be examined in the context of connected vehicles.

- **Security**

The security and privacy of modern technologies is one of their main worries. The success of any service depends on the privacy of its users, and any operator's top priority is to safeguard that privacy [14–16]. More security risks are related to MEC because it is more open than traditional mobile networks and allows for diverse suppliers, servers, and third-party apps. Moreover, since wireless networks are more susceptible to hacking, compute offloading is carried out over them. Therefore, it is necessary to use application-specific encryption mechanisms in order to improve security, but doing so will result in worse application performance. Therefore, a cryptographic method that balances performance and overhead is needed.

- **Concept Corroboration**

Through simulations, the work that has been done thus far is validated. Generally speaking, the work is based on straightforward hypotheses, which are fine at the beginning of the process but require real-time assumptions for advanced study. These notions must be tested under hardware limitations or at the very least realistic assumptions in order for MEC to be realized. To further bring the MEC concept to life, rigorous tests of the suggested architectures had to be conducted.

12.9 Challenges and Issues

There are numerous challenges associated with the MEC. Some of the challenges are discussed as follows:

- **Task Offloading:** Mobile electronics have speed, memory, and data storage restrictions. The multiple applications needed fast internet connectivity and powerful computers, which are not features that can be provided by mobile devices. Therefore, task offloading—also known as the move of resource-intensive jobs to other systems like the cloud, grid, or MEC server—is possible in such circumstances. The task can be delegated in two different ways. Fully Offload—In this case, the entire task or process is moved to a server, which may be a cloud server or an MEC server. Fully offloading is sometimes referred to as coarse offloading. Partial Offload—In this case, the tasks or processes are split into various portions, some of which are then transferred to the server. The tiny portion of the code that was chosen is the chunks. Fine-grained offloading is another name for the partial offload.

- **Congestion Control:** Latency occurs when there are a large number of queries and the data flow grows significantly due to the MEC server's inadequate resources. The server's limited resources are the primary cause of the bottleneck. The ability of each server and workstation is the cause of the bottleneck. When more data can be obtained in-network but insufficient resources are accessible, network layer issues are primarily concerned. It is one of the major problems in the MEC network, and it may be fixed by (1) buffering traffic. When the network's resources have been utilized up and its capacity is full, this feature stores the packet. Until the buffer capacity is reached, it queued them. (2) Wisely allocated resources—This strategy makes use of constrained resources, awarding grants one at a time when assistance requests come in. The architecture is created for hierarchical smart homes by design to offer edge network smart devices low-latency services.
- **Resource Allocation:** In terms of storage and processing power, MEC servers are superior to portable devices. Due to its constraints (i.e., limited compute and network resources), allocating computational and network resources under various/variable conditions is a challenging issue. Numerous researchers have offered numerous strategies to address this problem.
- **Security and Privacy issue:** The MEC network faces significant obstacles with regard to data security and privacy. Applications from user devices are transmitted wirelessly to MEC. Data are then made available to a trespasser and turned

into data theft. It decreases by utilizing various encryption methods that take longer and impair the functionality of the application. Data security is a crucial concern when some jobs are moved from the edge to the cloud at that time. The security of the infrastructure, the edge servers, the edge network, and the edge devices are some privacy considerations.

- **Mobility:** With respect to data security and privacy, the MEC network confronts considerable challenges. Wirelessly transmitted applications from personal computers are sent to MEC. Then, data are made accessible to a trespasser and are stolen. Implementing different encryption techniques that take more time and limit the app's functionality reduces it. When some jobs are migrated from the edge to the cloud at that time, data security is a critical problem. Some concerns regarding privacy include the security of the infrastructure, the edge servers, edge network, and edge devices.
- **Standardization:** The MEC network lacks any standardized frameworks or protocols. Due to a lack of an environment, the researchers always make assumptions when dealing with MEC. Therefore, the major problem with MEC is its standardization. If this problem is solved, all of the researchers' presumptions will be shown false and MEC will remain uniform.

12.10 Conclusion

MEC is a potential technology that extends cloud capabilities to the cellular network's edge in order to enhance a variety of performance measures, including latency, bandwidth, quality of service, and optimized wireless resources. MEC is a great combination of cellular network and IT infrastructure that adds compute capabilities to the cellular network. Future-oriented applications like video streaming, gaming, augmented reality, connected cars, and e-health, which demand high compute and low latency, have a significant chance of being provided by the MEC. This chapter covers mobile edge computing (MEC), including its definition, traits, benefits, architectures, models, and applications. This chapter also contains some ideas for future research topics and application cases to improve comprehension. The MEC concept, however, is not yet developed and needs to be further validated with real-time considerations in realistic circumstances.

References

1. Vhora F, Gandhi J. A comprehensive survey on mobile edge computing: challenges, tools, applications. In *2020 fourth international conference on computing methodologies and communication (ICCMC)* 2020 Mar 11 (pp. 49-55). IEEE.
2. Rasheed A, Chong PH, Ho IW, Li XJ, Liu W. An overview of mobile edge computing: Architecture, technology and direction. *KSII Trans. Internet Inf. Syst. (TIIS)*. 2019;13(10):4849-64.
3. Huda SA, Moh S. Survey on computation offloading in UAV-Enabled mobile edge computing. *J. Netw. Comput. Appl.* 2022 Feb 1:103341.
4. Elgendy IA, Zhang WZ, Zeng Y, He H, Tian YC, Yang Y. Efficient and secure multi-user multi-task computation offloading for mobile-edge computing in mobile IoT networks. *IEEE Trans. Netw. Serv. Manage,* 2020 Aug 28;17(4):2410-22.
5. Haibeh LA, Yagoub MC, Jarray A. A survey on mobile edge computing infrastructure: Design, resource management, and optimization approaches. *IEEE Access* 2022 Feb 18;10:27591-610.
6. Zaman SK, Jehangiri AI, Maqsood T, Umar AI, Khan MA, Jhanjhi NZ, Shorfuzzaman M, Masud M. COME-UP: computation offloading in mobile edge computing with LSTM based user direction prediction. *Appl. Sci.* 2022 Mar 24;12(7):3312.
7. Shakarami A, Ghobaei-Arani M, Shahidinejad A. A survey on the computation offloading approaches in mobile edge computing: A machine learning-based perspective. *Comput. Networks* 2020 Dec 9;182:107496.
8. https://enterprisersproject.com/article/2021/2/what-mobile-edge-computing-mec, available online 2023.
9. https://thenewstack.io/what-is-mobile-edge-computing/, Available online 2023.
10. Cisco White Paper, "Cisco Visual Networking Index: Forecast and Methodology, 2014-2019", May 2015.
11. Islam A, Debnath A, Ghose M, Chakraborty S. A survey on task offloading in multi-access edge computing. *J. Syst. Archit.* 2021 Sep 1;118:102225.
12. https://portal.etsi.org/portals/0/tbpages/mec/docs/mobile-edge_computing_-_introductory_technical_white_paper_v1%2018-09-14.pdf, Available online 2023.
13. Kaiqiang Wang and Minwei Shen and Jung-Yon Cho and Arijit Banerjee and Jacobus E. van der Merwe and Kirk Webb, "A Fast Moving Personal Cloud in the Mobile Network" *Proceedings of the 5th Workshop on All Things Cellular: Operations, Applications and Challenges*, 19-24, 2015. https://api.semanticscholar.org/CorpusID:364384
14. J. Pathak, S. L. Tripathi, "A Novel Model for Resisting Side Channel Attack by Masking of Gates" *J. Eng. Res.* (April 2022) https://doi.org/10.36909/jer.ICMET.17165

15. J. Pathak, S. L. Tripathi, Abhishek Kumar Singh "High Level Transformation techniques for designing Reliable and Secure DSP Architectures" *AI Techniques for Reliability Prediction for Electronic Components* IGI Global Publishers
16. J. Pathak and S. L. Tripathi, "Novel Obfuscated Secure Architecture for Baugh Wooley Multiplier," *2023 International Conference on Artificial Intelligence and Applications (ICAIA) Alliance Technology Conference (ATCON-1)*, Bangalore, India, 2023, pp. 1-4, doi: 10.1109/ICAIA57370.2023.10169781

13

Descriptive Analysis of the Potential of 5G Technologies in Empowering Decentralized Systems

Megha Agarwal[1]*, Shahnawaz Alam[2], Dewansh[2], Avni Singh[2] and Snehil Singh[2]

[1] Department of CSIS, Shri Ramswaroop Memorial University, Lucknow, India
[2] MCA, Department of CSIS, IoT, Shri Ramswaroop Memorial University, Barabanki, India

Abstract

Networks have evolved from time to time as 1G, 2G, 3G, 4G, and currently 5G, which is in the testing phase in many countries, with researchers experiencing security issues. First, we will tell the evolution of technology, how it occurs, and then explain the networks, how the network works, and how the internet connects the whole world. How it impacts offline businesses, the role of 5G in the growing online business, its role in social media, and so on shall also be discussed. This chapter also focuses on how the 5G network impacts modern technologies such as artificial intelligence (AI), Internet of Things (IoT), and self-driving cars.

Finally, if all current problems with 5G networks are resolved, it could become the foundation of the Smart City concept, which is expected to solve many ongoing problems associated with infrastructure, public transport, environment, etc. This work goes a step further and introduces a Smart Resource and Service Discovery application, based on intent recognition and machine learning-assisted techniques, to enable the automated discovery of resources and services exposed over the marketplace. Evaluation results demonstrate the accuracy of our approach and its suitability to provide decentralized Telecom Marketplaces with intelligent and data-driven discovery.

Keywords: 5G, IoT, artificial intelligence, smart city, decentralization

**Corresponding author*: meghaaggarwal.csis@srmu.ac.in

Sandhya Avasthi, Suman Lata Tripathi, Namrata Dhanda and Satya Bhushan Verma (eds.) Decentralized Systems and Distributed Computing, (263–276) © 2024 Scrivener Publishing LLC

13.1 Introduction

5G is the 5th-generation mobile network. It is a new global wireless standard after 1G, 2G, 3G, and 4G networks. 5G enables a new kind of network that is designed to connect virtually everyone and everything together including machines, objects, and devices. 1G, 2G, 3G, and 4G all led to 5G, which is designed to provide more connectivity than was ever available before [1].

5G is a unified, more capable air interface. It has been designed with an extended capacity to enable next-generation user experiences, empower new deployment models, and deliver new services.

5G chipsets with process nodes in the range of 1028 nm primarily include baseband processors for 5G infrastructure and RFIC components. Qualcomm's Snapdragon X55 5G modem pairs with a single-chip 14-nm RF transceiver for 5G sub-6 GHz for smartphone and other 5G devices.

Things have changed a lot since the first generation of mobile technology. The 1G era was defined by briefcase-sized phones and short conversations between a relatively small numbers of professional people [2].

In the lead up to 2G, the demand for mobile services grew and never slowed down. Phones that could fit in your pocket, SMS, and mobile internet access were hallmarks of the 3G world.

Thanks to 4G, we have smartphones, app stores, and YouTube.

Now, 5G is completely reshaping both our professional and personal lives by enabling new use cases like connective vehicles, augmented reality, and enhanced video and gaming.

5G is designed to deliver peak data rates up to 20 Gbps based on IMT-2020 requirements. Qualcomm Technologies' flagship 5G solutions, the Qualcomm Snapdragon X65, is designed to achieve up to 10 Gbps in downlink peak data rates.

However, 5G is about more than just how fast it is. In addition to higher peak data rates, 5G is designed to provide much more network capacity by expanding into a new spectrum, such as mm wave.

Broadly speaking, 5G is used across three main types of connected services, including enhanced mobile broadband, mission-critical communications, and the massive IoT. A defining capability of 5G is that it is designed for forward compatibility with the ability to flexibly support future services that are unknown today.

- Enhanced mobile broadband

In addition to making our smartphones better, 5G mobile technology can usher in new immersive experiences such as VR and AR with faster, more uniform data rates, lower latency, and lower cost-per-bit.

- Mission-critical communications

5G can enable new services that can transform industries with ultra-reliable, available, low-latency links like remote control of critical infrastructure, vehicles, and medical procedures.

- Massive IoT

5G is meant to seamlessly connect a massive number of embedded sensors in virtually everything through the ability to scale down in data rates, power, and mobility providing extremely lean and low-cost connectivity solutions.

13.1.1 Evolution of 5G Networks

The term evolution means something changes time to time, either changes in network or in gadgets or in anything. Every new generation of wireless networks delivers faster speed and more functionality to our phones. The world has so far experienced four generations with the fifth in few places.

13.1.2 Generations of 5G

- The first generation was used first in 1980s in the USA. Its wave frequency spectrum is between 824 and 894 MHz. 1 MHZ can make one million operations per second with a data transfer speed of 144 kbps; it can only be used to make voice calls.
- The second generation was launched in 1990 in Finland, with the wave frequency spectrum of 380–1800 MHz. The data transfer speed is 144 kbps; it support voice calls, text messages, SMS, picture messages, and simple internet browsing.
- The third generation was launched in Japan in 2003. Its wave frequency spectrum is between 900 and 1800 MHz; the data transfer speed is 14 Mbps. It supports voice, video calls, web browsing, and the use of chat applications.
- In 2009, the fourth generation came from South Korea. Its wave frequency spectrum is between 900 and 2000 MHz; the data transfer speed is 100 Mbps. With more increased internet speed, the fourth generation added LTE (Long-Term Evolution) and then LTE+ (also called LTE advanced).

- The evolution of 5G, the fifth generation of wireless cellular networks, has been a complex process involving multiple stages and technologies.

13.1.3 Phases of 5G

- 2015 - The International Telecommunication Union (ITU) sets the first official standards for 5G networks, including minimum data transfer speeds of 20 Gbps.
- 2016 - The first 5G test networks are deployed, including Verizon's 5G trial in the United States and KT Corporation's trial in South Korea.
- 2017 - The 3rd Generation Partnership Project (3GPP), the organization responsible for developing wireless standards, releases the first set of 5G specifications.
- 2018 - The first commercial 5G networks are launched in South Korea, followed by the United States, China, and other countries.
- 2019 - The first 5G-enabled smartphones are released by major manufacturers such as Samsung, Huawei, and LG.
- 2020 - The COVID-19 pandemic accelerates the adoption of 5G as more people rely on remote work, online education, and telemedicine. Figure 13.1 shows the evolvement phases of mobile communication techologies over time.

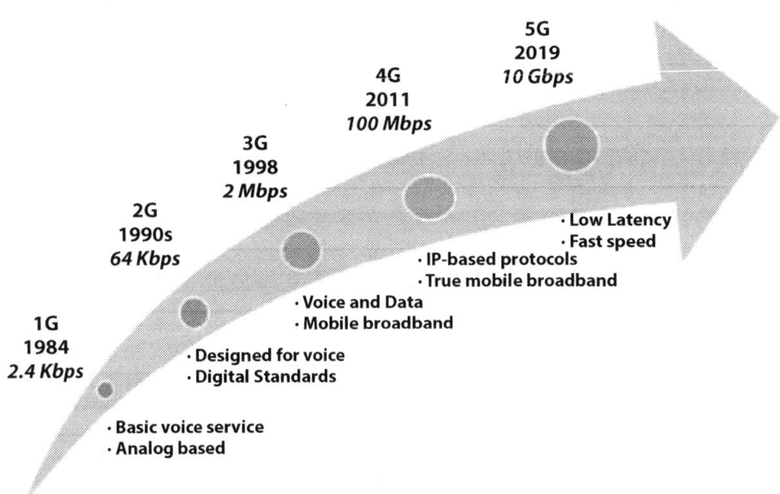

Figure 13.1 Phases of 5G network [14].

- 2021 - 5G networks continue to expand, with some operators beginning to roll out 5G standalone (SA) networks that operate independently of existing 4G infrastructures [3].

13.2 Building 5G Network

Chips play a critical role in 5G networks, as they are responsible for the processing of data and the transmission of signals. There are different types of chips used in 5G networks, including:

- Modem chips:

These chips are used to convert digital signals into radio waves that can be transmitted over the air.

- Radio frequency (RF) chips:

These chips are used to amplify and filter the radio signals that are transmitted over the air.

- Antenna tuning chips:

These chips are used to optimize the performance of the antenna by adjusting its impedance to match the frequency of the signal.

- Baseband chips:

These chips are responsible for processing the digital signals received from the modem and preparing them for transmission over the air. Figure 13.2 shows the architecture of 5G network. It discusses about key technologies used in 5G network, along with its internal architecture and industries involved.

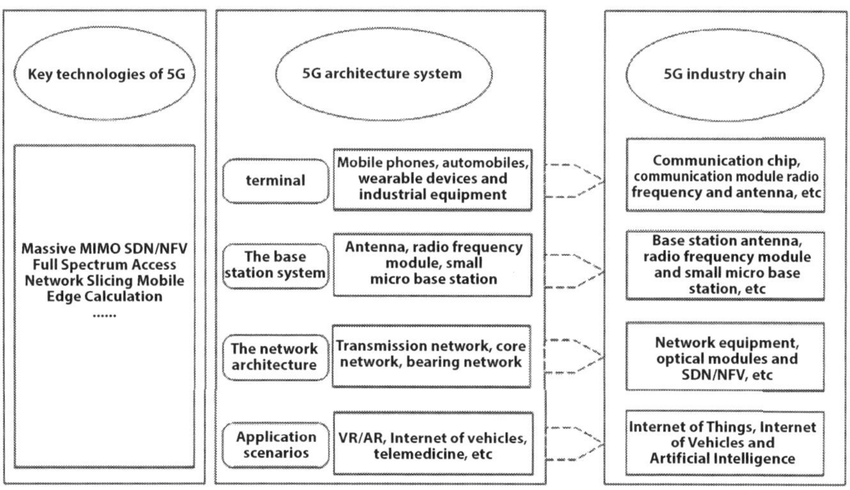

Figure 13.2 Architecture of the 5G network [15].

- Application processors:

These chips are used to run the applications and software that are used on 5G devices.

In addition to these chips, there are also specialized chips used in 5G network infrastructure, such as beam forming chips, which are used to direct signals towards specific users, and encryption chips, which are used to secure the transmission of data over the network.

13.3 Blockchain on 5G Networks

Blockchain technology has the potential to transform various industries, including telecommunications and 5G networks. Some possible applications of blockchain in 5G networks include the following:

- Decentralized network management:

Blockchain technology can be used to enable decentralized network management, where nodes in the network can collectively manage network resources and make decisions based on a consensus mechanism. Figure 13.3 shows the how blockchain technology works on 5G network. Blockchain is a popular technology used for decentralized system and providing security infrastructure's.

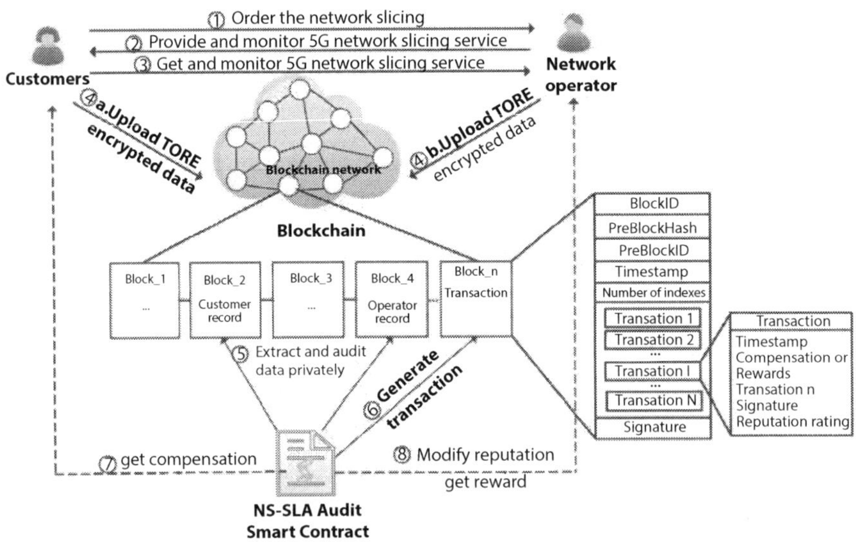

Figure 13.3 Blockchain technology on 5G network [13].

- Network security:

Blockchain can be used to enhance the security of 5G networks by providing a tamper-proof ledger that can be used to store critical network information, such as device identities and access permissions.

- Smart contracts:

Smart contracts can be used to automate the management of network resources and services, such as billing and payment processing.

- Identity management:

Blockchain technology can be used to create a decentralized identity management system for 5G devices, which can improve security and privacy.

- Supply chain management:

Blockchain can be used to create a transparent and tamper-proof supply chain for 5G equipment and devices, which can help reduce the risk of counterfeit products and ensure the authenticity of network components.

However, implementing blockchain technology in 5G networks requires overcoming technical and regulatory challenges. For example, the high throughput and low latency requirements of 5G networks may pose challenges for blockchain scalability, while regulatory requirements may need to be addressed to ensure compliance with data protection and privacy regulations.

13.4 Real-World Impact of 5G

5G technology will not only usher in a new era of improved network performance and speed but also new connected experiences for users. In healthcare, 5G technology and Wi-Fi 6 connectivity will enable patients to be monitored via connected devices that constantly deliver data on key health indicators, such as heart rate and blood pressure. In the auto industry, 5G combined with ML-driven algorithms will provide information on traffic, accidents, and more; vehicles will be able to share information with other vehicles and entities on roadways, such as traffic lights. These are just two industry applications of 5G technology that can enable better, safer experiences for users [12].

5G runs on the same radio frequencies that are currently being used for your smartphone, on Wi-Fi networks, and in satellite communications, but it enables technology to go a lot further. Beyond being able to download a full-length HD movie to your phone in seconds (even from a crowded stadium), 5G is really about connecting things everywhere reliably, without lag so people can measure, understand, and manage things in real time [4]. This has enormous potential, and together, we will take it to the next level.

As an AI language model, my real-world impact lies in the ability to assist individuals and organizations in various industries to improve their decision-making processes and outcomes. I can help with tasks such as natural language processing, data analysis, and language generation, which can have a significant impact on business operations, research, and communication.

For example, in the field of finance, I can help analyze large datasets and provide insights that can inform investment decisions, leading to better financial outcomes. In the healthcare industry, I can help identify patterns in patient data, leading to more accurate diagnoses and treatment plans, which can improve patient outcomes.

In the education sector, I can help with automated grading and feedback on assignments, which can save time for educators and provide more personalized feedback to students, improving their learning outcomes. In journalism, I can assist with writing articles and reports, improving the quality and speed of news production.

Overall, my real-world impact is in providing efficient and accurate solutions to complex problems across different industries, ultimately improving productivity, outcomes, and decision-making processes.

13.4.1 Impact on Business

The 5G network has the potential to bring significant benefits to businesses across various industries. Here are some of the ways in which 5G could impact business:

Faster and more reliable internet speeds: With 5G, businesses can expect significantly faster internet speeds, reducing latency and providing a more seamless experience for customers. This can improve the speed of online transactions and enhance the customer experience.

Improved connectivity: 5G networks are designed to support a large number of connected devices. This could enable businesses to connect more devices to the internet and enable more automation, creating new efficiencies and opportunities for innovation.

New business models: 5G could enable new business models, such as edge computing, where computing resources are moved closer to the end-user. This could enable businesses to provide faster and more personalized services, as well as new business models that rely on real-time data analysis.

Enhanced mobile experiences: 5G will provide enhanced mobile experiences, allowing businesses to offer more advanced mobile services to customers. This could include virtual and augmented reality applications, which could be used in fields such as entertainment, education, and healthcare.

Improved logistics and supply chain management: 5G could enable more efficient logistics and supply chain management, allowing businesses to track inventory and shipments in real time. This could lead to cost savings, improved customer satisfaction, and greater operational efficiency.

Overall, the impact of 5G on business is likely to be significant, providing new opportunities for innovation and growth. However, businesses will need to invest in upgrading their infrastructure and skills to take full advantage of the benefits of 5G.

13.4.1.1 Impact on Business After and Before 5G

The introduction of 5G networks has the potential to significantly impact businesses in a number of ways, both before and after the adoption of the technology. Before the adoption of 5G, businesses may experience slower connectivity and data transfer speeds, which can impact their ability to process and analyze data quickly. This can lead to delays in decision-making, reduced productivity, and decreased customer satisfaction. Additionally, businesses may need to invest in additional infrastructure and equipment to support their operations.

After the adoption of 5G, businesses may see significant improvements in their connectivity and data transfer speeds, which can enable them to process and analyze data more quickly and efficiently. This can lead to improved decision making, increased productivity, and enhanced customer satisfaction. Additionally, businesses may be able to take advantage of new technologies that require high-speed connectivity, such as virtual and augmented reality, which can enable them to create new and innovative products and services.

Overall, the impact of 5G on businesses will depend on a variety of factors, including the specific industry, the size of the business, and the extent to which it relies on high-speed connectivity and data transfer. However, it is clear that businesses that invest in 5G technology and infrastructure will likely have a competitive advantage over those that do not.

13.4.2 Impact on Infrastructure

The rollout of 5G networks is likely to have a significant impact on infrastructure in a number of ways.

Firstly, the deployment of 5G networks will require significant upgrades to existing infrastructure, such as cell towers and base stations, to support the increased data transfer speeds and capacity that 5G offers [5]. This will require substantial investment by telecommunication companies and governments to build new infrastructure and upgrade existing ones.

Secondly, 5G networks will require the installation of a much larger number of small cells, which are low-powered base stations that provide wireless connectivity to smaller geographic areas. This will require the installation of a large number of new small cells on existing infrastructure such as utility poles and buildings. This will also require coordination with municipalities to secure access to locations for installation.

Thirdly, the deployment of 5G networks will require the expansion of fiber optic networks to connect cell towers and small cells, which will provide the necessary backhaul to carry the increased data traffic generated by 5G. This will require additional investment in fiber optic infrastructure, including laying new fiber optic cables and upgrading existing ones.

Lastly, the deployment of 5G networks may also require changes to local regulations and policies, such as zoning and permitting requirements, to allow for the installation of new infrastructure in public and private spaces.

Overall, the deployment of 5G networks will require significant investment in infrastructure and changes to existing regulations and policies to support the rollout of the new technology. However, the benefits of 5G, such as increased data transfer speeds, capacity, and connectivity, will make these investments worthwhile in the long term.

13.4.3 Impact on Humans as Well as on Nature by the 5G Network

The deployment of 5G networks has the potential to impact both humans and nature in various ways, both positive and negative. Here are some potential impacts:

13.4.3.1 Impact on Humans

Increased connectivity and faster data transfer speeds: 5G networks have the potential to provide much faster data transfer speeds than previous generations of wireless networks. This could enable new applications such as remote surgery, augmented reality, and autonomous vehicles.

13.4.3.1.1 Improved Healthcare

5G networks could enable remote patient monitoring and other telemedicine applications, which could improve access to healthcare services.

13.4.3.1.2 Increased Electromagnetic Radiation Exposure

The deployment of 5G networks could result in increased exposure to electromagnetic radiation, which could have potential health implications [6]. However, the World Health Organization has stated that there is no evidence to suggest that exposure to electromagnetic radiation from wireless networks is harmful to human health.

13.4.3.1.3 Digital Divide

The deployment of 5G networks may exacerbate the digital divide by leaving some communities without access to high-speed internet.

13.4.3.2 Impact on Nature

Increased energy consumption: 5G networks require more infrastructure and energy than previous generations of wireless networks, which could have an impact on the environment [7].

13.4.3.2.1 Increased Electromagnetic Radiation Exposure

As mentioned above, the deployment of 5G networks could result in increased exposure to electromagnetic radiation, which could have potential environmental implications [8]. However, research on the impact of electromagnetic radiation on wildlife is still ongoing.

13.4.3.2.2 Potential for E-Waste

As 5G networks require new hardware and infrastructure, there is a risk of generating electronic waste if older equipment is not properly recycled or repurposed.

Overall, while the deployment of 5G networks has the potential to bring significant benefits to humans, it is important to carefully consider and address any potential negative impacts on both humans and nature [10].

13.5 Future of 5G

The fifth-generation (5G) network technology is set to revolutionize the way we connect with the world. With its superfast speeds, lower latency, and higher bandwidth, 5G will enable new applications and services that were not possible before. This new technology promises to transform not only the telecommunications industry but also other sectors such as healthcare, transportation, and entertainment. 5G networks promise faster internet speeds, lower latency, and better connectivity for devices [11]. This is made possible by high-frequency radio waves, also known as millimeter waves, which have a larger bandwidth than the radio waves used in 4G networks.

The benefits of 5G technology are numerous. For instance, faster internet speeds can facilitate the development of new technologies such as augmented reality and virtual reality, which require a high-speed, low-latency network to function effectively.

The network is expected to be at least 20 times faster than the current 4G technology. This means that users can download and upload files, at lightning-fast speeds. Additionally, this means that data will be transmitted almost instantaneously, making it ideal for real-time applications such as autonomous driving, remote surgery, and virtual reality.

Additionally, the improved connectivity of 5G networks will enable more devices to be connected simultaneously, opening up new possibilities for the Internet of Things [9]. This could lead to advancements in fields such as healthcare, transportation, and logistics.

Another advantage of 5G is its increased bandwidth, which allows more devices to connect to the network simultaneously [10]. This is particularly important in the era of the Internet of Things, where millions of devices are expected to be connected to the internet. With 5G, there will be no more network congestion, and devices will be able to communicate with each other seamlessly.

However, the deployment of 5G networks also poses some challenges. For one, the use of millimeter waves means that 5G networks have a shorter range than 4G networks and can be easily disrupted by obstacles such as buildings and trees.

Furthermore, the rollout of 5G networks requires significant infrastructure upgrades, such as the installation of new base stations and antennas. This means that the cost of implementing 5G networks is likely to be high, at least in the short term.

Despite these challenges, the benefits of 5G networks are expected to outweigh the costs. In fact, some experts believe that 5G networks could drive significant economic growth, creating new jobs and opening up new markets for businesses.

13.6 Limitations

Technology becomes obsolete quickly and so 5G may also be soon superseded by a 6G network. 6G network refers to the next generation of wireless communication technology, which is expected to provide faster, more reliable, and more secure connectivity than 5G. While 5G technology is still being rolled out in many countries, research and development on 6G technology have already begun.

One of the key features of 6G is its ability to handle extremely high data rates, which could be up to 1 terabyte per second. This would enable high-speed data transfer, streaming, and communication for applications such as augmented reality, virtual reality, and holographic communication.

Another important feature of 6G is its ability to provide low-latency communication, which means faster response times and improved real-time connectivity. This could have significant applications in areas such as autonomous vehicles, robotics, and industrial automation.

In addition, 6G is expected to offer improved security and privacy features, such as better encryption and authentication methods, to protect user data and prevent cyber threats.

While 6G technology is still in its early stages of development, it has the potential to revolutionize the way we communicate, work, and interact with technology in the future. However, it will likely take several years before 6G networks become widely available and are adopted by consumers and businesses around the world.

References

1. https://www.qualcomm.com/5g
2. https://www.sciencedirect.com/science/article/abs/pii/S1084804520301673
3. https://www.nasdaq.com/articles/the-future-of-5g-should-be-decentralized-201910-21
4. Kaifeng Yue, Yuanyuan Zhang, Yanru Chen, Yang Li, Lian Zhao, Chunming Rong, Liangyin Chen *IEEE Commun. Surv. Tutor.*, 23 (4), 2191-2217, 2021.
5. Dinh C Nguyen, Pubudu N Pathirana, Ming Ding, Aruna Seneviratne *J. Netw. Comput. Appl.*, 166, 102693, 2020.
6. https://www.ericsson.com/en/5g
7. https://timesofindia.indiatimes.com/blogs/voices/5g-a-future-for-everyone/
8. Yatong Wang, Gang Feng, Yao Sun, Shuang Qin, Ying-Chang Liang *IEEE Trans. Veh. Technol.*, 69 (10), 12124-12135, 2020.
9. https://ieeexplore.ieee.org/abstract/document/9024627
10. https://libraries.catholic.edu/research-help/how-do-i/google-scholar.html
11. https://timesofindia.indiatimes.com/blogs/voices/5g-network-the-technological evolution-of-india/
12. https://www.ncbi.nlm.nih.gov/pmc/articles/PMC8747744/

13. Xiao, K., Geng, Z., He, Y. et al. A blockchain-based privacy-preserving 5G network slicing service level agreement audit scheme. *J. Wirel. Commun. Netw.*, 2021, 165 (2021). https://doi.org/10.1186/s13638-021-02037-8
14. M. Attaran, "The impact of 5G on the evolution of intelligent automation and industry digitization," *J. Ambient Intell. Humaniz. Comput.*, 2021. https://link.springer.com/article/10.1007/s12652-020-02521-x
15. Jianjun Hou, Lijun Xu," Management of university and artificial intelligence statistics for 5G edge computing", *Wireless Networks*, Springer 2021, https://doi.org/10.1007/s11276-021-02689-w.

14

Merging Blockchain and Deep Learning for Authentication and Security Architectures

Abhay Kumar Yadav[1*] **and Satya Bhushan Verma**[2]

[1]*Guru Gobind Singh Indraprastha University, New Delhi, India*
[2]*Shri Ramswaroop Memorial University, Barabanki, India*

Abstract
Deep learning (DL) has attracted a lot of attention recently due to its ability to make informed decisions. Features such as dependability, operational transparency, security, reliable data provenance, and traceability are missing in centralized servers based on DL systems. This chapter highlights the need to merge DL and blockchain technology for eliminating the drawbacks of centralized servers. Generally, DL algorithms are employed to authenticate, spot anomalies, and offer system security in cryptography and biometric systems. Blockchain, on the other hand, provides decentralized and secured storage of data. It is important for DL algorithms and the blockchain-based security system to establish balance between confidentiality and efficacy as network sensors are energy-constrained devices. This paper also highlights security models presented for providing authentication and security features using Blockchain, DL, or both based on different systems based on clouding type, models, consensus protocols, applications, services, and deployment goals.

Keywords: DL, blockchain, multi-layered security architecture, healthcare, IoT, SCM

**Corresponding author*: abhaybbdnitm01@gmail.com

Sandhya Avasthi, Suman Lata Tripathi, Namrata Dhanda and Satya Bhushan Verma (eds.)
Decentralized Systems and Distributed Computing, (277–294) © 2024 Scrivener Publishing LLC

14.1 Introduction

Deep learning (DL) idea has been present in nearly every industrial sector. Making prediction based on the present data or situation is the fundamental concept of DL. In the healthcare sector, clinicians use DL models for identifying patient's diseases and problems by analyzing their symptoms. A recent popular application of DL is shown in predicting the spread rate of the COVID-19 pandemic. DL-based techniques have helped health practitioners to successfully identify COVID-19 disease using CT scan and X-ray images [1]. Blockchain's increasing popularity can help in the sustainability of global supply networks. Blockchain is an open and distributed technology that employs secure and unchangeable transaction records to provide novel techniques for protection and distribution of data in a peer-to-peer setting. The technology is said to hold great promise for supply chain sustainability by improving security, accountability, and efficiency [2].

Using biometric and facial recognition, DL models can assist authorities in detecting physical real-time threats [3]. The efficacy and efficiency of a DL system are determined by the data values used in model training phase. In the context of model training, many deep learning algorithms heavily rely on centralized storage and analysis, which exposes them to risks such as single point failures and data manipulation by malicious actors. Any alteration made to the data utilized in deep learning operations can potentially compromise the integrity of the training model. However, by leveraging blockchain, a decentralized system, it becomes possible to effectively address concerns related to data integrity, security, and confidentiality [4]. In this regard, blockchain technology is one such possible Sustainability-Oriented Innovation (SOI). DL combined with blockchain has the potential to provide several benefits, including autonomous and reliable decision-making processes, effective management of data markets, enhanced data security measures, improved model development for accurate predictions, seamless model sharing, and increasing flexibility of systems built on DL principles.

14.1.1 Motivation and Structure

Unethical hackers encounter significant difficulties when attempting to modify, manipulate, or erase data stored within the blockchain. This technology functions as a decentralized system, employing a peer-to-peer (P2P) structure to store and process transactions and data [5]. The network comprises multiple nodes that carry out the validation and storage of transactional data in block form. A block contains a collection of transactional data and is securely connected with preceding blocks, forming a chain-like structure. To ensure data integrity, whenever a user mining a block adds a new block to their local chain, it is distributed to all participating nodes across the network.

DL encompasses artificial intelligence and machine learning, extending its capabilities. Within existing DL systems, models have the ability to learn intricate representations of diverse data types, including images, text, and audio signals. DL empowers hardware to execute a wide range of tasks with remarkable precision, surpassing human performance in certain scenarios. Various industries have leveraged DL for applications such as image classification, object recognition, autonomous driving, disease prediction, and voice identification. This paper's structure is as follows: Section 14.2 will primarily delve into the integration of blockchain and DL, discussing their notable advantages and disadvantages. Sections 14.3 and 14.4 will highlight models given by different researchers and discusses applications of blockchain and DL across different domains. Section 14.5 will address open issues related to authentication and the implementation of multi-layered security architectures. In Section 14.6, the authors will conclude with discussing the future scope of integrating blockchain and DL in emerging technologies; the paper ends with an emphasis on conclusory statements and future directions.

14.2 Merger of Blockchain and DL

DL models are employed and trained by different entities, and blockchain technology can be useful in establishing the authenticity and trustworthiness of these models [6]. By leveraging blockchain, the entire lifecycle of DL models can be captured, including their development, modification, and usage stages. The blockchain works as a ledger, recording various transactions associated with the model, thereby facilitating the identification of the model's owner, participants, data sources, datasets, and the underlying algorithms. This ensures transparency and accountability by providing an immutable record of the model's generation procedures and base model (see Figure 14.1).

DL can be integrated with blockchain technology to create systems that are decentralized, protected, and unchangeable. A possible use case is creating decentralized networks for DL where data are spread across various points on a blockchain, and training and predictions are done on these distributed points. Another possible use case is creating transparent and

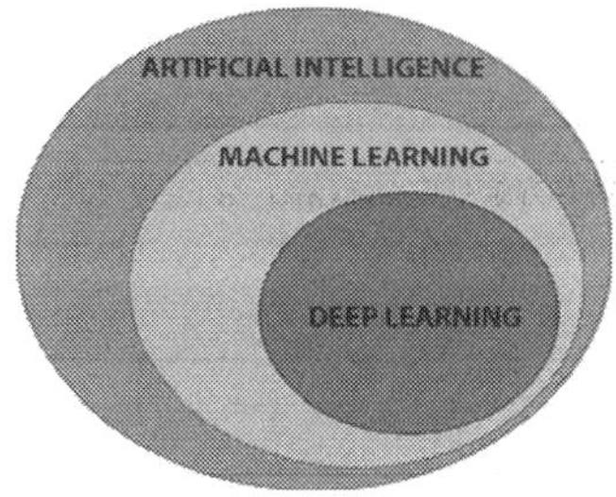

Figure 14.1 Relation between DL, machine learning, and artificial intelligence.

protected systems for sharing and confirming the results of DL research, where data and model elements are stored on a blockchain and can be conveniently examined and confirmed.

The blockchain serves as a worldwide, decentralized, and validated database where all nodes in the network can store and share data. When combining DL and blockchain, they form a powerful and resilient infrastructure for the data collected, analyzed, and utilized by DL-driven applications [7]. This integration ensures the durability and decentralization of the data infrastructure, providing a reliable foundation for the operations of DL-driven systems (see Figure 14.2).

Table 14.1 presents an outline of the notable characteristics of blockchain and DL. It offers a concise summary of how these technologies

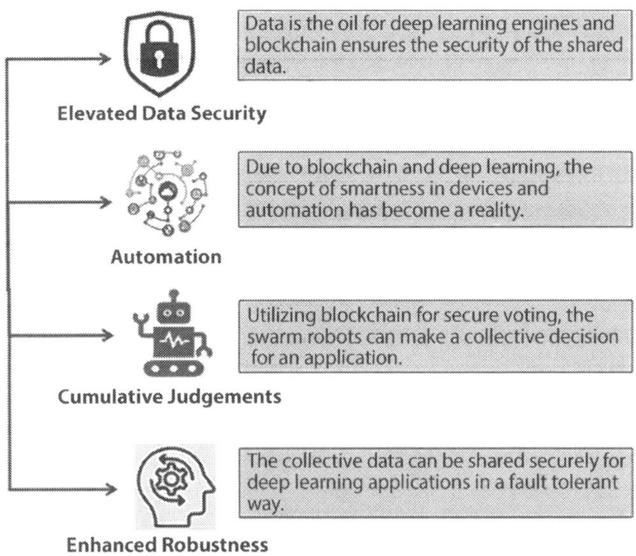

Figure 14.2 DL techniques and benefits from blockchain integration.

can be effectively employed to address various challenges across different domains. By leveraging the specific features of blockchain and DL, organizations can effectively tackle existing obstacles and enhance their operations in diverse fields.

Most developers of security providing software rely on signature-based digitized methods to safeguard network traffic from cyber-attacks.

Table 14.1 Characteristics of blockchain and DL.

DL	Blockchain	Outcome
Layered	Transparent	Collaborative Model Update
Merging big data analytics and other data-intensive approaches	Cybersecurity	Immutable ledger helps in tackling hacking issues. It validates integrity in data management and reduces IP-based DDoS attacks and single point of failure vulnerability.
Modeling using different layers and implementation of resource-intensive property	Insurance, IoT	Used across different sectors such as transportation, efficient supply chain management claim settlement, stream risk management along with increasing transparency.
Supervised, unsupervised, and semi-supervised	Forecasting	Using ML algorithms along with some features of DL and blockchain can be used for an efficient decentralized distributed forecasting model
Diagnostic and scalable	Legality immutability	Managing smart contracts with previously agreed rules, termination, and accessibility across different parties. Blockchain features such as immutability can provide a highly reliable system for legal and prioritized works.
Optimized, prioritized, and corner parameters	FinTech, Banking	Payment can be streamlined with increased security. Reduces auditing complexity and provide globalized transactions [8].

Signature, which is a unique sequence of different characters associated with each identified malware, enables efficient and effective classification of future samples. However, malware attackers can easily evade this technology by utilizing techniques like encryption, polymorphism, and obfuscation. Moreover, the sheer volume of malicious files being distributed daily renders the signature-based approach ineffective. To combat malware attacks, there is a need for intelligent malware detection solutions. Researchers are exploring the construction of models that can identify malware using various feature representations, employing classification algorithms [9]. Many of these methods are created on learning-based models. Although a shallow learning model is somehow successful in malware detection, they still fall short when it comes to addressing the challenges posed by false malware detection. Implementing blockchain can solve the issues due to its potential advantages along with DL techniques.

- **Data Security:** As blockchain is a decentralized system, data stored on it are secure. Private blockchain systems are used to process and store privacy and confidentiality. Securing the private keys of nodes is crucial as they serve as the sole means to access blockchain data [10]. By leveraging DL algorithms on the stable data stored in the blockchain, decision-making processes can benefit from enhanced reliability, accuracy, and dependability.
- **Automated Decision-Based Processes:** Blockchain is a widely recognized technology that ensures peer-to-peer transactions. The traceability aspect of blockchain provides a convenient way to verify the judgments made by DL models. Additionally, it ensures the integrity of documents by preventing tampering during the human-assisted auditing process.

14.2.1 Limitations of DL and Blockchain

14.2.1.1 Limited Utility of Blockchain Nodes

The issue of splitting an entire chain is limitation of blockchain. Endpoints that continue to use the old software would not accept transactions from derived chain. The fork is responsible for making the chain follow a similar path as previous blocks.

14.2.1.2 *High Dependency on Energy*

Blockchain consumption of energy presents another disadvantage. A real-time ledger necessitates the use of energy. Whenever a newer node is made, it creates a connection with the already present nodes at the same time [11].

14.2.1.3 *Providing Guarantee*

Blockchain decision trees have lower predictive capabilities, causing overfitting. On the other hand, logistic regression may suffer from multicollinearity issues. k-NN classifier models face challenges in handling mixed-type data, making them susceptible to interdependencies among inputs and irrelevant features.

14.2.1.4 *Computational Scalability*

SVM exhibit limitations in terms of real-time optimization of changing complexity and associated computational scalability. The Naive Bayes approach can introduce bias when the number of training datasets is increased.

14.2.1.5 *Neural Networks*

They have strong predictive capabilities and the ability to handle coupled inputs [12]. They effectively utilize the predictive value of various input combinations. However, neural networks can be sensitive to relevant features, and managing large datasets and complex models can become challenging.

14.2.1.6 *DL Classification Framework*

Issues regarding lower flexibility of server architecture exist in Pytorch [13]. TensorFlow has limitation with code complexity, limited input channels, and frequent changing interfaces.

14.3 Application of Blockchain-Based DL Models

This section presents a sectional categorization for listing present-day literature based on combined attributes of blockchain and DL techniques.

Today, blockchain and DL technology are gaining popularity and appreciation across the globe. Blockchain, a revolutionary technology, caused quite a stir with the introduction and trading of cryptocurrencies. ML has been advancing rapidly in leveraging available data to identify patterns and extract insights. The integration of ML and blockchain technologies has the potential to enhance their disruptive capabilities. Together, they can accelerate exploratory data analysis and enhance transaction security. Additionally, the distributed nature of blockchain can provide a robust and reliable data source for ML, which relies on large datasets for accurate predictions.

DL applications and services often have time limitations, and the time-specific features of blockchain can help enhance these services. Depending on their specific design, features, and governing rules, blockchain platforms integrated with DL frameworks can be called public, private, or consortium/federated [14].

14.3.1 Smart Contract-Based Security Analysis [15]

Whenever a smart contract code is breached, the ratiocinating and recorded data are blown. Gateway-based threats may be used in stealing data from blockchain. Other major research issues include detection of errors and faults, tackling with malicious smart contracts, and securing with appropriate coding without disrupting the network.

14.3.2 Cybersecurity

Intrusion detection systems are used to detect distributed DoS attacks, [8]. AI-based systems have other issues; false-positive detection is much more prevalent than genuine threats. The algorithms may also be compromised due to imbalanced data and comparative attributes provided in datasets that are useful in training the algorithm. Incorrect and compromised IDS have a lower probability of protecting patient information [16].

14.3.3 Healthcare Sector

DL models are trained on healthcare data for forecasting [1] disease presence and recommend preventive measures. The healthcare data include disease information and patient profiles, helping in the prediction of the patient's conditions. The centralized storage of disease and patient profile data may result in a failure.

14.3.4 Cloud Storage

Cloud storage services have various performance-related issues because of load imbalances and background disruption from different processes [17]. This creates a significant impact on a different application with large work sets and limited time constraints. Using blockchain distributed storage solutions, the client's information is divided into small chunks. Each piece of information is stored in its own location [20].

14.3.5 Banking and FinTech Segments

A ledger is a transaction holding data based on previously executed transactions. Each user in blockchain has a copy of the ledger. The ledger consists of ever-modifying data blocks of data containing information that has been completed [18]. Blockchain's popularity is expected to continue in the near future; it can be expected that its limitation would be eliminated soon [19]. These constraints include higher computational and electricity-related costs.

14.3.6 IoT in Data Analytics

IoT sensors generate massive data for a wide range of applications, successfully processing the data for improving decision-making capability and enhancing productivity as well as accuracy, which is a critical step in establishing IoT as a newer business concept improving people's lives [20].

14.3.7 Internet of Vehicles (IoV)

The use of IoT in the automobile industry has improved modern vehicles' networking capabilities. Cars nowadays are often outfitted with various trending gadgets, collision detectors, and smart software that permit them to efficiently transfer data among them [21]. By incorporating blockchain into IoV, data security can be ensured and data are safely transferred between entities [22]. Different components include the network. Next-generation vehicle networks bring cybersecurity issues with them. To properly address these issues, identifying the corrupting offending entities is necessary. IoV provides multiple services-based activities in real-time data [32].

14.3.8 Law Enforcement and Associated Things

As AI advances, law enforcement agencies are implementing data analysis techniques based on symbolic AI and DL on a larger scale [23]. This includes face, fingerprint, and violence detection. Although these factors may appear unrelated, digitization is related with computer vision and is rapidly expanding as a result of DL. These algorithms may be performed well on datasets with fixed parameters, but fails miserably on real-world applications [24].

14.4 Study of Existing Integrated Models

In pharmacogenomics research, DL models with hospital information management are becoming increasingly important. The research paper [25] presents an ovarian cancer prediction model based on DL techniques. Blockchain is used to share patient information along with modeling ovarian cancer predictions with cooperating organizations. The preexisting datasets are valuable in assessing the effectiveness of encryption and decryption techniques by analyzing the time required for these processes. Nevertheless, it is crucial to acknowledge that the model introduced in this research has solely been examined on a restricted number of instances per category and its capability to forecast unseen data within the learned categories [31].

14.4.1 Arrhythmia Classification

The research [26] discussed DL models for categorizing arrhythmia illness using a blockchain platform. The study described in [28] utilizes a multi-layered denoising stacked auto-encoder (SDA) for extracting features and identifying irregular heartbeats associated with arrhythmia. During the testing phase, the SDA model is trained to minimize false-positive results in arrhythmia classification. Once trained, the relevant information is stored in hardware and accessed through pointers in a ledger. To ensure privacy and protect user identity, the Hyperledger blockchain is employed in the network. This is particularly beneficial for distinguishing patients from clinicians and establishing secure connections for electronic signatures. In the proposed architecture, the actual data are stored in a cloud drive, while the blockchain network maintains only pointers or links to the data's location. This approach raises scalability by minimizing storage-related requirements of the system.

For authenticating patient access to read/write, smart contracts are utilized to establish access policies, enabling secure execution of logic within the Hyperledger blockchain framework.

14.4.2 Blockchain-Based DL Model in IoT

The IoT comprises a vast network of interconnected nodes, including network nodes, sensing nodes, and digital nodes [27]. Within this network, sensor data assembled by nodes are shared between peers and a centralized server device. However, the IoT faces significant security challenges, including vulnerabilities and the risk of denial-of-service (DoS) attacks. To address these concerns and enhance security, blockchain technology is employed, eliminating the need for a central authority. This study focuses on the development of an adaptive healthcare system that tailors user experiences based on their information and behaviors. DL techniques are utilized, specifically to know the exchange of public keys among interconnected parties. The proposed strategies demonstrate superior efficiency compared to existing benchmark models. DL models, inspired by the structure of human neurons, simulate the intricate network of interconnected layers found in the human brain. Multi-layer NNs are leveraged to create DL solutions that prioritize privacy and data protection [28].

14.4.3 Currency Supervision

With the growing popularity of digital currencies like Bitcoin, traditional methods of extracting valuable insights from large volumes of data and evaluating the worth of digital assets have become increasingly challenging [29, 30]. This study focuses on digital assets and explores the application of deep learning (DL) technology to develop a model for analyzing their value. To enhance the research, a permission mechanism is integrated with delegated PoS and PBET algorithms, which leads to the proposal of a transient delegated practical Byzantine fault tolerance (DDPBFT) mechanism. Ultimately, an upgraded blockchain technology is employed to establish a controlled digital economy system eliminating the need for physical currencies.

14.4.4 Ovarian Cancer Detection

By implementingpharmacogenomic and DL methodologies, an algorithm was created with the objective of detecting ovarian cancer during earlier stages [25]. The study was used to identify the genetic fingerprints that contribute to the disease's susceptibility. The training model will predict possible mutation causing cancer based on the image. An 86% efficiency was achieved after many iterations.

14.5 Challenges and Issues

The merger of blockchain and DL technology is an increasing field of research, but it still faces several challenges and unresolved issues that require careful consideration in future efforts. Furthermore, there are other factors and strategies that impact the development of both blockchain and DL technologies, and their interplay can have a significant influence. This section will delve into the critical unresolved issues and research challenges that must be addressed to improve the performance of blockchain and DL. Additionally, we will explore research opportunities in related domains within a broader framework.

The smart contracts utilize a Deep Autoencoder Neural Network to identify abnormalities and establish a mutual network for controlling traffic by programming. This innovative design provides a platform for creating a secure platform that facilitates and executes interconnected transactions within critical infrastructure networks without relying on a centralized authority. It represents a transformative approach to integrating AI into the blockchain network, not merely as a supplementary framework that enhances its capabilities, but as an integral component essential for its functioning. However, the addition of newer blocks in the network creates a more permanent record, leading to a decrease in the network's efficiency as its size expands.

DL techniques can be used for compressing data along with assisting in data reduction. Because recorded data are irreversible, the growing size of the ledger poses a significant challenge to be addressed appropriately.

The increasing utilization of blockchain is expected to present several challenges, particularly regarding user requirements for internet access, data speed, transaction speed, and the volume of transactions carried out by participants. To address the growing storage and computing demands of the blockchain ledger, it becomes crucial to significantly decrease the existing nodes and transactions incorporated into blockchain, aiming to fulfill the anticipated user demands.

The DL-based methods necessitate the preservation of both privacy and security. If any corrupted node gets into a unified system to modify the instructions, DL systems are vulnerable to hacking at the moment.

The compromise of data can lead to concerns regarding the privacy of personal information since training data often include a substantial amount of personal data. It is critical to consider methods for protecting training data of DL models from hackers.

As a critical component of today's networks, big data extraction [29] is useful in many sectors. In the near future, data processing could make collected data extremely useful. Big data are widely used in blockchain and DL applications, as a place where data are gathered, sorted, and stored in blocks before being presented as training data.

When multiple IoT devices share the same connection state, blockchain can improve spectral efficiency and 5G traffic optimization along with maintaining privacy in 5G application communications networks and beyond [31]. There is currently no existing framework that can offer a complete and secure solution encompassing the CIA triad, while also preserving privacy and providing remote authentication [32–34]. Despite the advantages of blockchain technology and the proposed frameworks based on it to enhance security components, there are still significant challenges that researchers are expected to encounter when it comes to securing blockchain-based IoT systems in the future.

14.6 Conclusion

The merger of blockchain, DL, and IoT provides a number of potential opportunities and benefits, and it is critical in a number of ways. IoT is needed to connect various types of DL applications, and it will be even more important in providing large-level information in real time for learning and decisions. Meanwhile, as blockchain-based apps gained popularity, the size of data on DL increased exponentially. This paper focused on and discussed the significance of integrating blockchain and DL, making it a desired approach for establishing safe and decentralized model sharing, along with efficient operation. The paper began with a concise introduction to blockchain and DL, encompassing essential concepts, classifications, and popular use cases. Additionally, the authors focused on examining prominent models that have aimed to leverage the integration of these technologies or proposed architectural frameworks with specific parameters. Moreover, the paper examines the interplay between blockchain and DL, highlighting their mutual usability and the advantages they offer each other. Additionally, the paper identifies several issues, challenges, and problems in this domain, which can serve as potential areas for future research by various scholars in the field. After in-depth analysis and evaluation of various domains, this chapter explores the application of blockchain-based DL model systems. Therefore, researchers who are interested and motivated to address the aforementioned challenges are encouraged to pursue further research in these areas.

References

1. Ahmad, R. W., Salah, K., Jayaraman, R., Yaqoob, I., Ellahham, S., & Omar, M. The role of blockchain technology in telehealth and telemedicine. *Int. J. Med. Inf.*, 148, 104399. 2021.
2. Ismaeel Al Ridhawi, Moayad Aloqaily, and Yaser Jararweh. An Incentive-based Mechanism for Volunteer Computing Using Blockchain. *ACM Trans. Internet Technol.*, 21, 4, Article 87, 2021; DOI: https://doi.org/10.1145/3419104
3. Shafay, Muhammad & Hassan, T aimur & Velayudhan, Divya & Damiani, Ernesto & Werghi, Naoufel. Deep Fusion Driven Semantic Segmentation for the Automatic Recognition of Concealed Contraband Items. 2021; 10.1007/978-3-030-73689-7_53.
4. Cao, X., Zhang, J., Wu, X. *et al.* A survey on security in consensus and smart contracts. *Peer Peer Netw. Appl.,* 2022; https://doi.org/10.1007/s12083-021-01268-2
5. Webb, M.E., Fluck, A., Magenheim, J. *et al.* Machine learning for human learners: opportunities, issues, tensions and threats. *Educ. Technol. Res. Dev.,* 69, 2109–2130, 2021; https://doi.org/10.1007/s11423-020-09858-2
6. Abhishek Gupta, Alagan Anpalagan, Ling Guan, Ahmed Shaharyar Khwaja, DL for object detection and scene perception in self-driving cars: Survey, challenges, and open issues, *Array*, Volume 10, 2021; 100057, ISSN 2590-0056, https://doi.org/10.1016/j.array.2021.100057.
7. Fran Casino, Thomas K. Dasaklis, Constantinos Patsakis, A systematic literature review of blockchain-based applications: Current status, classification and open issues, *Telemat. Inform.*, Volume 36, 2019; Pages 55-81, ISSN 0736-5853, https://doi.org/10.1016/j.tele.2018.11.006.
8. K. Arulkumaran, M. P. Deisenroth, M. Brundage and A. A. Bharath, "Deep Reinforcement Learning: A Brief Survey," in *IEEE Signal Process. Mag.*, vol. 34, no. 6, pp. 26-38, 2017; doi: 10.1109/MSP.2017.2743240.
9. Maldonado, A.D.; Morales, M.; Navarro, F.; Sánchez-Martos, F.; Aguilera, P.A. Modeling Semiarid River–Aquifer Systems with Bayesian Networks and Artificial Neural Networks. *Mathematics,* 2022; 10, 107. https://doi.org/10.3390/ math10010107
10. Ziaee Adib, Seyed Morteza & pour, Amir & Ghahfarokhi, Farzaneh. The impact of blockchain mega-trend on the tourism industry. 2019; doi: 10.13140/RG.2.2.25803.34083.

11. A. K. Yadav and V. P. Vishwakarma, "Adoptation of Blockchain of Things(BCOT): Oppurtunities & Challenges," in *2022 IEEE International Conference on Blockchain and Distributed Systems Security,* Institute of Electrical and Electronics Engineers Inc., 2022; doi: 10.1109/ICBDS53701.2022.9935985.
12. L. Račić, T. Popović, S. čakić and S. Šandi, "Pneumonia Detection Using DL Based on Convolutional Neural Network," *2021 25th International Conference on Information Technology (IT)*, 2021; pp. 1-4, doi: 10.1109/IT 51528.2021.9390137.
13. Zhang, Dehai & Cui, Menglong & Yang, Yun & Yang, Po & Xie, Cheng & Liu, Di & Yu, Beibie & Chen, Zhibo. Knowledge Graph-Based Image Classification Refinement. *IEEE Access* 2015; 10.1109/ACCESS.2019.2912627.
14. Jabbar, S., Lloyd, H., Hammoudeh, M. *et al.* Blockchain-enabled supplychain: analysis, challenges, and future directions. *Multimedia Syst.,* 27, 787–806, 2021; https://doi.org/10.1007/s00530 -020-00687-0
15. Huang, Yongfeng & Bian, Yiyang & Li, Renpu & Zhao, J. & Shi, Peizhong. Smart Contract Security: A Software Lifecycle Perspective. *IEEE Access* 7. 1-1. 2019; 10.1109/ACCESS.2019.2946988.
16. P. K. Ghosh, A. Chakraborty, M. Hasan, K. Rashid, and A. H. Siddique, "Blockchain Application in Healthcare Systems: A Review," *Systems*, vol. 11, no. 1, p. 38, Jan. 2023; doi: 10.3390/systems11010038.
17. Sheth, H.S.K., Tyagi, A.K. Mobile Cloud Computing: Issues, Applications and Scope in COVID-19. In: Abraham, A., Gandhi, N., Hanne, T., Hong, TP., Nogueira Rios, T., Ding, W. (eds) *Intelligent Systems Design and Applications*. ISDA 2021. Lecture Notes in Networks and Systems, vol. 418. Springer, Cham. 2022; https://doi.org/10.1007/978-3-030-96308-8_55
18. P. K. Ghosh, A. Chakraborty, M. Hasan, K. Rashid, and A. H. Siddique, "Blockchain Application in Healthcare Systems: A Review," *Systems*, vol. 11, no. 1, p. 38, Jan. 2023; doi: 10.3390/systems11010038.
19. M. Kumar and A. K. Singh, "Distributed Intrusion Detection System using Blockchain and Cloud Computing Infrastructure," *2020 4th International Conference on Trends in Electronics and Informatics (ICOEI) (48184)*, 2020, pp. 248-252, doi:10.1109/ICOEI48184.2020.9142954.
20. A. Mughal and A. Joseph, "Blockchain for Cloud Storage Security: A Review," *2020 4th International Conference on Intelligent Computing and Control Systems (ICICCS)*, 2020, pp. 1163-11f69, doi: 10.1109/ICICCS48265.2020.9120930.
21. Prakash, A., Avasthi, S., Agarwal, K., & Hussain, M. (2024). Modern Healthcare Systems: Unveiling the Possibility of AIoT for Remote Patient Monitoring, in: *Developments Towards Next Generation Intelligent Systems for Sustainable Development* (pp. 180-203). IGI Global.

22. Alladi, Tejasvi & Kohli, Varun & Chamola, Vinay & Yu, F. Securing the Internet of Vehicles: A DL-Based Classification Framework. *IEEE Networking Lett.,* 2021; PP. 10.1109/LNET.2021.3058292.
23. S. Raaijmakers, "Artificial Intelligence for Law Enforcement: Challenges and Opportunities," in *IEEE Secur. Priv.*, vol. 17, no.5, pp. 74-77, Sept.-Oct. 2019; doi: 10.1109/MSEC.2019.2925649.
24. Mishra S., Tyagi A.K. The Role of Machine Learning Techniques in Internet of Things-Based Cloud Applications. In: Pal S., De D., Buyya R. (eds) *Artificial Intelligence-based Internet of Things Systems. Internet of Things (Technology, Communications and Computing).* Springer, Cham. 2022; https://doi.org/10.1007/978-3-030-87059- 1_4
25. Abraham, Misha & Am, Hima & Srinivasan, Chungath & Namboori, Dr. Krishnan. Healthcare security using blockchain for pharmacogenomics. *J. Int. Pharm. Res.,* 2019; 6. 529-533.
26. Kumar, R., Badwal, L., Avasthi, S., & Prakash, A. (2023, January). A Secure Decentralized E-Voting with Blockchain & Smart Contracts. In *2023 13th International Conference on Cloud Computing, Data Science & Engineering (Confluence)* (pp. 419-424). IEEE.
27. Avasthi, S., Sanwal, T., Sharma, S., & Roy, S. (2023). VANETs and the Use of IoT: Approaches, Applications, and Challenges. *Revolutionizing Industrial Automation Through the Convergence of Artificial Intelligence and the Internet of Things*, 1-23.
28. A. Juneja and M. Marefat, "Leveraging blockchain for retraining DL architecture in patient-specific arrhythmia classification," *2018 IEEE EMBS International Conference on Biomedical & Health Informatics (BHI)*, 2018, pp. 393-397, doi: 10.1109/BHI.2018.8333451.
29. Ali, Aitizaz, Muhammad F. Pasha, Jehad Ali, Ong H. Fang, Mehedi Masud, Anca D. Jurcut, and Mohammed A. Alzain. "DL Based Homomorphic Secure Search-Able Encryption for Keyword Search in Blockchain Healthcare System: A Novel Approach to Cryptography" *Sensors* 22, 2022; no. 2: 528. https://doi.org/10.3390/s22020528
30. Huiling Fan, The digital asset value and currency supervision under DL and blockchain technology, *J. Comput. Appl. Math.*, 2022, 114061, ISSN 0377 -0427, https://doi.org/10.1016/j.cam.2021.114061.
31. E. Hegland Hjort Kure, P. Engelstad, S. Maharjan, S. Gjessing and Y. Zhang, "Distributed Uplink Offloading for IoT in 5G Heterogeneous Networks Under Private Information Constraints," in *IEEE Internet Things J.*, vol. 6, no. 4, pp. 6151 -6164, Aug. 2019; doi: 10.1109/JIOT.2018.2886703.
32. Sheth, H.S.K., Tyagi, A.K. Mobile Cloud Computing: Issues, Applications and Scope in COVID-19. In: Abraham, A., Gandhi, N., Hanne, T., Hong, TP., Nogueira Rios, T., Ding, W. (eds) *Intelligent Systems Design and Applications.* ISDA 2021. Lecture Notes in Networks and Systems, vol 418. Springer, Cham. 2022; https://doi.org/10.1007/978-3-030-96308-8_55.

33. SB Verma, Brijesh P., and BK Gupta, Containerization and its Architectures: A Study, *ADCAIJ: Adv. Distrib. Comput. Artif. Intell. J.*, Vol. 11 N. 4 (2022), 395-409, eISSN: 2255-2863, DOI: https://doi.org/10.14201/adcaij.28351
34. Satya B Verma, Shashi B V, Data Transmission in BPEL (Business Process Execution Language), *ADCAIJ: Adv. Distrib. Comput. Artif. Intell. J.* Regular Issue, Vol. 9 N. 3 (2020), 105-117 eISSN: 2255-2863 DOI: https://doi.org/10.14201/ADCAIJ202093105117 105

15

5G Embedded Decentralized Systems for Secure and Efficient Data Sharing

Asjad Suhail Akhtar, Garima Srivastava*, Shikha Singh and Sachin Kumar

Amity University Lucknow Campus, Lucknow, India

Abstract
The emergence of decentralized systems, powered by blockchain applications and networks and other distributed system technologies, is revolutionizing the way we store, manage, manipulate, and share. However, the growth and scalability of these systems have been faced by challenges related to speed, connectivity, and security threats. The advent of 5G technology promises to solve these challenges by enabling faster and more efficient speed for data transmission, improving connectivity and reliability, and enhancing security encryption and privacy threats. This chapter will provide an overview of 5G technology in decentralized systems, covering topics such as overview of decentralized systems, 5G technology, 5G for decentralized systems, use cases, and applications. Studies done address potential challenges and opportunities that arise with interoperability of the technologies. Further work also includes security and privacy considerations that arise, and regulatory issues related to government.

***Keywords*:** 5G architecture, network slicing, distributed databases, blockchain technology, peer-to-peer networks, internet of things (IoT), edge computing

15.1 Overview of Decentralized Systems

A decentralized system is a type of network architecture that distributes control among many nodes or participants instead of relying on a single centered authority [1]. In contrast to centralized systems, where a single entity that presents a node controls the network and the flow of data,

**Corresponding author*: gsrivastava1@lko.amity.edu

Sandhya Avasthi, Suman Lata Tripathi, Namrata Dhanda and Satya Bhushan Verma (eds.) Decentralized Systems and Distributed Computing, (295–316) © 2024 Scrivener Publishing LLC

decentralized systems operate in a peer-to-peer (P2P) fashion, allowing for greater transparency, security, and resilience.

One of the key benefits of decentralized systems is their ability to operate without a single point of failure [2]. Because control is distributed across multiple nodes, even if one or more nodes fail or are compromised, the system can continue to function. This makes decentralized systems highly resilient to cyber-attacks, natural disasters, and other disruptions. Another advantage of decentralized systems is their potential for greater privacy and security [3]. By distributing control and limiting the amount of information that any one node can access, decentralized systems can provide a higher level of protection against data breaches and other security threats. This is particularly important in contexts where sensitive or personal data are being transmitted or stored.

Decentralized systems have a wide range of applications, including in finance, healthcare, logistics, and more. One of the most well-known examples of a decentralized system is blockchain technology, which uses a distributed ledger to record and verify transactions. Other examples include peer-to-peer networks, distributed databases, and decentralized file storage. However, decentralized systems are not without their challenges. One major issue is scalability, as the size and complexity of the network can make it difficult to maintain performance and efficiency. Another challenge is interoperability, as different nodes may use different protocols or standards, making it difficult to ensure seamless communication and integration.

Despite these challenges, decentralized systems are likely to play an increasingly important role in the future of technology. With the emergence of new technologies such as 5G, edge computing, and the Internet of Things (IoT), decentralized systems will be able to take advantage of new opportunities for connectivity, security, and innovation.

15.1.1 Peer-to-Peer (P2P) Network

A peer-to-peer (P2P) network as in Figure 15.1 is a type of decentralized network that allows individual devices or nodes to communicate directly with each other without the need for a centralized server or authority [1]. In a P2P network, all nodes have equal status and can act as both a client and a server at the same time, exchanging data and resources with each other. P2P networks have been widely used for file sharing, content distribution, and other applications that require fast and efficient data transmission. One of the most well-known examples of P2P networking is the BitTorrent protocol, which is used for peer-to-peer file sharing.

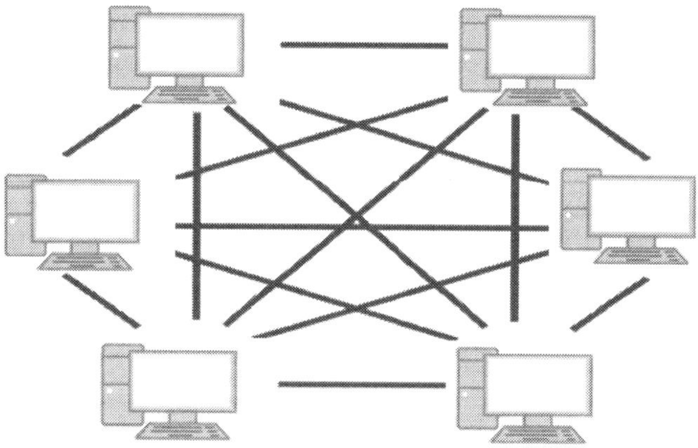

Figure 15.1 Peer-to-peer (P2P) network.

P2P networks offer several advantages over traditional client-server architectures, including:

- Scalability: P2P networks can be easily scaled up by adding more nodes, which can share the workload and provide redundancy in case of failure.
- Cost-effectiveness: P2P networks are typically less expensive to operate than centralized networks since they do not require a dedicated server infrastructure.
- Resilience: P2P networks are more resilient to failure and attack than centralized network since there is no point of attack or failure.
- Anonymity: P2P networks can provide a greater degree of anonymity than centralized networks since nodes can communicate directly with each other without the need for intermediaries.

Furthermore, P2P networks also have some disadvantages, including:

- Performance: P2P networks can suffer from performance issues, particularly when the network becomes congested or when nodes are not well connected.
- Security: P2P networks are vulnerable to attacks and security breaches, particularly if nodes are not properly authenticated or if malicious nodes are introduced into the network.

- Governance: P2P networks can be more difficult to govern and regulate than centralized networks, since there is no central authority to enforce rules or policies.

15.1.2 Distributed Databases

A distributed database as in Figure 15.2 is a database system that stores data across multiple physical locations or nodes in a network, rather than on a single centralized server [3]. In a distributed database system, each node can process queries and transactions independently, and the results are synchronized across the network to maintain consistency and data integrity [4].

Distributed databases are used in a variety of applications, including supply chain management, online transactions, and cloud computing. Some of the key benefits of using a distributed database include:

- Scalability: Distributed databases can handle big volumes of data and can be effortlessly scaled up by increasing the numbers of nodes to the network.
- High availability: Distributed databases are designed to make sure high availability and fault tolerance, since the data are duplicated across multiple nodes.
- Reduced latency: Distributed databases can reduce the latency of queries and transactions by storing data closer to the point of use.
- Security: Distributed databases can provide greater security than centralized databases since the data exist on multiple nodes and can be encrypted and secured more effectively.

However, there are also some challenges associated with using a distributed database, including:

- Complexity: Distributed databases can be more difficult to achieve, implement, and manage than centralized databases, due to the need to ensure data consistency and integrity across multiple nodes.
- Performance: Distributed databases can suffer from performance issues, particularly when nodes are not well connected or when there is a high volume of data traffic.

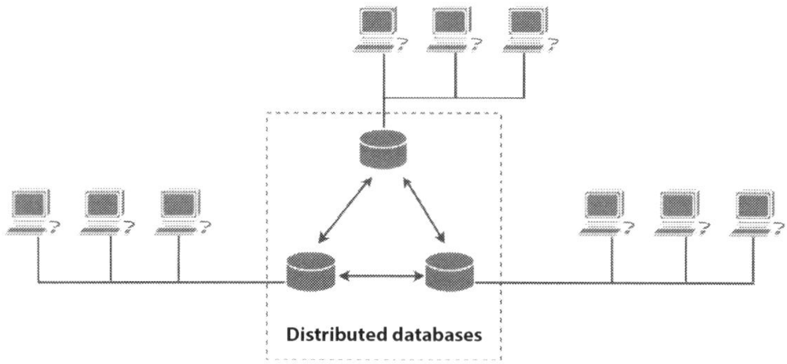

Figure 15.2 Distributed database.

15.1.3 Blockchain

Blockchain as in Figure 15.3 is a decentralized digital ledger technology that enables secure and transparent transactions without the need for intermediaries such as banks or governments [4]. It was first introduced in 2008 as a part of the cryptocurrency Bitcoin, but its potential applications have since expanded to various industries. In a blockchain network, each block of data (or transaction) is cryptographically secured and linked to the previous block, forming a chain of blocks that cannot be modified or deleted without the consensus of the network. This makes it virtually impossible for any individual or entity to manipulate the data or tamper with the integrity of the system [5].

Blockchain technology offers several advantages over traditional database systems, including:

- Decentralization: Blockchain networks are decentralized, which means there is no single influence controlling the system. This makes it more secure to any attack and failure and provides greater trust and transparency.
- Security: Blockchain networks use cryptographic algorithms to secure and protect the data, making it nearly unfeasible for any breach of the system.
- Transparency: Blockchain networks offer greater transparency, as all participants have some kind of entry level to the same information and can validate the authenticity of the block [16, 17].

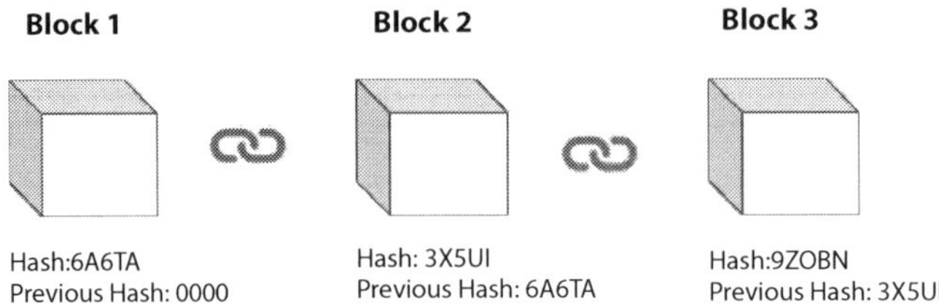

Figure 15.3 Visual representation of blockchain.

- Immutability: Once a block is added to the blockchain, it is immune to any modification or deletion, making sure of the integrity of the data.

Blockchain has several applications, including:

- Cryptocurrencies: Blockchain is the backbone of many cryptocurrencies, such as Bitcoin, Ethereum, and Litecoin [4].
- Supply chain management: using blockchain, it is possible to keep an eye on the movement of supply and keep the authenticity and integrity of the supply chain.
- Voting systems: Blockchain technology can be used to create transparent and secure voting systems that eliminate the possibility of fraud.
- Profile management: Blockchain technology can be used to generate safe and decentralized profiling management systems that protect personal information and prevent identity theft [18, 19].

15.2 5G Technology

5G technology is the 5th gen of wireless communication technology that pledges rapid data transfer, minimize latency, and greater connectivity. It is designed to be more efficient and faster than its predecessor, 4G/LTE. 5G technology uses radio waves at higher frequency, which allows for quick data transfer rates, lower latency, and greater network capacity [6].

One of the features of 5G technology is its possibility to support many devices simultaneously. This is possible due to its advanced network architecture, which uses small cells and millimeter wave frequencies to connect devices to the internet [7]. This technology is predicted to enable the deployment of more advanced apps, such as VR and AR, autonomous vehicles, and smart cities [2].

The 5G technology architecture is designed to support three different types of services: improved mobile broadband (eMBB), ultra-reliable and low-latency communications (URLLC), and massive machine-type communications (mMTC) [8]. The eMBB service provides faster data transfer rates, while the URLLC service offers reliable, low-latency communications for applications such as autonomous vehicles and remote surgery. The mMTC service is designed for the Internet of Things (IoT) devices, which require low-power, low-data-rate connectivity.

5G technology also offers several other benefits, such as improved energy efficiency, enhanced security features, and improved network slicing capabilities, which allow operators to provide specific services tailored to the needs of their customers. Furthermore, the implementation of 5G technology also faces several challenges, such as the need for greater infrastructure investments, the availability of spectrum, and the need for new business models to support the deployment of advanced applications [9].

15.2.1 Key Futures of 5G Technology

- High Speed: 5G technology provides much faster data transfer speeds correlated to previous generations of wireless technology. 5G networks can provide top speed of data rates up to 20 Gbps, which is almost 20 times more than the 4G LTE network speed. This high-speed connectivity can enable a range of new applications, such as VR, AR, and UHD video streaming.
- Low Latency: 5G networks have reduced the latency compared to previous generations of wireless technology. Latency is the time it takes for a signal to travel from the device to the network and back again. 5G networks have a latency of less than 1 ms, which is almost 10 times faster than 4G LTE networks. This low latency is crucial for real-time applications like gaming, autonomous vehicles, and remote surgeries.
- Massive IoT Connectivity: 5G technology provides massive connectivity for the Internet of Things (IoT) devices. It can add up to one million devices per square kilometer, which is a hundred times more than 4G LTE networks. This massive IoT connectivity can enable the deployment of smart cities, smart homes, and industrial automation.

- Network Slicing: 5G technology allows network operators to add many virtual networks within a single physical network. This feature is called network slicing [9]. Each network slice can be enhanced for a specific use case, such as low-latency, high-speed, or huge IoT connectivity. Network slicing enables operators to provide modified services to different users and industries.
- Edge Computing: 5G technology enables edge computing, which is the processing of data near the source rather than passing it to a centralized location for processing. This reduces latency and impacts the efficiency of applications that need real-time data processing, such as autonomous vehicles, remote surgeries, and VR.
- Security: 5G technology provides enhanced security features compared to previous generations of wireless technology. It supports advanced encryption algorithms and authentication mechanisms to secure user data and block unauthorized access to the network.

15.2.2 5G Architecture

5G architecture as in Figure 15.4 is engineered to be more adaptable and flexible compared to previous generations, enabling new levels of automation, interconnectivity, and efficiency. It consists of three key components as in Figure 15.5: the user plane, the control plane, and the management plane.

1. User Plane: The user plane is responsible for transporting user data between the end-user device (e.g., laptop or smartphone) and the application server. It consists of the Radio Access Network (RAN) and the core network [10].

    - Radio Access Network (RAN): The RAN is responsible for obtaining wireless connectivity between the end-user device and the core network. It includes the base station and the radio access network controller.
    - Core Network: The core network is responsible for managing the data traffic with the end-user device and the application server. It includes several components such as the User Plane Function (UPF), the Session Management Function (SMF), and the Access and Mobility Management Function (AMF).

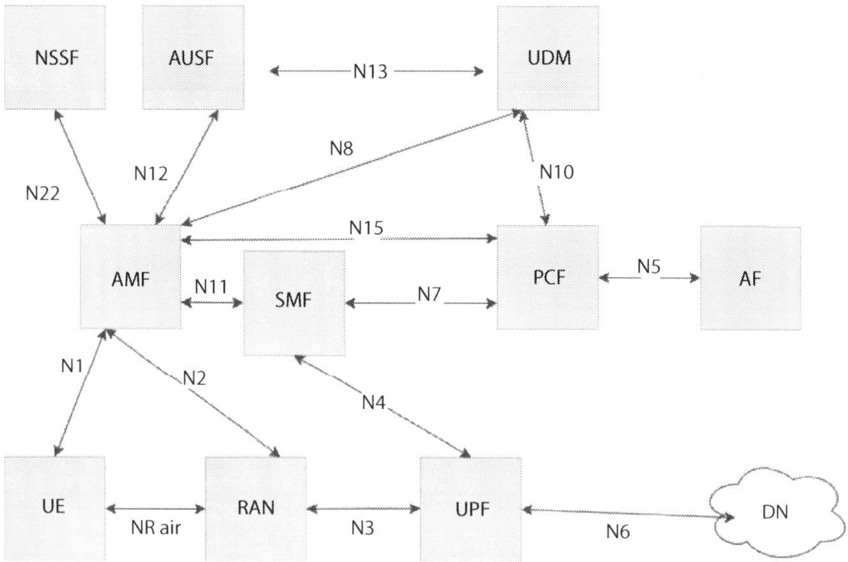

Figure 15.4 Visual representation of 5G architecture.

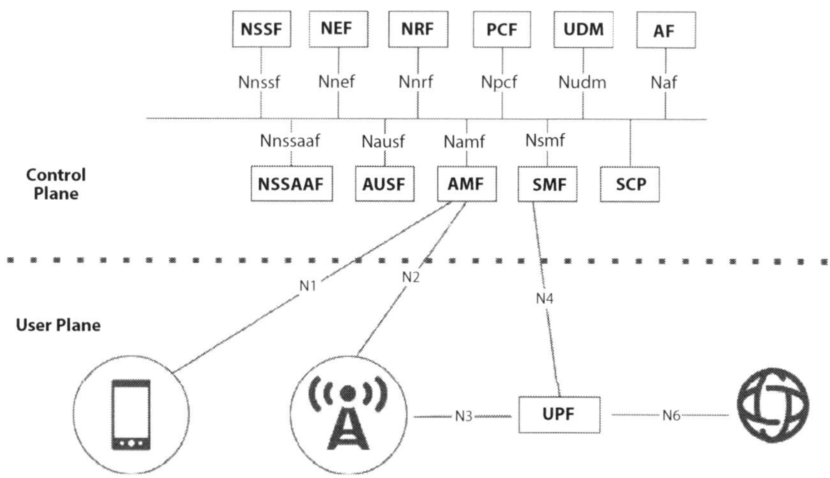

Figure 15.5 The user plane and the control plane of 5G architecture.

2. Control Plane: it is responsible for managing the signaling between the end-user device and the network. It includes several components such as the Control and User Plane Separation (CUPS), the Policy Control Function (PCF), and the Network Slice Selection Function (NSSF) [8].

- Control and User Plane Separation (CUPS): CUPS separates the user data traffic from the control signaling, enabling more flexible network design and management.
- Policy Control Function (PCF): PCF is responsible for controlling the quality of service (QoS) for different network slices and applications.
- Network Slice Selection Function (NSSF): NSSF is responsible for choosing the proper network portion for a specific application or user.

3. Management Plane: The management plane is responsible for managing and monitoring the network, ensuring efficient network performance and reliability. It includes several components such as the Network Function Virtualization (NFV), the Software-Defined Networking (SDN), and the Operation, Administration, and Maintenance (OAM) functions [7].

    - Network Function Virtualization (NFV): NFV enables the network functions to be virtualized and run on commodity hardware, improving network flexibility and scalability.
    - Software-Defined Networking (SDN): SDN enables network configuration and management through software, improving network agility and flexibility.
    - Operation, Administration, and Maintenance (OAM): OAM functions enable network operators to monitor and manage network performance and reliability.

15.2.3 5G Advantages Over Previous Generations

- Increased data transfer speeds: 5G offers data transfer rates up to 100 times more than 4G, with the capability for even faster speeds in the future. This means faster download and upload speeds for streaming, downloading large files, and browsing the internet.
- Lower latency: 5G offers way less latency than 4G, meaning the time it needs for data to be transmitted between devices is significantly reduced. This is particularly important for applications like online gaming, video conferencing,

and self-driving vehicles where even a small delay can be problematic.

- Greater capacity: 5G offers significantly greater capacity than previous generations, meaning more devices can be connected simultaneously without a drop in speed or performance. This is particularly important for the Internet of Things (IoT) and other applications that require multiple devices to be connected at once.
- Enhanced energy efficiency: 5G technology is developed to be extra energy-efficient than past gens, which is important for devices with limited battery life, such as smartphones, wearables, and IoT devices.
- Network slicing: 5G networks allow for "network slicing," which means that different parts of the network can be dedicated to precise applications or services [6]. This allows for extra efficient use of network resources and can improve work for specific applications.
- Increased security: 5G offers improved security features over previous generations, including stronger encryption, more secure authentication, and better protection against hacking and cyber-attacks.

15.3 5G for Decentralized Systems

5G technology has the potential to address some of the challenges and problems associated with decentralized systems, but it is not a silver bullet solution for all issues. The next points will address how 5G can potentially solve some of these challenges [2].

15.3.1 Scalability

Scalability is a significant challenge in decentralized systems, particularly in blockchain networks, where the number of users and transactions can rapidly increase, leading to slow transaction processing times and high transaction fees [11]. Here are some ways that 5G technology could potentially help address the scalability problem in decentralized systems:

- High-Speed Connectivity: 5G networks provide significantly quicker data transfer speeds than past generations of wireless networks. With 5G, data can be transmitted at rates of up to 20 Gbps, which is approximately a hundred times faster than 4G. This high-speed connectivity can help decentralized systems process transactions more quickly, reducing transaction times and improving overall network performance.

- Low Latency: 5G networks have less latency than previous gens of wireless networks, which means that there is lower delay between when data are sent and received. This low latency can help improve the performance of decentralized systems, particularly those that require real-time processing, such as financial transactions or gaming applications.
- Edge Computing: 5G networks enable edge computing, which deals with processing data closer to the edge of the network, rather than in a centralized data center. This can help decrease network congestion and improve the scalability of decentralized systems [11]. By processing data closer to the edge of the network, 5G can help reduce the burden on centralized nodes and enhance the overall efficiency of the network.
- Network Slicing: 5G networks support network slicing, which allows different applications and services to be isolated in dedicated network slices. This can help improve the scalability of decentralized systems by isolating different parts of the network and preventing network congestion [8]. For example, a blockchain network could be isolated in its network slice, ensuring that the network is not impacted by other applications or services.

15.3.2 Security

5G technology has the potential to enhance the security of decentralized systems in several ways. Here are some of the ways that 5G could potentially help address security concerns in decentralized systems:

- Encrypted Communication: 5G networks support end-to-end encryption, which means that data transmitted over the network are encrypted and can only be decrypted by the intended recipient [9]. This can help protect data from unauthorized access and ensure that it remains secure throughout the transmission process.

- Authentication and Identity Management: 5G networks support advanced authentication and identity management capabilities, which can help prevent unauthorized access to decentralized systems. With 5G, users can be authenticated using multiple factors, such as biometric data, smart cards, and passwords, to establish that only permitted users can get to the network.
- Enhanced Network Visibility: 5G networks provide greater visibility into network traffic, which can help identify potential security threats and enable faster responses to security incidents. This enhanced visibility can help detect and prevent attacks such as DDoS attacks and network intrusions.
- Enhanced Device Security: 5G networks provide enhanced device security features such as secure boot, device attestation, and secure firmware updates. These features can help prevent malicious actors from tampering with devices and can ensure that devices are running the latest and most secure firmware.
- Enhanced Network Resilience: 5G networks are engineered to be more resilient to network disruptions and attacks, which can help ensure the availability and security of decentralized systems. For example, 5G networks support network slicing, which can isolate different parts of the network and prevent network disruptions from spreading to other parts of the network [9].

15.3.3 Interoperability

Interoperability is a major challenge in decentralized systems, as different systems may use different protocols and standards that may not be compatible with each other. 5G technology has the potential to address interoperability challenges in decentralized systems in several ways:

- Standardization: 5G networks are built on standardized protocols and specifications, which can help ensure that different systems are able to communicate with each other seamlessly. This standardization can help ensure that decentralized systems are able to interoperate with each other, even if they use different protocols and standards.
- Service Orchestration: 5G networks support service orchestration, which allows for the management and coordination

of services across different networks and systems. This can help ensure that services in decentralized systems are able to interoperate with each other and that data can be exchanged between different systems seamlessly.

- Network Slicing: 5G networks support network slicing, which allows for the making of VN (Virtual Network) that can be customized for different applications and services. This can help ensure that different systems in decentralized systems are able to interoperate with each other by providing a common platform for communication and data exchange.
- Edge Computing: 5G networks back edge computing allows for processing and storage of data to be moved closer to the edge of the network. This can ensure that data can be processed and analyzed in real time, which is critical for many decentralized systems.

15.3.4 Internet of Things (IoT)

In a decentralized system, the Internet of Things (IoT) devices need to communicate with each other in a secure and efficient manner. 5G technology can help solve the IoT connectivity problem in a decentralized system by providing several key features:

- High Bandwidth: 5G technology provides high bandwidth, which allows for faster data transfer between IoT devices [12]. This is important for actual-time applications that require less latency and more data rates.
- Low Latency: 5G technology offers low latency, which reduces the delay between receiving and sending data. This is important for real-time applications that require immediate response times.
- Massive Machine Type Communications (mMTC): 5G technology can support many IoT devices through mMTC [12]. This enables many IoT devices to communicate with each other simultaneously.
- Ultra-Reliable Low Latency Communications (URLLC): 5G technology provides URLLC, which ensures reliable and timely delivery of critical data. This is crucial for applications like autonomous vehicles and industrial automation.

- Network Slicing: 5G technology enables network slicing, which allows for the creation of VN (Virtual Network) that can be customized for specific applications. This ensures that IoT devices receive the required network resources to perform their tasks.

15.3.5 Low Latency

It is a critical condition for many applications that operate on decentralized systems, including real-time gaming, augmented/virtual reality, and autonomous vehicles. 5G technology offers several solutions to address the low latency problem in decentralized systems.

One of the key solutions that 5G provides is edge computing, which involves processing data at the edge of the network, closer to the end-users or devices [11]. This reduces the amount of data that need to be sent back and forth between the devices and centralized servers, thus reducing latency. With 5G's high bandwidth and low latency, edge computing can be seamlessly integrated with decentralized systems to provide faster and more efficient processing.

Another solution that 5G offers is network slicing, which involves creating virtual networks within the physical network infrastructure. Each virtual network can be customized to meet the specific requirements of the application, including latency, bandwidth, and security. By creating dedicated slices for latency-sensitive applications, 5G can ensure that these applications get the necessary network resources to operate smoothly. 5G also supports advanced radio technologies, such as massive MIMO (multiple-input multiple-output) and beamforming. Massive MIMO allows for multiple antennas to be used to communicate with multiple devices simultaneously, improving the efficiency and capacity of the network. Beamforming focuses the transmission of the radio signal in a specific direction, reducing interference and improving signal strength, which can improve the reliability and latency of the network.

Finally, 5G's network architecture supports distributed processing, which allows processing to be distributed across multiple network nodes. This can reduce the amount of data that needs to be transmitted across the network and improve the overall latency of the system.

In summary, 5G technology offers several solutions to address the low latency problem in decentralized systems, including edge computing, network slicing, advanced radio technologies, and distributed processing. These solutions can provide faster and more efficient processing, making it possible to support latency-sensitive applications on decentralized systems.

15.3.6 Edge Computing

Edge computing refers to the processing and analysis of data at or near the source of data generation, which can help reduce latency and improve overall system performance. In a decentralized system, edge computing is particularly important for ensuring efficient and secure data processing, especially in scenarios where large volumes of data need to be processed in real time [11].

5G networks have the potential to enable a range of edge computing applications, such as real-time video analytics, smart traffic management, and industrial automation. Data are generated, reducing the need to transmit data over long distances to centralized data centers for processing. This not only helps reduce latency and improve system performance, but also helps reduce the load on centralized data centers, making them more scalable and cost-effective [13].

Additionally, 5G technology enables more advanced and sophisticated edge computing capabilities, such as network slicing, which allows different virtual networks to be created within a single physical network, each with unique characteristics and performance requirements. This can help ensure that critical applications and services receive the necessary network resources and priority to operate effectively, while also allowing for more efficient use of network resources overall.

15.4 Use Cases and Applications

5G technology is still relatively new, and its adoption in decentralized systems is still in its early stages. However, there are already several use cases where 5G is being used to enhance the capabilities of decentralized systems, and there are many potential future applications as well.

15.4.1 Current Use Case

- Decentralized applications (dApps): Some decentralized applications, such as those based on blockchain technology, are starting to use 5G networks to improve their speed and scalability. With 5G, dApps can handle more transactions per second, reducing the time and cost of verifying transactions [5]. For instance, the 5G-enabled blockchain-based platform called "Streamr" aims to provide decentralized real-time data delivery and monetization. The platform leverages 5G networks to improve the efficiency of data processing, increase security, and provide reliable and efficient data monetization.

- Edge computing: 5G networks are being used to support edge computing, which allows for processing and storage of data to be moved closer to the edge of the network [12]. This can be particularly useful for decentralized systems that require actual-time data processing, such as those used in smart cities or industrial IoT applications. For example, a 5G-enabled edge computing platform called "Mobile Edge Computing (MEC)" can provide local data processing and storage resources to support real-time and low-latency applications like video analytics, traffic management, and predictive maintenance.
- Internet of Things (IoT): 5G networks are being used to support IoT devices, which can be integrated into decentralized systems to provide real-time data and control capabilities [12]. This is particularly useful for applications that require monitoring and control of distributed systems, such as in agriculture or environmental monitoring. For example, a 5G-enabled IoT platform can provide reliable and efficient connectivity to enable real-time monitoring of the environment, smart irrigation, and precision agriculture.
- Decentralized Cloud Computing: 5G networks can also be used to support decentralized cloud computing platforms like "Fog Computing" [13]. Fog computing provides a distributed computing infrastructure that enables data processing at the edge of the network to reduce the latency and bandwidth requirements of cloud computing. With 5G, Fog computing can provide a more reliable and efficient computing infrastructure that can support decentralized applications like blockchain, IoT, and machine learning.

15.4.2 Future Use Case

- Decentralized Finance (DeFi): Decentralized finance is a speedily growing field that uses blockchain technology to provide financial services without intermediaries. With 5G, DeFi applications can become even more accessible and user-friendly, allowing for faster and cheaper transactions [5]. 5G can also improve the scalability of DeFi applications by providing high-bandwidth and low-latency connections that can support the processing of many transactions.

- Smart Cities: 5G technology can be used to create smart cities that leverage the power of IoT devices and decentralized systems to enhance the quality of life for citizens. For example, smart traffic management systems can be implemented using 5G-enabled edge computing, which can process real-time data from sensors and cameras to optimize traffic flow [14]. 5G can also enable smart energy systems that can balance energy supply and demand in real time, reducing waste and costs.
- Healthcare: Decentralized systems can be an important role in the healthcare department by improving patient data management and medical research. With 5G, healthcare providers can use decentralized systems to securely store and share patient data, lowering the risk of data breaches and improving the accuracy of diagnoses. Decentralized systems can also be used to improve medical research by facilitating data sharing between researchers and medical institutions.
- Gaming: The gaming industry is expected to benefit greatly from 5G technology, as it can provide high-speed, low-latency connections that can support immersive gaming experiences. Decentralized gaming platforms can use 5G to provide real-time multiplayer gaming, enabling players to connect from anywhere in the world.
- Supply Chain Management: Decentralized systems can be used to improve supply chain management by providing actual-time visibility of goods and reducing the risk of fraud. With 5G, supply chain managers can use decentralized systems to track the movement of goods in real time, enabling them to optimize logistics and reduce waste.

15.5 Challenges and Opportunities

The implementation of 5G technology in decentralized systems is expected to bring many benefits, such as higher data rates, lower latency, and increased capacity. However, there are also several challenges that need to be addressed to ensure the successful implementation of 5G in decentralized systems. Some of these challenges are as follows:

- Infrastructure: The deployment of 5G infrastructure requires many small cells, base stations, and other components to achieve full coverage, which can be costly and challenging, especially in rural or remote areas. In addition, upgrading existing infrastructure to support 5G can be time-consuming and expensive [15].
- Security: The security of decentralized systems is critical, and 5G technology presents new security risks. For instance, as 5G networks are more distributed, they may be more open to attacks. Hence, it is vital to have powerful security measures to make sure the protection of data and transactions on decentralized networks.
- Interoperability: Decentralized systems and 5G networks may use different protocols and standards, making interoperability a significant challenge. Developers need to ensure that their systems are compatible with 5G networks, and that data can be exchanged seamlessly.
- Cost: Implementing 5G in decentralized systems can be expensive. For example, organizations need to contribute to new hardware and software to support 5G, and upgrading existing infrastructure can be costly.
- Regulatory challenges: Implementing 5G in decentralized systems may also face regulatory challenges, such as spectrum allocation and licensing, which can impact the deployment of 5G networks.
- Skill and knowledge gap: 5G technology is relatively new, and the skill and knowledge gap in the market could pose a significant challenge for the implementation of 5G in decentralized systems [15]. Organizations need to train their staff to ensure that they have the required skills and knowledge to operate and manage 5G infrastructure effectively.

As it is a new technology, there is a lot of potential to it in the future as it is promising. Here are few of the opportunities for implementing 5G in decentralized systems and networks:

- Increased Network Capacity: With 5G, decentralized systems can handle many devices and data traffic, providing more bandwidth and faster data transfer rates [9]. This increased network capacity can support new applications and services and enable more productive use of network resources.

- Improved Efficiency: 5G's network slicing feature allows network resources to be partitioned and specialized for specific use cases, increasing network efficiency, and reducing latency. This feature also enables decentralized systems to prioritize certain types of traffic, such as critical data, over non-critical traffic.
- Enhanced Security: 5G offers improved security features, such as enhanced authentication and encryption, which can assist in protecting decentralized systems from cyber-attacks and unauthorized access [12]. The use of blockchain technology in conjunction with 5G can also provide additional security measures.
- Innovative Applications: The high-speed and low-latency capabilities of 5G can help the development of innovative applications and services for decentralized systems, such as autonomous vehicles, smart homes, and telemedicine.
- Integration with Emerging Technologies: 5G can be integrated with new technologies such as artificial intelligence (AI), machine learning (ML), and the Internet of Things (IoT), to establish more intelligent and efficient decentralized systems.

Overall, the implementation of 5G technology in decentralized systems provides opportunities for improved network capacity, efficiency, security, innovative applications, and integration with emerging technologies. However, there are also challenges associated with this implementation, such as the lavish cost of infrastructure upgrades and ensuring compatibility with existing systems.

15.6 Conclusion

The advent of 5G technology promises to solve these challenges by enabling faster and more efficient speed for data transmission, improving connectivity and reliability, and enhancing security encryption and privacy threats. Chapter will provide an overview of 5G technology in decentralized systems, covering topics such as - overview of decentralized systems, 5G technology, 5G for decentralized systems, use cases and applications. Study done addresses potential challenges and opportunities that arises with interoperability of the technologies. Further work also includes security and privacy considerations that arise, and regulatory issues related to government. Chapter provides a comprehensive overview of how 5G technology can be embedded with decentralized blockchain platform, recent developments in diversified applications of blockchain in 5G is surveyed in a wide range of popular use-case domains, The main findings derived from include possible research challenges with open issues. Lastly, we complete this survey by shedding new light on future directions of research on this newly emerging area.

References

1. X. Z. Y. F. J. L. Z. Zhao, "5G Technology and Its Application in Decentralized Systems: A Review," *IEEE Access*, pp. 74525-74538, 2020.
2. E. Gelenbe, "Blockchain and 5G: The Impact on Business," *IEEE Technol. Soc. Mag.*, pp. 70-75, 2018.
3. O. S. a. B. D. Larry Peterson, *Private 5G: A Systems Approach*, Systems Approach LLC, 2022.
4. I. MORRIS, "FRONTHAUL/C-RAN," 3 8 2018. [Online]. Available: https://www.lightreading.com/mobile/fronthaul-c-ran/vran-tech-hits-resistance-at-sk-telecom/d/d-id/741230.
5. K. S. C. L. P. P. R. ,. T. S. A. Chaer, "Blockchain for 5G: Opportunities and Challenges," *IEEE Globecom Workshops*, pp. 1-6, 2019.
6. VIAVI , "What is 5G Architecture?," 2022. [Online]. Available: https://www.viavisolutions.com/en-us/what-5g-architecture.
7. 5g americas, "White Paper Archive," 2022. [Online]. Available: https://www.5gamericas.org/white-paper-archive/.
8. K. Mori, "Autonomous decentralized systems: Concept, data field architecture and future trends," *Proceedings ISAD 93: International Symposium on Autonomous Decentralized Systems*, pp. 28-34, 1993.
9. J. E. Allchin, An Architecture for Reliable Decentralized Systems, Atlanta, Georgia: Georgia Institute of Technology, 1983.
10. 5G Public Private Partnership (5G PPP), "5G-Architecture WG – Open Call for White Paper Inputs," 2016. [Online]. Available: https://5g-ppp.eu/.
11. G. W. T. a. R. B. Zhou, "Deep reinforcement learning-based methods for resource scheduling in cloud computing: A review and future directions," *arXiv preprint*, 2021.
12. D. H. S. Z. N. K. H. W. a. K. R. C. Jinglin Zou, "Integrated Blockchain and Cloud Computing Systems: A Systematic Survey, Solutions, and Challenges.," *ACM Comput. Surv.*, p. 36, 2021.

13. E. U. A. M. A.-M. a. A. M. K. Ogbodo, "A Survey on 5G and LPWAN-IoT for improved smart cities and remote area applications: From the aspect of architecture and security," *Sensors*, 2022.
14. P. Y. N. V. S. &. G. P. William, "Blockchain Technology for Data Privacy using Contract Mechanism for 5G Networks," in *2022 3rd International Conference on Intelligent Engineering and Management*, 2022.
15. A. A. A. C. S. K. A. &. G. M. Filali, "Multi-access edge computing: A survey.," *IEEE Access*, pp. 197017-197046, 2020.
16. F. A. A. M. A. A. A. M. &. A. H. Khan, "Blockchain technology, improvement suggestions, security challenges on smart grid and its application in healthcare for sustainable development.," in *Sustain. Cities Soc.*, 2020.
17. S. J. H. P. a. H. C. Rathore, "Deep learning and blockchain-empowered security framework for intelligent 5G-enabled IoT," in *IEEE Access 9*, 2021.
18. Ferrag, Mohamed Amine, and Lei Shu. "The performance evaluation of blockchain-based security and privacy systems for the Internet of Things: A tutorial." *IEEE Internet Things J.*, 8.24 (2021): 17236-17260.
19. Leng, Jiewu, *et al.* "Blockchain security: A survey of techniques and research directions", *IEEE Trans. Serv. Comput.*, 15.4 (2020): 2490-2510.

16

Artificial Intelligence in a Distributed System of the Future

Kadambri Agarwal[1*], Ojasw Khare[1], Aastha Sharma[1], Ayushi Prakash[2] and Abhishek Kumar Shukla[1]

[1]Department of CSE, ABES Engineering College, Ghaziabad, India
[2]Department of CSE, AKGEC, Ghaziabad, India

Abstract

The integration of artificial intelligence (AI) approaches with distributed systems has emerged as a game-changing paradigm, transforming how we build, administer, and optimize large-scale networks. This chapter emphasizes the significant influence of AI on distributed systems, highlighting how it improves productivity and scalability. Due to their capacity to manage massive volumes of data and provide services to a variety of users, distributed systems, such as cloud computing platforms, Internet of Things (IoT) networks, and edge computing infrastructures, have taken the place of central processing units in modern computing. However, the difficulties in managing distributed systems have created substantial problems that call for clever solutions. By offering automation, self-optimization, and adaptive decision-making capabilities, AI plays a crucial role in tackling these issues. Distributed systems may analyze large datasets in real time using machine learning algorithms, resulting in intelligent resource allocation, load balancing, and fault tolerance decisions. Additionally, security breaches and system failures are detected and mitigated with the use of AI-powered anomaly detection systems, assuring the ongoing availability and dependability of services.

Keywords: IoT, autonomous, distributed system, AI, reinforcement learning, OpenAI, chatbots

**Corresponding author*: kadambriagarwal@gmail.com

Sandhya Avasthi, Suman Lata Tripathi, Namrata Dhanda and Satya Bhushan Verma (eds.) Decentralized Systems and Distributed Computing, (317–336) © 2024 Scrivener Publishing LLC

16.1 Introduction

Artificial intelligence (AI) has been one of the prominent topics in recent times. It can promote the fastest and the most intelligent way to solve problems and behaves like human intelligence. It can create an atmosphere that can deal with all sorts of data that can lead to encouraging the critical thinking skills of inventors to make discoveries and meet project deadlines. AI is trained to fetch a large amount of data and tries to recognize the patterns, judgment, and the previous propagation dataset, and make decisions like humans do. It is one of the greatest discoveries by which it encourages the world to be creative and innovative in all areas. Whether it would be in the atmospheric, environment, or industrial sectors, AI leads in all areas to enhance the capability of computers that require human intelligence. For artificial intelligence (AI), the next gradation is to use it as a distributed system [1, 2]. In AI (artificial intelligence), a distributed system is mainly used as a collection of phenomenon processes by which it can be physically separated but in a central way connected to share the common goals and the same messages to the phenomenon processes occurring or to be executed at a running time. Distributed systems in AI can be one of the hottest topics in the world to execute the process more smoothly and in the fastest way to achieve goals. This can share the same messages and the same type of aim to achieve a small running time in local learning and intelligence. It can work as an automated machine that performs all the tasks by distributing them individually with a shared goal and the same centralized computer sets. There are various branches of AI:

1. Machine learning, a branch of artificial intelligence, is the process of teaching computers to learn from data without being explicitly programmed. Machine learning algorithms are created to find patterns in data and generate predictions. Supervised learning, unsupervised learning, and reinforcement learning are the three categories of machine learning. In supervised learning, the computer is trained using a dataset that has been labeled and contains the desired output. Unsupervised learning involves training the computer on an unlabeled dataset and letting it search for patterns on its own.
2. NLP is an area of artificial intelligence that focuses on teaching computers to comprehend and interpret human language. Applications like chatbots, virtual assistants, and

machine translation all make use of NLP. Sentiment analysis, named entity identification, and part-of-speech tagging are just a few of the techniques used in NLP. While named entity recognition recognizes entities like persons, locations, and organizations, sentiment analysis determines the tone of a text. The function of each word in a sentence is identified via part-of-speech tagging.

3. Robotics is a branch of artificial intelligence that focuses on creating robots that can work independently or under human supervision. Numerous industries, including manufacturing, healthcare, and space exploration, use robotics. Robotics uses a variety of methods, including control systems, motion planning, and computer vision. Robots can observe their surroundings thanks to computer vision, and motion planning chooses the most efficient route for them to take. Control systems are used to manage a robot's behavior and make sure it completes tasks safely.
4. Expert systems are computer programs that simulate a human expert's decision-making processes in a given sector. Applications like financial planning and medical diagnosis use expert systems. Expert systems can offer suggestions or solutions to a problem by using a knowledge base and a set of rules. Expert systems are made to be transparent, allowing users to know how a suggestion or solution was arrived at.

16.2 AI in the Industrial Sector

For over three decades, the realm of artificial intelligence has been a subject of intensive research and innovation. Within the industrial sector, significant technological strides have been achieved during this time, marked by robust hardware and software developments, amplified computing power, and enhanced data transfer capabilities [3]. The integration of artificial intelligence has unlocked a realm of unprecedented possibilities, especially in the production of intricate and personalized items in small batch runs, fostering flexibility and efficiency like never before. The transformative potential of AI in the industrial sector goes beyond mere advancements; it is revolutionizing the entire landscape. By enabling businesses to streamline operations, boost productivity, and curtail costs, AI is ushering in a new era of efficiency and optimization. This has been substantiated by the remarkable surge in investments in generative AI firms, soaring to a staggering $2.6 billion through

110 acquisitions in 2022, according to Insights research. As AI continues to play a momentous role in the industrial domain, it is reshaping various facets of operations and production, ushering in a future where innovation and automation seamlessly converge to propel industries toward unprecedented heights. Predictive Maintenance leverages AI algorithms to analyze sensor data and other sources, enabling the anticipation of machinery maintenance needs, reducing downtime, and enhancing overall efficiency. Quality Control is bolstered by AI, which monitors product quality in real time, promptly identifying defects and implementing adjustments to elevate production quality. In Supply Chain Management, AI optimizes the entire supply chain by analyzing data from suppliers, logistics providers, and customers, unearthing opportunities for cost savings and heightened efficiency. Production Planning is streamlined with the aid of AI, which optimizes scheduling, inventory management, and resource allocation. Finally, in Energy Management [4, 5], AI optimizes energy usage in factories and industrial settings, leading to reduced energy costs and enhanced sustainability. Embracing AI in the industrial realm propels innovation and augments productivity, revolutionizing the future of the sector (Figure 16.1).

16.2.1 Applications

Figure 16.1 shows application of AI in different area of manufacturing.

16.2.2 AI in the United States

In the realm of AI in the United States, two notable advancements stand out: ChatGPT and Google's Bard. ChatGPT, a sophisticated language

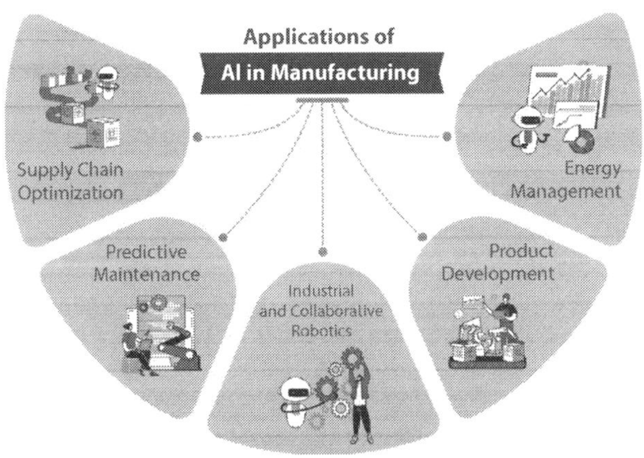

Figure 16.1 AI applications in manufacturing (source: www.compunneldigital.com).

model developed by OpenAI, operates through chat-based interactions and is continuously evolving on a 2021 database, showcasing an impressive display of AI capabilities. Utilizing the GPT (Generative Pre-trained Transformer) architecture, ChatGPT is designed to generate human-like responses to queries in natural language. Its extensive training on a vast corpus of text data, comprising books, papers, and websites, empowers it with a comprehensive understanding of linguistic nuances and patterns. Renowned for its versatility, ChatGPT can provide answers to a wide range of questions, making it a favored choice in chatbots and virtual assistants for its ability to produce responses closely resembling human speech. On the other hand, Google's Bard, similar to ChatGPT and Bing Chat, is a conversational and generative AI chatbot that showcases its prowess in various domains. Bard offers assistance in creative work, clarifying complex concepts, and gathering knowledge. It handles intricate inquiries, including personalized recipe suggestions based on available ingredients, which is traditional. One of Bard's significant strengths lies in its ability to identify and examine connections between seemingly unrelated data pieces, allowing it to uncover correlations even when there is no apparent direct link between specific genes and pharmacological targets.

16.2.3 AI in India

AI in India has witnessed notable developments, with the introduction of Lexi, an Indian version of ChatGPT by the financial technology company Velocity. This innovative model offers users a personalized and user-friendly chat experience, integrated into Velocity Insights, the company's existing analytics solution. Moreover, Lexi leverages the WhatsApp platform to deliver daily business reports to e-commerce brands, enhancing their operational efficiency and decision-making processes. In a significant stride towards the advancement of AI in India, Finance Minister Nirmala Sitharaman, during the Union Budget 2023, announced the establishment of three centers of excellence for AI in prestigious educational institutions. This marks a crucial step towards realizing the nation's vision of harnessing the potential of AI in various domains. Chandrasekhar, in his speech at the event, emphasized the transformative impact of the India AI ecosystem on future tech segments, alongside finance and other sectors. He highlighted the remarkable adoption rates of India's fintech ecosystem, reaching an impressive 87% compared to the global average of 67%, with the Unified Payments Interface (UPI) playing a pivotal role in this transformation. The confluence of Lexi's advanced capabilities and the government's commitment to nurturing AI expertise through centers of excellence sets the stage for a promising AI landscape in India, poised to drive innovation and propel the nation toward a technology-driven future.

16.3 AI in the Economic Sector

In the economic sector, artificial intelligence (AI) is wielding a profound impact, transforming organizational processes, decision-making, and customer engagement. This transformative technology is redefining business practices, and here are some illustrative examples of AI applications in the corporate world. Process automation takes a significant leap with AI, as it can handle monotonous tasks such as data entry and analysis, freeing up human resources to focus on strategic responsibilities, leading to reduced costs and increased productivity. AI-powered chatbots are revolutionizing customer service, offering personalized advice [6] and support around the clock. This not only enhances customer satisfaction but also alleviates the burden on human customer support representatives.

Marketing and advertising strategies benefit from AI's data analysis capabilities, enabling the development of tailored marketing campaigns based on customer trends and preferences. This optimized approach often leads to increased sales and more effective marketing initiatives. Supply chain management is experiencing an AI-driven optimization process by analyzing inventory, production, and shipping data. Machine learning algorithms identify inefficiencies and bottlenecks, making valuable recommendations for improvements. AI's application in financial analysis has become indispensable for predicting trends and guiding investment decisions. By analyzing financial data, including stock prices and economic indicators, machine learning algorithms uncover patterns and connections that might elude human analysts. Product development receives a boost with AI, as it leverages consumer feedback, market trends, and competitor behavior to streamline and expedite the creation of new products and services. Machine learning algorithms predict customer demand and identify opportunities for innovation. Business intelligence [7] takes a significant leap forward with AI, as it analyzes diverse data sources, producing insights and recommendations. Machine learning algorithms uncover patterns, connections, and anomalies that human analysts might miss, empowering businesses with deeper insights. Autonomous AI is driving the development of autonomous cars, revolutionizing logistics, and transportation by reducing costs and enhancing efficiency and safety. This groundbreaking advancement paves the way for a future where autonomous vehicles operate without human assistance. In the economic landscape, AI is proving to be a game-changer, reshaping the way businesses operate and positioning them to thrive in an increasingly technology-driven world.

16.3.1 Applications

JPMorgan Chase analyzes credit data and determines credit risk using AI algorithms. Additionally, they employ AI to spot financial transaction fraud. Ford is utilizing AI to create self-driving vehicles, which have the potential to completely transform the transportation sector. Based on a customer's browsing and purchase history, Amazon employs AI algorithms to tailor product recommendations for them. They also employ AI to forecast product demand and improve supply chain operations. In India, the National Payment Corporation of India (NPCI) introduced Pi, an AI-based Chatbot, to provide real-time information on services like Fastag, AePs, and RuPay. By leveraging technology to transform real samples into digital images, Pi can automatically review visual data and offer insightful information, enhancing user experiences and streamlining services.

Furthermore, Niti Ayog established the Atal Mission Innovation in 2016 to support business owners in embracing AI technologies, fostering a climate of innovation and growth. Amidst the challenges posed by the pandemic, EdTech companies have played a crucial role in the education

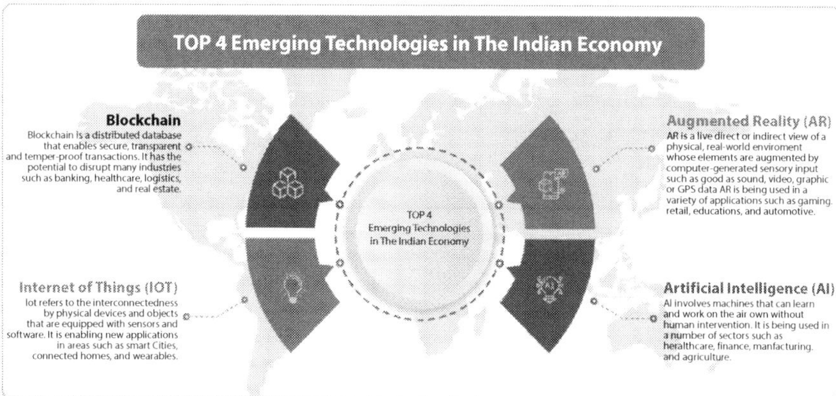

Figure 16.2 Use of technology in India (source: www.aeologic.com).

sector by offering the latest AI-driven technology to students. This has encouraged students to delve into AI technologies, explore creative projects, and utilize flexible design layouts, promoting a future-ready generation adept at harnessing the power of AI for transformative advancements in the nation's progress (Figure 16.2).

16.4 Environment Sector

Artificial intelligence (AI) offers a diverse array of applications to address environmental challenges, ranging from designing energy-efficient buildings to monitoring deforestation and optimizing the use of renewable energy sources. At a global scale, AI can be employed for satellite monitoring of worldwide emissions, while at a local level, it can enable smart homes to efficiently manage heating and lighting by automatically turning them off after a predetermined time. One significant area where AI proves invaluable is climate modeling. By analyzing various factors such as greenhouse gas emissions, changes in land use, and natural disasters, AI helps forecast the future impact on the climate. This aids decision-makers in their endeavors to mitigate and adapt to climate change effectively.

Furthermore, AI plays a crucial role in energy management. Utilizing data from sensors, weather predictions, and energy networks, AI optimizes energy consumption in buildings and cities. This not only leads to cost savings but also reduces carbon emissions and minimizes energy waste. AI also has significant contributions to waste management. Through the use of sensors, drones, and computer vision, AI can accurately determine the type, quantity, and location of waste. This optimization enhances waste management practices, leading to decreased trash, increased recycling rates, and reduced environmental impact from landfills. In the context of conservation, AI is instrumental in monitoring forests. With the aid of drones, satellite photography, and machine learning techniques, AI enhances conservation efforts by facilitating the detection of deforestation, illegal logging, and forest fires. Similarly, wildlife conservation benefits from AI-driven solutions [8, 9]. Video traps, acoustic sensors, and machine learning algorithms are employed to track and safeguard wildlife. This advanced technology assists in combating poaching, monitoring animal populations, and reducing human–wildlife conflicts. Specific applications of AI in the environmental sector include real-time analysis guidance, monitoring methane emissions, assessing air quality, measuring environmental footprints, and reducing ICT emissions. In conclusion, AI's potential in the environmental sector is vast and promising. Its ability to

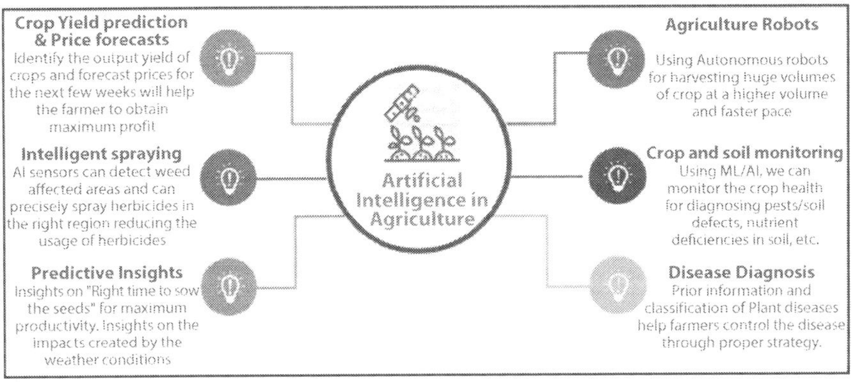

Figure 16.3 Applications of AI in agriculture (image source: www.wipro.com).

analyze vast amounts of data and provide actionable insights allows for more effective and sustainable approaches to address environmental challenges, thus fostering a greener and more sustainable future (Figure 16.3).

16.5 Robotics Skill

A robot is a machine capable of sensing and interacting with its environment. AI gives robots a computer vision to navigate, sense, and calculate their reaction accordingly. Robots learn to perform their tasks from humans through machine learning, which again is a part of computer programming and AI.

In the realm of AI-driven robotics, three main types of artificial intelligence come into play:

- Weak Artificial Intelligence: This form of AI is designed to simulate human thought and behavior but with a limited scope. Robots equipped with weak AI follow pre-programmed responses and commands to complete tasks. While they may appear to understand and respond to user input, their comprehension is superficial. Examples of weak AI in action include popular virtual assistants like Alexa and Siri. These devices rely on pre-defined algorithms to provide responses and execute tasks requested by users.
- Strong Artificial Intelligence: In contrast to weak AI, strong AI enables robots to operate with a higher level of autonomy and intelligence. Once trained for specific tasks, robots with

strong AI can function without constant human monitoring. Prominent examples include self-driving vehicles, which utilize advanced AI algorithms to navigate roads and make real-time decisions based on the environment. Additionally, humanoid robots equipped with strong AI can perceive their surroundings, interact with objects, and carry out tasks with impressive adaptability and autonomy.

- Specialized Artificial Intelligence: This type of AI is employed when robots are designed to perform specific, specialized tasks. Specialized AI [10] is tailor-made for particular functions, limiting the robot's capabilities to a narrow range of activities. Industrial robots often utilize specialized AI to execute repetitive and well-defined tasks with precision and efficiency. Examples include robots used in manufacturing processes for tasks like painting, welding, and tightening. The application of AI in robotics offers a multitude of benefits, revolutionizing the way robots operate and interact with their surroundings. Here are some examples of AI-driven enhancements in robotic capabilities:
- Computer Vision: AI-powered computer vision algorithms enable robots to perceive and understand their environment through visual data. This empowers them to recognize objects, identify patterns, and navigate obstacles, making them more versatile and adaptive in complex environments.
- Object Recognition: With the help of AI, robots can accurately identify and analyze objects based on their features, such as shape, color, texture, and size. This capability is particularly valuable in tasks like sorting items, assembling components, and quality control in manufacturing processes.
- Motion Planning: AI-driven motion planning algorithms enable robots to determine the most efficient and safe path to perform tasks. By analyzing the environment and potential obstacles, robots can navigate smoothly and avoid collisions, ensuring a higher level of safety and efficiency in their operations.
- Machine Learning: AI-powered machine learning algorithms allow robots to learn and improve from their experiences. Through continuous data analysis and pattern recognition, robots can adapt to changing conditions, optimize their performance, and even handle unforeseen situations more effectively. Robots equipped with reinforcement

learning algorithms can learn and improve through trial and error. By receiving rewards for correct actions and penalties for mistakes, they can fine-tune their behavior and decision-making, making them more versatile and adaptable to new situations.

- Natural Language Processing (NLP): AI-driven NLP empowers robots to comprehend and respond to human commands and inquiries. This capability is instrumental in developing interactive and user-friendly interfaces, allowing robots to interact seamlessly with humans and cater to their needs in various applications, such as home automation and customer service.
- The integration of AI into robotics has revolutionized the capabilities of robots, making them more efficient, autonomous, and versatile across various industries. From manufacturing and logistics to healthcare and education, AI-driven robotics is reshaping how robots interact with the world, paving the way for a future where intelligent machines play an increasingly vital role in our daily lives.

16.5.1 Applications

Boston Dynamics is at the forefront of developing dynamic, intelligent, and adaptable robots by employing sensor-based controls, enabling these robots to navigate various situations and terrains with ease. They have made significant strides in creating advanced robotic systems capable of impressive mobility and agility. UI Path, a leading player in robotic process automation, offers an enterprise software platform equipped with robot licensing, provisioning, scheduling, monitoring, and alerting capabilities. Their robots take on time-consuming tasks, streamlining communication across legacy applications, thereby liberating developers to focus on creating new AI-oriented applications. With its cutting-edge solutions, UI Path has significantly contributed to the automation revolution, facilitating increased productivity and efficiency across industries.

Qualcomm, renowned for its powerful smartphone chips, plays a pivotal role in driving the AI revolution. Their signal processors enable devices to handle complex tasks related to image and sound capabilities, ushering in a new era of AI-powered functionalities in smartphones and other devices. Qualcomm's contributions have accelerated the integration of AI into everyday devices, enhancing user experiences and opening up possibilities for new applications and services.

16.6 Interaction Between the Human and the Machine

In AI (artificial intelligence), the interaction between the human and the machine is a phenomenon process by which a lot of information is exchanged level by level (conscious state). A number of patterns can be read by the computer with the help of the AI-updated computer system; reading the neural subconscious to conscious state entails the interaction of humans to the machine-level language translator to read and recognize the patterns and make the decision based on the transmitted signal via AI (artificial intelligence). Interaction with humans means the computer system is designed and maintained, trained, and stored in such a way to behave like a human. Similarly, they show the same gestures and critical thinking like humans do. Humans also show acceptance regarding the numerous types of discovery related to interaction with the machine. Interaction with the machine also shows that the machine interacts frequently and it generally promoted the introvert generation and builds a great cycle to establish creation and discovery. Some of the features during interaction can build deep learning to enhance the capability, and more operations are performed using the trained dataset to make the best decision as per the previous interaction to the computer [11]. It can translate a large amount of training data into the trained dataset and make future predictions on thinking skill to behave like human intelligence. The aim is to establish an AI sequence to build a more interactive skilled computer system to give the best output at a running time of execution, and it takes a small amount of time to complete all the processes in the real-time execution. Interacting with the computer in AI form shows the social life of the machine, to give the correct gestures similar to humans. It is a great achievement to work on this and promote great discoveries to create the best platform for the machine and humans to work on continuously. Interactive AI applications involve direct communication between people and AI-powered machines. To provide consumers with engaging and dynamic experiences, these applications combine a variety of AI techniques, including machine learning, computer vision, and natural language processing.

16.6.1 Applications

Interactive AI applications encompass various instances that demonstrate the versatility and utility of artificial intelligence. Chatbots, for example, are AI-powered conversational agents capable of communicating through text or speech, finding applications in personal assistants, customer service, and other domains requiring a conversational interface. Virtual assistants, like Siri, Alexa, and Google Assistant, leverage AI to perform tasks such as event reminders, reservations, and suggestions. Moreover, AI drives personalized user experiences in e-commerce, social networking, and online advertising through recommendation engines that offer tailored goods, content, and services based on user preferences and past actions. Additionally, AI-powered augmented reality applications enable real-world object recognition and provide detailed information and interactions through smartphone apps. In the realm of education, AI facilitates individualized learning experiences with adaptive learning systems that adjust according to students' pace and preferences, while AI-driven teaching programs offer valuable guidance and feedback to learners. The breadth of interactive AI applications continues to expand, revolutionizing various industries and domains with enhanced efficiency and user-centric experiences. Examples include the Google Assistant feature on mobile phones, Alexa in Amazon tools, Light Switch (prominent discovery to show the interaction between humans and the machine), automated car driving, and Chat gpt Business contract. A few recent projects are cycle and loops of data links, human and the digital life, and chatbot of AI lovers.

16.7 Recent Projects in AI

AI (artificial intelligence) in the next generation marks an ongoing large discovery in the real world, which promotes creativity, knowledge, and critical thinking of the knowledge database. The next generation of AI in a distributed system can share the same goal and the same type of pattern that promotes the distributed system analysis of the informative database with the centralized computer system. It can share the same memory to fetch the data and distribute it individually to perform each task to promote the nature of AI (artificial intelligence) in distributed preformatted centralized tasks to build and create the patterns of the fetched dataset. Several notable recent AI projects have emerged as groundbreaking advancements in the field:

- ***AlphaFold:*** Revealed by DeepMind, a Google-owned research firm, in 2020, AlphaFold represents an AI system of remarkable accuracy in predicting the three-dimensional (3D) structure of proteins. This revolutionary technology holds great promise in accelerating drug development and transforming pharmaceutical research.

- ***GPT-3:*** Unveiled by OpenAI in 2020, GPT-3 is a versatile language model capable of generating text closely resembling human language. Its widespread adoption in applications like chatbots, content generation, and language translation has redefined interactions between humans and AI.
- ***Automated Machine Learning (AutoML):*** Developed by Google in 2018, AutoML streamlines the process of creating machine learning models. This powerful framework automatically generates models that match or even surpass those developed by human experts, democratizing AI application development.
- ***Deepfake Detection:*** The proliferation of deepfakes, manipulated videos or images depicting fictitious events, has posed significant challenges in combatting misinformation. Nonetheless, recent advancements in AI algorithms have led to highly accurate deep fake detection, offering potential solutions to curbing the dissemination of false information.
- ***Reinforcement Learning for Gaming:*** OpenAI's 2019 publication introduced OpenAI Five, an AI system achieving superhuman performance across various games, including Dota 2. Leveraging reinforcement learning, OpenAI Five agents collaborate seamlessly as a team, demonstrating outstanding strategic prowess, such as in agriculture cropping patterns, weather forecasting, geographical patterns, patterns of walls, social life prediction, neural patterns, financial sector, AI cryptocurrencies, military pipeline products, fixed platform economy security, data integration firewall, detection of fraud users, human interaction, human emotions, AI skill computing patterns, big data analysis protection, AI paradigm, satellite projection robot, and robotic skill share. This process of evolution and creation of discovery related to AI and deep learning patterns occurs all around the world and shows the creativity and the generalized patterns of the analysis of data.

16.8 Terminologies Regarding AI for the Next Generation

An AI pattern can trigger the creation of discoveries and shows some of the prominent analyses to some of the issues, and the terminologies can promote AI in the next generation of pattern analysis creation. There are areas by which it can promote and also translate a large number of patterns as well as analysis regarding future prediction [11] for an AI (artificial intelligence).

1. An approach to AI called explainable AI (XAI) emphasizes the necessity for AI systems to give justifications for their choices and actions. This is crucial in fields like healthcare and finance where accountability and transparency are essential.
2. A method of AI called "human-in-the-loop" (HITL) entails human supervision and input into the decision-making process of the AI. In addition to helping the AI system perform better over time, this can ensure that it makes morally sound and accurate decisions.
3. Federated Learning: In federated learning, models are trained using data that are dispersed over several different devices or systems without the data being moved to a centralized location. Regarding security and privacy, this can be helpful.
4. Edge AI: Edge AI describes AI programs that are installed locally, as opposed to in the cloud, or on IoT devices, smartphones, and tablets. This can help with performance enhancement, latency reduction, and privacy problems.
5. Synthetic Data: Artificially produced data are referred to as synthetic data and are used to train AI algorithms. This can be especially helpful in applications where it is difficult or expensive to gather real-world data.
6. Transfer Learning: A machine learning method that involves transferring knowledge acquired from one task to another is called transfer learning. This can increase the precision and effectiveness of the model.
7. Generative AI: This term describes AI programs that can create new and unique content, such as writing, photos, music, and videos. Applications like content development, advertising, and entertainment can all benefit from this.
8. Artificial intelligence (AI) applications that can function autonomously and without human supervision are referred to as autonomous AI. Applications for this include robots, drones, and self-driving cars.

9. Droids Agent: They are the type of continuous process by which it can perform the creation of the analysis of data followed by the aim and the reasoning skills to perform all the tasks related to the patterns recognized in a distributed execution. It can also be directed by two or more terminology goals or the assistant.
10. Robotic nature skill: It is a process by which it can promote the null touch operation performed by the machines. Again, it shows the interaction between the human and the machine to perform the task within a limited period. It is free from external control and controls itself independently and executes all the tasks.
11. Backpropagation learning: It is based on a method of creating an estimation of the result if it is not shown as expected. Then, all the patterns are detracted; in other words, it backpropagates all the methods to learn all the patterns linking and again predict the decision as per the expected result.
12. Memory boosting: It is the creation of the periodic skills in which it shows the shared memory communication to the related skills and the creation of the AI, boosting the execution process.
13. Rule-based: It follows up all the related searches and shows the rule-based system to perform all the tasks related to the creation of the patterns via critical skills to solve the problems.

16.9 Real-Time Project and Future Implementation

Real-time AI projects exemplify the capabilities of automated skills and knowledge, enabling tasks to be efficiently performed in a distributed manner while sharing the same space and memory allocation. Depending on individual interests and competence levels, numerous real-time AI projects offer exciting possibilities for exploration. Among these projects are building chatbots, which utilize natural language processing and machine learning techniques to communicate with users, provide support, and respond to queries in various fields such as customer service, healthcare, and education.

Additionally, image and video recognition projects hold significant popularity, employing deep learning techniques to train models capable of identifying objects, people, emotions, and other visual features, with applications ranging from security and surveillance to entertainment. Sentiment analysis, another prominent AI endeavor, involves analyzing the emotional tone of text using machine learning and natural language processing, with data sourced from social media posts, consumer reviews, and various other platforms. Furthermore, AI's pivotal role in fraud detection cannot be overlooked, with machine learning techniques employed to detect patterns and abnormalities in financial transactions [12] and credit card usage, safeguarding against fraudulent activities. As the field of AI research into autonomous vehicles rapidly advances, engaging in projects that create algorithms for self-driving cars, drones, and other autonomous systems presents exciting opportunities for reshaping transportation and enhancing vehicle automation. The future of real-time AI projects is brimming with innovation and potential, offering boundless opportunities for advancements in technology and its applications across various domains and many more like IOT, robotic skill, cryptocurrencies, face recognition, plant patterns, species evolution, human-to-machine intelligence, the evolution of AI, dataset training in AI, financial dataset, number of transaction patterns, natural language processing, image detection, business interaction, car drivers, mobile automation share, AI networking system, AI computing methods, deep learning, neural setup detection, creation to analyze the different pattern sets, nature of natural raw facts, AI mining, energy utilities, public safety, transportation implementation, logistic utility, 6G network [13–15], etc.

16.10 Conclusion

In conclusion, the integration of artificial intelligence (AI) in a distributed system of the future holds immense potential across various sectors. Industries are experiencing the transformative power of AI in process automation, quality control, supply chain management, and production planning, leading to increased efficiency and reduced costs. The economic sector benefits from AI-driven advancements in customer service, marketing, financial analysis, and product development, fostering growth and innovation. The environment sector leverages AI for climate modeling, energy management, waste reduction, and wildlife conservation, promoting sustainability and environmental preservation. The robotics skill in AI enhances automation and autonomy, paving the way for intelligent systems that operate seamlessly without human intervention. AI's interaction with humans has evolved through conversational agents, sentiment analysis, and emotion recognition, improving user experiences and service delivery.

Recent AI projects, such as AlphaFold and GPT-3, showcase groundbreaking achievements in protein structure prediction and language generation, reflecting the rapid progress in AI research. Real-time AI projects, encompassing applications like chatbots, image recognition, and fraud detection, offer practical solutions to complex challenges. As AI continues to advance, its distributed system of the future promises to unlock new possibilities, reshape industries, and create a more intelligent and interconnected world.

References

1. Girija, D. K., Rashmi, M., William, P., & Yogesh, N. (2023, May). Framework for Integrating the Synergies of Blockchain with AI and IoT for Secure Distributed Systems. In *International Conference on Data Analytics and Insights* (pp. 257-267).
2. Facchini, S. D. (2022). Decentralized Autonomous Organizations and Multi-agent Systems for Artificial Intelligence Applications and Data Analysis. In *IJCAI* (pp. 5851-5852).
3. Trivedi, R., & Khadem, S. (2022). Implementation of artificial intelligence techniques in microgrid control environment: Current progress and future scopes. *Energy AI*, 8, 100147.
4. Lynggaard, P. (2014). Artificial intelligence and Internet of Things in a "smart home" context: A Distributed System Architecture, 5(1), 22-25.
5. Dzitac, I., & Bărbat, B. E. (2009). Artificial intelligence+ distributed systems= agents. *Int. J. Comput. Commun. Control*, 4(1), 17-26.
6. Parunak, H. V. D. (2006). Applications of distributed artificial intelligence in industry. Foundations of distributed artificial intelligence, 2(1), 18.
7. Bond, A. H., & Gasser, L. (Eds.). (2014). *Readings in distributed artificial intelligence*. Morgan Kaufmann.
8. Huhns, M. N. (2012). *Distributed Artificial Intelligence: Volume I* (Vol. 1). Elsevier.
9. Raman, S. (2023). A Virtual World for Troubleshooting Distributed Systems
10. Paramasivam, K., Prathap, M., & Sharif, H. (2020). Heterogeneous Large-Scale Distributed Systems on Machine Learning. In *Deep Neural Networks for Multimodal Imaging and Biomedical Applications* (pp. 47-68). IGI Global.
11. Fisch, D., Jänicke, M., Kalkowski, E., & Sick, B. (2012). Learning from others: Exchange of classification rules in intelligent distributed systems. *Artif. Intell.*, 187, 90-114.
12. Verbraeken, J., Wolting, M., Katzy, J., Kloppenburg, J., Verbelen, T., & Rellermeyer, J. S. (2020). A survey on distributed machine learning. *ACM Comput. Surv. (CSUR)*, 53(2), 1-33.

13. Goyal, N., Sharma, S., Kumar Rana, A., & Tripathi, S.L. (Eds.). (2022). *Internet of Things: Robotic and Drone Technology* (1st ed.). CRC Press. https://doi.org/10.1201/9781003181613
14. Dasari L. Prasanna; Suman Lata Tripathi, "Machine and Deep-Learning Techniques for Text and Speech Processing," in *Machine Learning Algorithms for Signal and Image Processing*, IEEE, 2023, pp. 115-128, doi: 10.1002/9781119861850.ch7
15. Arora, K., Kumar, V., Prashar, D., & Tripathi, S.L. (Eds.). (2023). *Modern Computational Techniques for Engineering Applications* (1st ed.). CRC Press. https://doi.org/10.1201/9781003407409

17

The Metaverse and Mental Health: The Future

Vandana Yadav[1]* **and Namrata Dhanda**[2]

[1]*Amity Institute of Information Technology, Amity University, Lucknow, U.P., India*
[2]*Department of Computer Science & Eng., Amity University, Lucknow, U.P., India*

Abstract

Using Google Trends search revealed that Intransmissible tokens and the metaverse were some of the most popular tech terms in 2022. Our review's objective is to explain the metaverse and Intransmissible tokens in the context of how they might be used to treat mental health conditions. Technology development has been transforming human life at an accelerating rate. The metaverse, commonly referred to as the three-dimensional (3D) web, is an online environment where physical and virtual reality coexist. With Intransmissible tokens serving just as primary establishment elements in the new, expansive digital world, it may transform the internet as we now know it. With the additional capacity to alter the surrounding environment in accordance with specific demands and requirements, this realistic 3D virtual environment has all the qualities of the actual world. For the past ten years, numerous mental health illnesses have been treated with the application of mixed reality, augmented reality, and virtual reality. Promising results have emerged from studies on its use in the identification and management of mental health conditions. Many have praised virtual reality/augmented reality/mixed reality as a solution to the severe death of access to mental health care and mental health professionals. On the other hand, young adults frequently engage in social media use and 3D immersive gaming, which can result in behavioral addiction, insecurities, anxiety, and despair. Constantly browsing through social networking sites also reduces people's attention spans and makes teens more likely to experience the signs of attention deficit hyperactivity disorder.

Keywords: Tokens, 3D, augmented reality, immersive, Intransmissible, metaverse, virtual reality, mixed reality

**Corresponding author*: vandanayadav771@gmail.com

Sandhya Avasthi, Suman Lata Tripathi, Namrata Dhanda and Satya Bhushan Verma (eds.)
Decentralized Systems and Distributed Computing, (337–354) © 2024 Scrivener Publishing LLC

17.1 Introduction

Web 1.0, also known as the "read-only" Internet, was the first version of the Internet. It was succeeded by Web 2.0, a more interactive iteration that allowed users to send data to the network. After that, the digital world involved virtual, augmented, and mixed reality shortly thereafter. All of these technological advancements have now coalesced to form a new world that is possible by only the metaverse, which is proclaimed as the Web 3.0 of the future [1].

Virtual reality has existed in its most primitive forms with limited immersion and interaction since the 1960s [1]. Wearing a helmet-mounted camera that combines stereo displays that use specialized optics and motion-tracking devices to generate immersive virtual reality simulates a physical presence in a virtual world [3]. Augmented reality adds virtual content to the real world by superimposing digital images on actual objects, whereas virtual reality substitutes a virtual environment for the real one. Consequently, users can interact with virtual things in the real environment via augmented reality. Users of VR can get psychologically absorbed in a virtual setting. Real and virtual elements can interact in mixed reality, giving users more control over digital things than is possible with augmented reality [4]. A virtual universe called the "metaverse" was developed using the Internet, cryptocurrencies, virtual reality, and other cutting-edge technologies. Neal Stephenson's science fiction book *Snow Crash* [10], from which the words "meta" and "universe" were taken, is where the term "metaverse" first appeared [11]. As interest in the metaverse has recently increased, Google Trends predicts that "metaverse" and "non-fungible tokens (Intransmissible tokens)" will be among the most prominent technology terms in 2022. By digitally engaging with other users, items, and environments in the enormous network of real-time 3D virtual simulations known as the metaverse, users can experience being "within" the internet [10]. The virtual world is equipped with all of the same conveniences as the actual one, including the capability for users to engage in different physical settings while donning personalized avatars for socializing, working, and playing. The metaverse already exists in its rudimentary form, and its evolution may be unpredictable [5].

Web 3.0 will alter our communication, collaboration, and social interactions. Unique digital assets known as Intransmissible tokens represent music, artwork, collectibles, and in-game objects. Intransmissible tokens are a form of data stored on a blockchain, a decentralized digital database that is frequently traded online alongside cryptocurrencies [1]. The technology underlying cryptocurrencies such as Bitcoin, Ethereum, and Intransferable tokens is blockchain. Blockchain maintains a secure, decentralized record of transactions, ensuring the confidentiality of data records free from outside interference [1]. The decentralized structure of the blockchain makes limitless commercial opportunities for metaverse social engagement possible. It is possible that nontransferable tokens will serve as the metaverse's building elements. By representing real-world identities, Intransmissible tokens' digital personas serve as access tokens to various areas within the metaverse.

In the metaverse, non-transferable tokens may be indispensable components of the economic ecosystem. Initiatives related to the metaverse of Intransmissible tokens would enhance the potential for using Intransmissible tokens to exchange virtual goods [6]. The identities, social interactions, and communal experiences of users of the metaverse would be significantly impacted by metaverse enterprises. By holding Intransferable tokens, users can voice their opinions or support specific causes. Therefore, proprietors of Intransmissible tokens who share similar interests could form communities for communicating, exchanging ideas, and working together.

What does the future hold for psychiatry and mental health, given the hype surrounding the metaverse and Intransmissible tokens? In spite of the fact that mental problems are the leading cause of sickness and disability [7] globally, a World Health Organization (WHO) study found that between 30 and 80% of people with mental health concerns never seek treatment [8]. Schedule conflicts, a lack of awareness, limited access to resources or therapy, financial hardship, and the inability to pay for ongoing care may all be to blame [1]. The metaverse is a three-dimensional amalgamation of social media, online commerce, virtual gaming, and the trading of digital art, real estate, and securities [1]. The metaverse will unquestionably have both positive and negative implications for mental health on our reality. According to Amara's law, we frequently exaggerate a technology's short-term influence while underestimating its long-term impact [23]. On the basis of this premise, it is possible to exaggerate the metaverse's capacity to predict future events while it is still in development. We plan to investigate the role the metaverse will play in the epidemiology of mental health disorders in light of the impending global mental health crisis [1].

17.2 How Has Technology and the Internet Already Impacted Mental Health?

The impact of technology and the internet on mental health has been both positive and negative [4]. Positively, technology has facilitated access to mental health-related information, resources, and support. Online support groups, forums, and counseling services have made it simpler for people to connect with others facing similar difficulties and to access mental health services from the comfort of their homes [3].

Nevertheless, technology and the Internet have had negative effects on mental health. The increase in cyberbullying and online harassment is one of the most significant negative effects. Social media platforms have facilitated the anonymous bullying and harassment of others, which can lead to melancholy, anxiety, and other mental health problems.

Constant connectivity enabled by technology can also induce feelings of tension and exhaustion. The pressure to continuously check emails, social media notifications, and other messages can make it difficult for people to disconnect and relax.

In addition, the use of technology and the Internet has been associated with sleep disturbances, particularly when electronic devices are used before bedtime. This can have an effect on mood and mental health, as sleep is necessary for physical and mental health.

Ultimately, technology and the Internet have had both positive and negative effects on mental health [10]. As technology continues to advance, it will be essential to consider how it can be utilized to promote mental health and well-being, as well as how to mitigate its negative effects [19].

17.2.1 These Concerns May Be Exacerbated by the Metaverse

One could assume that if we spend more time using technology, our risk of developing these issues would increase proportionally. Many metaverse designers acknowledge this fact. Important to their vision is the creation of a virtual world that is so immersive that it can enhance our sense of connection and work–life balance (e.g., via the sensation that you are walking to work or by taking the VR headset off).

However, it appears that the metaverse could also contribute to an increase in severe mental health issues if we dig deeper. A 2010 study, for instance, discovered that people were more likely to experience depersonalization symptoms after VR immersion. When we feel disconnected from our bodies and minds, we experience depersonalization. It can be a frightening experience, with symptoms including anxiety and panic attacks. Thus, while the metaverse may intend to create healthy boundaries between our private and personal lives, it may create problematic boundaries between our online and offline identities.

I find it difficult to imagine a universe where the metaverse does not exacerbate loneliness. Humans are tribal entities that have always flourished in communities. We are so accustomed to interacting with others that we have evolved to become ill without it. For instance, when we do not receive enough physical contact, a condition known as touch starvation occurs. When we partake in a pleasant touch experience, such as a hug, our skin sends a signal to our brain to release oxytocin. This will increase our contentment and social connections while decreasing our fear. In contrast, a dearth of physical contact increases cortisol production in the body (the stress hormone) [6]. This can put us at risk for a variety of problems, including burnout, cardiovascular issues, decreased productivity, and melancholy.

While the metaverse may include senses such as hearing and scent, it cannot include all of the intuitive and physical aspects of our physical interactions.

17.3 Importance of Researching the Metaverse's Effects on Mental Health

Studying the impact of the metaverse on mental health is important for several reasons:

1. The metaverse has the potential to significantly affect our way of life, including our mental health. As the metaverse becomes increasingly integrated into our daily existence, it is crucial to comprehend its potential effects on mental health in order to design it in a manner that promotes mental health [28].
2. The metaverse may present new mental health care and treatment opportunities. In the metaverse, virtual mental health services, immersive therapy, and virtual support groups could become more accessible. It is essential to investigate these potential advantages in order to determine how they can be utilized to promote mental health.
3. The metaverse may also present new mental health dangers and challenges. For instance, the metaverse may contribute to addiction and compulsive behavior or exacerbate preexisting mental health issues. It is essential to investigate these potential hazards in order to mitigate their negative effects.
4. The metaverse may also present new mental health dangers and challenges. For instance, the metaverse may contribute

to addiction and compulsive behavior or exacerbate preexisting mental health issues. It is essential to investigate these potential hazards in order to mitigate their negative effects.

17.4 Potential Positive Mental Health Effects of the Metaverse

17.4.1 Access to Virtual Mental Health Services

Access to virtual mental health services is one prospective mental health benefit of the metaverse [21]. The metaverse has the potential to significantly expand access to mental health services, especially for those who encounter barriers to traditional in-person services, such as transportation, time constraints, or stigma. Among the prospective advantages of virtual mental health services in the metaverse are the following:

1. **Comfort and accessibility:** Virtual mental health services can be accessed from anywhere, at any time, increasing access for individuals who may not have been able to utilize traditional mental health services. This can be especially advantageous for those who reside in rural or remote areas, or who have mobility or transportation issues [3].
2. **Anonymity and privacy:** Virtual mental health services in the metaverse can provide a greater sense of anonymity and privacy, which can be especially essential for individuals who are reluctant to seek help out of stigmatization or privacy concerns [25].
3. **Increased access to specialized care:** Virtual mental health services in the metaverse can increase access to specialized care, such as therapy for specific mental health conditions or services for specific populations, such as individuals or veterans.
4. **Cost-effectiveness:** Virtual mental health services in the metaverse may be more cost-effective than traditional in-person services, thereby increasing access for individuals who may not be able to afford conventional mental health services [4].

It is essential to note, however, that virtual mental health services may not be suitable for everyone, and that in some cases, in-person services may still be required. In addition, there are potential hazards and ethical concerns associated with virtual mental health services, such as the possibility of insufficient care and data privacy issues. As with any form of mental health care, it is essential that virtual mental health services in the metaverse are devised and implemented in an ethical and responsible manner.

17.4.2 Reduction of Stigma Surrounding Health

Another potential beneficial effect of the metaverse on mental health is the elimination of stigma associated with mental illness [2]. Mental health stigma can be a significant barrier to seeking assistance and gaining access to mental health services. In multiple ways, the metaverse has the potential to reduce stigma [33]:

1. **Anonymity and privacy:** As mentioned previously, the metaverse can provide a greater sense of anonymity and privacy, which can be particularly essential for individuals who may be hesitant to seek help out of stigmatization or privacy concerns [9]. Virtual access to mental health services can provide a sense of security and reduce feelings of humiliation and embarrassment [29].
2. **Normalization of mental health:** The metaverse can serve as a forum for open dialogue and mental health education. By incorporating mental health into the metaverse, mental health can become less stigmatized and more normalized.
3. **Representation and diversity:** The metaverse can provide a platform for diverse representation of individuals with mental health conditions, thereby decreasing stigma by increasing understanding of mental health issues.
4. **Community and support:** Individuals with mental health conditions can connect with others who share comparable experiences through the metaverse. This can foster a sense of community and support, thereby diminishing feelings of isolation and stigma.

It is essential to note, however, that stigma reduction surrounding mental health is a complex and continual process. While the metaverse has the potential to play a role in stigma reduction, it is essential that stigma be addressed in all aspects of society [12].

17.4.3 Opportunities for Social Connection and Support

The metaverse has the potential to facilitate social connection and support, which may have a beneficial effect on mental health. The metaverse can facilitate social connection and support in the following ways:

1. **Virtual social spaces:** The metaverse can provide virtual social spaces where people with shared interests, activities, or experiences can interact. These virtual spaces can be designed to nurture positive interactions and community spirit.
2. **Support groups:** The metaverse can provide a platform for virtual support groups where individuals can connect with others experiencing similar mental health difficulties. These support organizations can provide a sense of community, emotional support, and a safe environment for sharing experiences and coping mechanisms [18].
3. **Immersive experiences:** The metaverse can provide immersive experiences that facilitate meaningful human connections. Virtual reality experiences, for instance, can generate a sense of presence and help individuals feel physically present with others.
4. **Gaming communities:** Through online gaming communities, the metaverse can provide opportunities for individuals to interact with one another. These communities can foster a sense of social connection and support, as well as collaboration, communication, and problem-solving skills.

It is crucial to note, however, that virtual social connection should not completely replace in-person social connection. In-person social connection remains an essential component of mental health and well-being, and the metaverse should be viewed as a supplement rather than a replacement [14].

17.4.4 The Potential for Immersive Therapy

The metaverse may provide opportunities for immersive therapy, which may have a beneficial effect on mental health. Immersive therapy is therapy administered within a virtual environment, such as virtual reality or augmented reality. The metaverse can facilitate immersive therapy in the following ways:

1. **Exposure therapy:** Virtual reality environments can be used to simulate real-life situations that induce anxiety or dread. Individuals are progressively exposed to these situations in a safe and controlled environment as part of exposure therapy [30]. This can be especially beneficial for those with phobias or PTSD.
2. **Mindfulness and meditation:** The metaverse can provide immersive environments that encourage mindfulness and meditation. These environments can aid in the reduction of tension and anxiety, as well as the promotion of relaxation and well-being [13].
3. **Social skills training:** Virtual environments can be utilized to teach and practice social skills, including communication, problem-solving, and conflict resolution. This can be especially beneficial for those with social anxiety or autism spectrum disorder.
4. **Cognitive-behavioral therapy (CBT):** Virtual environments can be used to practice cognitive-behavioral therapy (CBT) techniques [2], such as cognitive restructuring and behavioral activation. These techniques can assist individuals in reframing negative thoughts and behaviors and fostering more constructive coping mechanisms [10].

It is essential to note, however, that immersive therapy is a relatively new field, and more research is required to determine its efficacy and best practices. Moreover, it may not be suitable for all individuals or mental health conditions. As with any form of mental health treatment, it is essential that immersive therapy in the metaverse be designed and implemented with responsibility and integrity.

17.5 Possible Detrimental Effects of the Metaverse on Mental Health

17.5.1 Risk of Addiction and Compulsive Behavior

The risk of addiction and compulsive behavior is a possible negative effect of the metaverse on mental health. The metaverse can contribute to addiction and compulsive behavior in the following ways [1]:

1. **Gaming and gambling:** Some individuals may develop an addiction to online gaming and wagering offered by the metaverse. When these activities are designed to be rewarding and immersive, with frequent reinforcement and stimulation, they can be particularly problematic [29].
2. **Addiction to social media and the internet:** The metaverse can serve as a platform for social media and internet addiction. Social media and other online activities can be designed to be highly engaging and addictive, with constant updates and notifications that can induce FOMO (fear of missing out) or compulsive monitoring [30].
3. **Escapism:** The metaverse can serve as a platform for escapism, in which people use virtual reality to avoid real-world problems or responsibilities. This is especially problematic when it interferes with essential facets of life, such as work, relationships, or physical health [28].
4. **Lack of Regulation:** The metaverse is a new and largely unregulated space, making it difficult to monitor and address problematic behavior. This can make it simpler for people to engage in addictive or compulsive behavior without experiencing negative consequences.

Individuals must be aware of the potential dangers of addiction and compulsive behavior in the metaverse and manage their use of technology and virtual reality. This can include setting utilization limits, taking breaks, and, if necessary, seeking assistance from mental health professionals [27]. It is also crucial for designers and developers of virtual reality technology to consider the potential risks and take measures to mitigate them, such as designing games and activities with the principles of responsible gaming in mind.

17.5.2 Impact on Physical Health

Despite the fact that the effects of the metaverse on physical health are primarily unknown, there are potential positive and negative effects to consider. The metaverse may impact physical health in the following ways [10]:

1. **Sedentary behavior:** The risk of sedentary behavior is one of the potential negative effects of the metaverse on physical health. A lack of physical activity can increase the risk of

obesity, cardiovascular disease, and other health concerns associated with prolonged virtual reality use [20].

2. **Increased physical activity:** However, the metaverse also offers opportunities for increased physical activity [25]. Some virtual reality experiences, such as fitness or athletics simulations, may motivate users to engage in physical activity.
3. **Improved accessibility:** The metaverse may also enhance the accessibility of physical activity for people with physical disabilities or limited mobility, for example. Virtual reality can offer an alternative way to engage in physical activity and promote overall health and well-being [26].
4. **Virtual health care:** The metaverse may also offer opportunities for virtual health care, such as telemedicine consultations and remote monitoring of health conditions [30]. This can enhance health care access for those who reside in rural or remote areas, or who have difficulty accessing traditional health care settings [14].

Ultimately, the metaverse has a complex and multifaceted effect on physical health. There are both potential negative and positive effects to consider, such as sedentary behavior and the risk of overuse injuries, as well as increased physical activity and enhanced access to health care [23]. It will be essential for designers and developers of virtual reality technology to consider the potential impact on physical health and to take measures to mitigate any adverse effects.

17.5.3 Concerns Regarding the Morality of Data Privacy and Mental Health Diagnoses

The use of technology and the metaverse in mental health raises significant ethical concerns regarding data privacy and the possible misuse of personally identifiable information [22]. In the metaverse, data privacy can influence mental health diagnoses in the following ways:

1. **Collection and use of personal data:** The metaverse may acquire sensitive health information, such as mental health diagnoses, from users as part of its collection and use of personal information. The use of these data must be strictly

governed in order to prevent the exploitation of personally identifiable information [9].

2. **Discrimination and stigmatization:** The collection of personal data, especially sensitive health information, can contribute to discrimination and stigmatization of people with mental health conditions. Employers and insurance companies, for instance, may use this information to deny employment or coverage [10].
3. **Informed consent:** Individuals must provide explicit consent after being completely informed about the collection and use of their personal data, particularly sensitive health information. This includes the data collection's purpose, how it will be used, and who will have access to it.
4. **Access to mental health care:** The use of technology and virtual reality in mental health care must ensure that all individuals, regardless of socioeconomic status or other personal characteristics, have access to high-quality mental health care [15].

To resolve these ethical concerns, it is crucial for virtual reality technology designers and developers to prioritize data privacy and implement stringent data protection measures. Professionals and researchers in mental health must also ensure that individuals are thoroughly informed and provide informed consent for the collection and use of their personal data. To prevent discrimination and stigmatization, policymakers and regulators must also establish clear guidelines and regulations regarding the use of personal data in mental health diagnoses in the metaverse [12].

17.6 Considerations for Designing the Metaverse with Mental Health in Mind

It is essential to design the metaverse with mental health in mind in order to provide individuals with access to virtual environments that promote mental health and well-being. Consider the following factors when designing the metaverse with mental health in mind:

1. **Accessibility:** Virtual environments should be accessible to individuals with mental health conditions, including those with sensory, cognitive, or physical impairments [22].

This may include closed captioning, audio descriptions, and voice recognition technology.

2. **User safety:** Virtual environments should prioritize user safety and mitigate the risk of harm, including the risk of addiction and other adverse mental health effects [21]. This may involve designing and implementing features that encourage breaks and limit excessive use, as well as tools for users to report detrimental or abusive behavior.
3. **Positive environments:** Virtual environments should promote mental health and well-being by providing opportunities for social support, positive reinforcement, and meaningful engagement [24]. This may include virtual support groups and team-based activities.
4. **Privacy and data protection:** Virtual environments should prioritize user privacy and data protection in order to prevent the misuse of personal information and reduce the likelihood of discrimination or stigmatization [20]. This can include implementing robust data protection measures and providing clear information to users about how their personal data will be collected and used.
5. **Evidence-based interventions:** Virtual environments should be designed to include cognitive-behavioral therapy and mindfulness meditation as evidence-based interventions for mental health [19]. This can include virtual reality therapy programs and other interactive interventions that have proven effective in treating mental health conditions.

By designing the metaverse with mental health in mind, virtual environments can provide individuals with novel access to mental health care, social support, and well-being-promoting positive experiences. It will be essential for virtual reality technology designers and developers to collaborate closely with mental health professionals and researchers to ensure that these considerations are completely integrated into the design process [26].

17.7 The Future of Mental Health and Metaverse

The future of mental health and the metaverse presents opportunities for innovation and advancement in mental health care, as well as obstacles and potential hazards to consider. The future of mental health and the metaverse will be significantly shaped by inter-disciplinary collaboration [7].

17.7.1 Innovation and Advancement Opportunities in Mental Health Care

1. **Accessible and affordable mental health care:** The metaverse can provide a cost-effective method to increase access to mental health care, especially in rural and underserved areas. The ability to conduct teletherapy and virtual support groups in the metaverse can reduce the barriers to accessing mental health care services [27].
2. **Customized mental health interventions:** The metaverse can provide personalized mental health interventions that are tailored to the specific requirements of each individual. Individuals with anxiety disorders can, for instance, receive exposure therapy through virtual reality therapy [17].
3. **Improved data collection and analysis:** The metaverse can provide rich data on users' mental health and behavior, allowing researchers and clinicians to better comprehend the causes of mental health disorders and devise more effective interventions [17].

17.7.2 Challenges and Potential Pitfalls to Consider

1. **Digital divide:** The digital divide, or the disparity in access to technology and the internet, could exacerbate existing health disparities, especially for low-income and rural residents [16].
2. **Privacy and data protection:** The accumulation and use of personal data in the metaverse may endanger the privacy and security of users. It will be essential to implement appropriate safeguards to secure the personal information of users.
3. **Addiction and compulsive behavior:** The immersive and engaging nature of the metaverse may result in addiction and compulsive behavior, which may have adverse effects on mental health [7].

17.7.3 The Influence of Interdisciplinarity on the Development of Mental Health and the Metaverse

1. **Collaborative research:** Researchers and clinicians from diverse disciplines should collaborate to develop and evaluate the efficacy of mental health interventions in the metaverse [8].
2. **Education and training:** Collaboration between mental health professionals and educators can aid in the development of educational resources and training programs for mental health professionals to use the metaverse effectively in their practice [18].

17.8 Conclusion

There is a possibility that certain mental health conditions could worsen if an individual were to reside in a metaverse world. It could even lead to the discovery and classification of novel mental diseases [5]. We cannot, however, dismiss the metaverse's promise of a wealth of readily available resources for individuals looking to better themselves. As a result of the internet's anonymity, more people could feel comfortable sharing their stories with professionals and others living through similar issues. Therefore, given the current severe scarcity of mental health practitioners, this quickly developing technology could dramatically increase access to mental health care. We might very soon encounter a vast Web 3.0 mental health arena that could revolutionize the mental health industry. For many facets of civilization, the metaverse offers amazing potential, but only time will tell how things turn out [2].

References

1. Businesswire. Revitalist is building virtual clinics called Revitaland in the Metaverse. Obtainable at: https://www.businesswire.com/news/home/20211112005581/en/Revitalist to Develop Revitaland Virtual Clinics in the Metaverse [Accessed February 8, 2022].Use Google Scholar

2. Smith, S. Will the metaverse only make things worse in terms of mental health online? The following link is available: https://www.dazeddigital.com/science-tech/article/55420/1/mental-health-online-internet-social-media-metaverse-illness-virtual-reality [Retrieved 15 February 2022].Use Google Scholar
3. Hajesmaeel-Gohari, S., S. Goharinejad, and others. An summary of pertinent studies on the effectiveness of virtual, augmented, and mixed reality technologies in the diagnosis and treatment of attention deficit hyperactivity disorder in children. pmid:http://www.ncbi.nlm.nih.gov/pubmed/34983446 *BMC Psychiatry* 2022;22:4.doi:10.1186/s12888-021-03632-1 PubMedUse Google Scholar
4. Rebenitsch L, Owen C. Review on cybersickness in applications and visual displays. *Virtual Real.* 2016 Apr 26;20(2):101-125 [FREE Full text] [CrossRef]
5. Park MJ, Kim DJ, Lee U, Na EJ, Jeon HJ. A literature overview of virtual reality (VR) in treatment of psychiatric disorders: recent advances and limitations. *Front. Psychiatry* 2019 Jul 19;10:505 [FREE Full text] [CrossRef] [Medline]
6. Baghaei N, Chitale V, Hlasnik A, Stemmet L, Liang HN, Porter R. Virtual reality for supporting the treatment of depression and anxiety: scoping review. *JMIR Ment. Health* 2021 Sep 23;8(9):e29681 [FREE Full text] [CrossRef] [Medline]
7. Kothgassner OD, Goreis A, Kafha JX, Van Eickels RL, Plener PL, Felnhofer A. Virtual reality exposure therapy for posttraumatic stress disorder (PTSD): a meta-analysis. *Eur. J. Psychotraumatol.* 2019 Aug 19;10(1):1654782 [FREE Full text] [CrossRef] [Medline]
8. Freeman D, Lambe S, Kabir T, Petit A, Rosebrock L, Yu LM, gameChange Trial Group. Automated virtual reality therapy to treat agoraphobic avoidance and distress in patients with psychosis (gameChange): a multicentre, parallel-group, single-blind, randomised, controlled trial in England with mediation and moderation analyses. *Lancet Psychiatry* 2022 May; 93 (5):375-388 [FREE Full text] [CrossRef] [Medline]
9. Number of social media users worldwide from 2018 to 2027. Statista. 2022. URL: https://www.statista.com/statistics/278414/number-of-worldwide-social-network-users/ [accessed 2022-03-28]
10. Weller K. Accepting the challenges of social media research. *Online Inf. Rev.* 2015;39(3):281-289 [FREE Full text] [CrossRef]
11. Persily N, Tucker J. *The Challenges and Opportunities for Social Media Research*. Stanford Cyber Policy Center. 2020. URL: https://cyber.fsi.stanford.edu/publication/challenges-and-opportunities-social-media-research [accessed 2022-10-24]
12. Orben A, Przybylski AK, Blakemore SJ, Kievit RA. Windows of developmental sensitivity to social media. *Nat. Commun.* 2022 Mar 28;13(1):1649 [FREE Full text] [CrossRef] [Medline]

13. Gayle D. Facebook aware of Instagram's harmful effect on teenage girls, leak reveals. The Guardian. 2021 Sep 14. URL: https://www.theguardian.com/technology/2021/sep/14/facebook-aware-instagram-harmful-effect-teenage-girls-leak-reveals [accessed 2022-10-24]
14. Valkenburg PM, van Driel II, Beyens I. The associations of active and passive social media use with well-being: a critical scoping review. *New Media Soc.*, 2021 Dec 31;24(2):530-549 [FREE Full text] [CrossRef]
15. Oakman J, Kinsman N, Stuckey R, Graham M, Weale V. A rapid review of mental and physical health effects of working at home: how do we optimise health? *BMC Public Health* 2020 Nov 30;20(1):1825 [FREE Full text] [CrossRef] [Medline]
16. De Sio S, Cedrone F, Nieto HA, Lapteva E, Perri R, Greco E, *et al.* Telework and its effects on mental health during the COVID-19 lockdown. *Eur. Rev. Med. Pharmacol. Sci.*, 2021 May;25(10):3914-3922 [FREE Full text] [CrossRef] [Medline]
17. Van Zoonen W, Sivunen AE. The impact of remote work and mediated communication frequency on isolation and psychological distress. *Eur. J. Work Organ. Psychol.*, 2022;31(4):610-621 [FREE Full text] [CrossRef]
18. Davis ML, Fletcher T, McIngvale E, Cepeda SL, Schneider SC, La Buissonnière Ariza V, *et al.* Clinicians' perspectives of interfering behaviors in the treatment of anxiety and obsessive-compulsive disorders in adults and children. *Cogn. Behav. Ther.* 2020 Jan;49(1):81-96. [CrossRef] [Medline]
19. Mann S, Holdsworth L. The psychological impact of teleworking: stress, emotions and health. *New Technol. Work Employ.*, 2003 Nov;18(3):196-211 [FREE Full text] [CrossRef]
20. Pelissier C, Paredes J, Moulin M, Bitot T, Fakra E, Fontana L. Telework and psychological health in hospital staff during the first wave of the COVID-19 epidemic in France. *Int. J. Environ. Res. Public Health,* 2021 Oct 03;18 (19): 10433 [FREE Full text] [CrossRef] [Medline]
21. Afonso P, Fonseca M, Teodoro T. Evaluation of anxiety, depression and sleep quality in full-time teleworkers. *J. Public Health (Oxf),* (forthcoming) 2021 May 25:fdab164 [FREE Full text] [CrossRef] [Medline]
22. Valkenburg PM, Schouten AP, Peter J. Adolescents' identity experiments on the internet. *New Media Soc.,* 2005 Jun; 7 (3): 383-402 [FREE Full text] [CrossRef]
23. Biener V, Kalamkar S, Nouri N, Ofek E, Pahud M, Dudley JJ, *et al.* Quantifying the effects of working in VR for one week. arXiv. Preprint posted online on June 8, 2022 2022 [FREE Full text]
24. Berryhill MB, Culmer N, Williams N, Halli-Tierney A, Betancourt A, Roberts H, *et al.* Videoconferencing psychotherapy and depression: a systematic review. *Telemed. J. E Health,* 2019 Jun;25(6):435-446. [CrossRef] [Medline]

25. Berryhill MB, Halli-Tierney A, Culmer N, Williams N, Betancourt A, King M, *et al.* Videoconferencing psychological therapy and anxiety: a systematic review. *Fam. Pract.* 2019 Jan 25;36(1):53-63. [CrossRef] [Medline]
26. Markowitz JC, Milrod B, Heckman TG, Bergman M, Amsalem D, Zalman H, *et al.* Psychotherapy at a distance. *Am. J. Psychiatry,* 2021 Mar 01;178(3):240-246. [CrossRef] [Medline]
27. Huckvale K, Venkatesh S, Christensen H. Toward clinical digital phenotyping: a timely opportunity to consider purpose, quality, and safety. *NPJ Digit. Med.,* 2019 Sep 6;2:88 [FREE Full text] [CrossRef] [Medline]
28. Foley GN, Gentile JP. Nonverbal communication in psychotherapy. *Psychiatry (Edgmont)* 2010 Jun;7(6):38-44 [FREE Full text] [Medline]
29. Craig TK, Rus-Calafell M, Ward T, Leff JP, Huckvale M, Howarth E, *et al.* AVATAR therapy for auditory verbal hallucinations in people with psychosis: a single-blind, randomised controlled trial. *Lancet Psychiatry,* 2018 Jan;5(1):31-40 [FREE Full text] [CrossRef] [Medline]
30. Valkenburg PM, Peter J. Preadolescents' and adolescents' online communication and their closeness to friends. *Dev. Psychol.,* 2007 Mar;43(2):267-277. [CrossRef] [Medline]

18

Decentralization in Education Governance: A Case Study of India

Mani Dublish[1]* and Anita Pati Mishra[2]

[1]*MCA Department, ABES Engineering College, Ghaziabad, India*
[2]*IT Department, IMS, Noida, India*

Abstract

The decentralization of educational systems is currently a household trend across the globe. The majority of nations are testing out or thinking about decentralizing education in some way. Decision-making authority is transferred through the process from central ministries of education to regional administrations, local managements, societies, and educational institutions. Yet, the possibility of the transmission might range from directorial decentralizing to a considerably wider transmission of economic responsibility to the local or regional level. The reasons for decentralization, the initial conditions of the education system, and the coalitions of actors involved all affect the direction, scope, and success of decentralization reforms. In an ecosphere, the mainstream of administrations has the disadvantages of central teaching facility supply, including impervious verdict-making, ineffective managerial and financial management, and deprived service excellence and accessibility. These points have made the supposed profits of reorganization quite alluring. Decentralized education delivery has the potential to be more effective, represent local objectives better, promote involvement, and, ultimately, increase coverage and quality. Decentralization has the potential to boost efficiency, which appeals to governments with tight budgets in particular. In addition to boosting efficiency, decentralization can also offload part of the financial burden of providing education services from central governments to local governments and communities. This can be done through cost recovery schemes, such as community funding, which allow local governments and communities to raise their revenue to support education. This chapter provides a detailed review of decentralization and distributed application in the field of education governance.

**Corresponding author*: mani.dublish@abes.ac.in

Sandhya Avasthi, Suman Lata Tripathi, Namrata Dhanda and Satya Bhushan Verma (eds.) Decentralized Systems and Distributed Computing, (355–370) © 2024 Scrivener Publishing LLC

***Keywords*:** Decentralization, education systems, virtual reality, distributed computing, online learning

18.1 Introduction

Decentralized education delivery has the potential to be more effective, represent local objectives better, promote involvement, and, ultimately, increase coverage and quality [1–5]. Education is a multifaceted endeavor that involves the facilitation of learning and the development of knowledge, skills, and good values. The principal objective of education is to equip students with the essential knowledge and skills required for their future endeavors, including life, employment, and civic engagement. Educators bear the responsibility of augmenting the qualifications, abilities, and skills of graduates throughout their educational journey. In general, academic courses are structured to encompass both theoretical and practical elements, which may include exercises, laboratory work, and apprenticeship experiences. Theoretical courses primarily involve the dissemination of knowledge through the delivery of large-group lectures, which may also incorporate opportunities for debate.

The decentralization of education in India was officially expressed through the 73rd Constitutional Amendment, enacted in 1992. This facilitated the widespread formation of local government units known as panchayats, which were established at various levels including the village, block, and district (with the village being the primary unit). These units were created through a democratic election process and were assigned specific functions and responsibilities. The allocation of educational obligations to Panchayati Raj organizations encompasses primary and secondary education, technical training and vocational education, adult and non-formal education, and the provision of libraries. The Panchayati Raj Act recommends the allocation of responsibilities about education among local authorities at different levels under several provisions. Over time, the educational system has undergone alterations in response to the evolving demands of students and the employment market [6]. The process necessitates that decision-makers rationalize and coordinate a wide range of complementary tasks, including curriculum development, instructional strategies, student evaluation, textbook production and distribution, teacher recruitment and compensation, school construction and renovation, education financing, and parent–teacher relationships. As a result, the debate is still unresolved. The decision-making process is made more difficult because each of these roles must be examined for primary, secondary, and university education, as well as frequently for preschools and adult literacy. This is a summary of several growing consensus areas. There is conflicting and scant information available about how decentralization affects educational services [7].

The sudden emergence of the pandemic and subsequent closure of educational facilities sparked a substantial effort by educators and educational institutions to adapt and implement innovative strategies. The aforementioned alterations were promptly implemented in response to the prevailing circumstances. Many educational institutions and systems quickly shifted to delivering remote education. Educational institutions have implemented a range of instructional approaches, employing diverse channels like television, radio, the Internet, and conventional postal services. Regardless of the outcomes, remote learning has emerged as the dominant method of delivering education for varying periods. Educational professionals demonstrated a proactive approach and displayed substantial endorsement for the modifications in instructional delivery.

The chapter is organized into six main sections. Section 18.1 introduces the need for a decentralized system in the education sector and the overall scenario. Sections 18.2, 18.3, and 18.4 provides an overview of centralized, distributed, and decentralized system on basic concepts.

18.2 Central Organizations

Centralized frameworks are a common and straightforward architecture in computing. In a centralized system, multiple client nodes are connected directly to a central server [8]. The clients initiate requests, and the server processes these requests and responds to the clients accordingly. This client/server model is prevalent in various organizations and has been widely used for many years.

Some key characteristics of centralized frameworks include the following:

Client–Server Architecture: Clients and a central server are the system's two basic building blocks. Clients ask the server for assets or offerings, which the website then analyzes and delivers the required solutions.

Communication: Communication in a centralized framework is typically achieved through well-defined protocols and communication channels. Clients send their requirements to the server, and the server replies with the required data or actions.

Resource Management: All resources, data, and computing power are concentrated in the central server. Clients rely on the server to perform most of the processing tasks and access shared resources.

Single Point of Failure: The central server becomes a single point of failure in the system. If the server goes down, clients may lose access to services and data until the server is restored.

Security: Centralized systems need robust security measures to protect sensitive data and prevent unauthorized access since all data are stored and managed in a central location.

While centralized frameworks have been widely used, they do have some limitations, especially in terms of scalability, fault tolerance, and potential performance bottlenecks. As a result, other architectural paradigms like distributed systems and decentralized systems have gained popularity to address these challenges [9, 10]. The limitations of the centralized framework are that it is difficult to move upwards and sometimes you get obstacles when demand increases. The components of the centralized framework are nodes, servers, and communication links as shown in Figure 18.1a, whereas Figure 18.1b shows a distributed connection of participating nodes.

The important feature of a centralized system is the presence of a global clock, one single central server, and dependent nodes or components. As the whole framework comprises a central node (a server/an ace) and numerous client nodes (a computer/a slave), all client hubs match up with

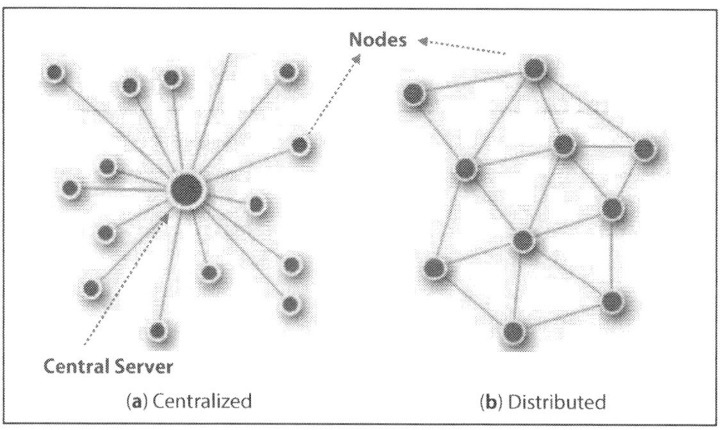

Figure 18.1 (a) Centralized and (b) distributed system network.

the worldwide clock(the clock of the central hub). All additional hubs within the structure are coordinated and served by a single central unit. The entire framework fails due to the central hub's failure. a lot of reasons because there is not anything else available to send/receive responses to/ requests from when the server is offline.

18.3 Decentralized Systems

The immense excitement surrounding Bitcoin has led to the rapid rise in popularity of other types of systems [11]. Many businesses are exploring uses for these technologies right now. In a decentralized system, every node makes independent decisions. The final behavior of the system is determined by the aggregate of the decisions taken from distinct nodes. Take note that the request is received by several entities (Figure 18.2).

18.3.1 Features of a Decentralized System

The main features of a decentralized system are the presence of a world-wide timer, fundamental components/nodes, and dependent components. Each node runs and adheres to a distinct schedule because they are distinct from one another. Many centralized devices can monitor interconnections from additional nodes. Part of the structure collapses as a consequence of one main node failure, but not the whole thing.

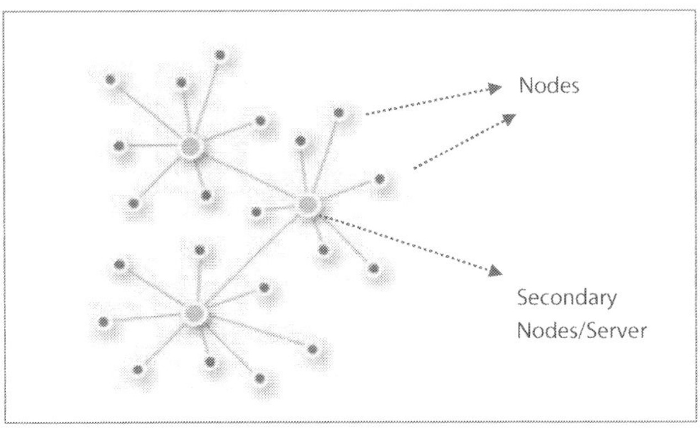

Figure 18.2 Decentralized framework.

18.3.2 Architecture of a Decentralized System

The decentralized system architecture is of two types: peer-to-peer (P-2-P) architecture and master–slave architecture. In P-2-P, every node is an ally of all other nodes. There is no hegemony of one node over another. On the other hand, in master–slave architecture, by voting, one node can coordinate a portion of the system and become the master, but this does not give that node superiority over the other nodes it is managing.

18.4 Distributed Systems

A system that is distributed is made up of several programming languages that divide computing power among numerous, autonomous computing hubs to accomplish one common objective. It is also known as distributed databases or distributed computer systems and uses multiple nodes to communicate and coordinate via shared networks. Other cyclical included systems or distinct software processes can likewise be symbolized by these nodes. They frequently represent various pieces of real gear. Distributed systems aim to get rid of individual points of breakdown and redundancies in the network. The components of distributed system are node (computer, mobile, etc.) and a communication link (cables, Wi-Fi, etc.).

18.4.1 Features of a Distributed System

- **Source allocation:** Software, equipment, and data can all be utilized by a distributed system. Parallel computation: multiple devices can carry out the same task at once.
- **Scalability:** If more machines are added, the computation and analyzing power can grow up as required.
- **Error detection:** Inconsistencies are easier to identify.
- **Transparency:** A node's position in the system is capable of communicating and interacting with other nodes.

18.4.2 The Architecture of a Distributed System

- **Peer-to-peer:** Each node is equal of one another and collaborates on a shared objective.
- **Client–server:** Certain nodes turn into server clusters to play coordinating, arbiter, and other responsibilities.

- **n-tier architecture:** Each of the parts of an application is stored by distinct system nodes, and these nodes work together to create an application for an individual or client.

18.4.3 Limitations of a Distributed System

It might be challenging to develop and fix programs for a system. These algorithms are challenging to utilize since there can be no temporal sorting of instructions or reports without a shared timer. Nodes may have different latencies, which must be considered while creating such algorithms. The level of intricacy increases as the number of interconnections does. The chronological sequencing of actions and occurrences is problematic since there is no uniform timestamp. It is challenging for a node to have a comprehensive understanding of the system and, consequently, make judgments based on the conditions of other networks in the system. It is not possible to track occurrences in this situation using the traditional method of recording absolute times.

18.4.4 Difference Between a Centralized Database and a Distributed Database

A consolidated database is essentially a sort of information that is kept, situated, and operated in just one place. From that place, this kind of database is controlled and amended. Therefore, a database system or consolidated computing system is primarily where this place is. A connection to the internet (LAN, a Wide Area Network, etc.) is utilized to access the central location [12]. Organizations or groups are the principal users of this integrated database.

Difference between Centralized Database and Distributed Database

S. no.	Basis of comparison	Centralized database	Distributed database
1.	**Meaning**	It is a database that is only kept in one place and is stored, located, and maintained there.	This database comprises multiple data that are linked to each other and scattered across different physical locations.

(Continued)

(Continued)

S. no.	Basis of comparison	Centralized database	Distributed database
2.	**Access time**	In a centralized database, the time it takes for several users to access data is greater.	This database is dispersed over numerous physical locations, making maintenance, modification, and backup very challenging.
3.	**Management of data**	Since all of the data are present in one place, managing, editing, and backing up this database is simpler.	The management, modification, and backup of this database are very difficult as it is spread across different physical locations.
4.	**Opinion**	The user is given a consistent and comprehensive perspective thanks to this database.	Since it is dispersed over several sites, it is challenging to give the consumer a consistent view.
5.	**Data consistency**	In comparison to dispersed databases, this database has higher data consistency.	Data consistency has decreased because this database might have some data copies.
6.	**Failure**	If a database failure occurs, users are unable to access the database.	In a database that is distributed, users have access to multiple databases in the event that one database fails.
7.	**Cost**	A centralized database is less expensive.	The cost of this database is high.

There are several different architectures for distributed database systems, including:

- **Client–server architecture:** In this approach, clients link to a centralized server, which administers the databases across the system. The responsibility for maintaining data storage, restricting access, and managing operations fall on the server that hosts the data.
- **Peer-to-peer architecture:** The spread-out database system's locations are all connected through this design. Each internet address is in charge of arranging commerce with other internet properties and managing its data management.
- **Joined architecture:** According to this layout, every site in the dispersed database architecture keeps a distinct database. However, the middleware component that connects the databases provides common access for finding information and querying.

Applications for distributed database systems include online shopping, banking, and telephony. Thoughtful thought must be given to issues like data dissemination, replication, and consistency when creating and administering a distributed database system.

18.5 Use of a Distributed Database in the Education System

A database that is distributed is a kind of repository that consists of several systems linked to each other and distributed over different geographical regions. As a result, data administration is possible regardless of additional tangible places for storage. Thus, databases that are spread across various geographical regions are connected via a computer network.

18.5.1 Functions of a Distributed Database System

- **Maintaining a record of information:** By extending the DDBMS catalogue, DDBMS's primary job is to keep track of data dispersion, breakdown, and replica.
- **Decentralized Query Analysis:** A DDBMS's primary role is essentially its capacity to connect to remote locations

and send queries as well as information between them via a channel of communication.

- **Duplicated Data Management:** The fundamental job of a DDBMS is to select which copy of a duplicated record to make available and to ensure that those copies are consistent.
- **Distributed Database Restoration:** The capability to recover from new sorts of failures, such as broken communication lines, as well as from individual site crashes.
- **Security:** The fundamental job of a DDBMS is to carry out distributed transactions while properly managing user access rights and data security.
- **Distributed Directories Administration:** In essence, a directory is a repository for information coming from a relational database. The directory can either be regional for every website or worldwide for the whole DDB. There could be architectural and ethical challenges with the directory's placement and distribution.
- **Dispersed Transaction Management:** The fundamental job of a database management system (DDBMS) is to design methods to manage inquiries and activities that utilize data from multiple locations, coordinate connections to dispersed data, and, in essence, preserve the authenticity of the entire database at all times.

18.6 Decentralized Education and Contemporary Solution

In order to remove entrance hurdles for players (expensive and lengthy development timetables) with technology that is simple to implement, it is important to concentrate on adoption across the industry. This calls for a set of decentralized tools, and the use of smart contracts, community-driven governance, and an open ecosystem may ensure independence from any outside interference. The certification of knowledge transfer in one or both ways is made possible by the use of a blockchain protocol that acts as a trusted, transparent, and legitimate middleman between two participants in the learning space. Education-related organizations concentrate on what they do best, which is teaching, and the majority are not already familiar with blockchain technology. Access is heavily constrained by technological expertise, research, development and deployment times, and cost constraints. Much greater influence is being exerted on the online learning environment by smaller education platforms like tutors, blogs, and teachers [13]. They react and decide more quickly than their more well-known competitors. These smaller education platforms need easily deployable, adaptable blockchain solutions to increase the credibility of their goods and expand their market reach if they are to compete head-to-head. We need to develop ready-to-use B2B open-source educational blockchain products in order to support the development of a decentralized education system.

Centralization and decentralization are key ideas in the management system concept that the public should think about since they have a big impact on how schools are operated and how well they educate our future leaders. Both ideas have to do with how decisions are made and who has authority in the management system. For many years, both centralization and decentralization have been practiced in numerous nations and regions throughout the world. They each have unique traits, benefits, and drawbacks that set them apart from one another. This essay seeks to describe the fundamental ideas of centralization and decentralization in the context of the educational system, as well as its benefits and drawbacks.

Additionally, centralization eliminates redundant equipment and other facilities, allowing for the utilization of these resources jointly rather than requiring them to be provided for various tasks. Coordination is enhanced, and because of the unity of command, coordination is made simpler. In contrast to centralization, decentralization refers to the degree of delegation of power to the individual schools at the lowest levels. According to this idea, the top grants local officials more liberty, which encourages inventiveness, resourcefulness, and personal enhancement particularly in the area of problem solving. The local communities or regions then have some influence over the policies that are made in their areas under the supervision of the central government. Decentralization of education involves the transfer of financial and decision-making power from higher to lower tiers of government and organizational entities. This may have an impact on how school systems decide how to allocate assets and spend money, as well as how to organize instruction (e.g., curriculum, books, ways to teach, and schedule), manage personnel (e.g., hiring/firing, salary ranges, responsibilities for teaching, and training), and plan and manage public schools [14–17].

There are four questions that can help us provide better solution of college and school-going students:

1. How can the problem of reaching these kids be resolved?
2. Are the tools currently available to educational organizations capable of eradicating gender discrepancies in tutoring and guaranteeing equivalent access to all stages of education and vocational training for the weak, including those with incapacities, original people, and kids in difficult circumstances?
3. Is the current system set up to benefit from the expected growth of the EdTech industry over the next five years, which is expected to double?
4. Can the current collection of educational tools be used to set up the Web 3.0 integration of the upcoming educational platforms?

18.6.1 Rebuilding Education through Decentralization

The distribution of resources in the area of education must follow a hierarchical structure, starting with the federal government and moving through state departments to district and municipal governments. To make sure that the benefits of national efforts successfully reach students, this sequential movement is essential. Early on in the COVID-19 pandemic, when there was little information available about the illness, politicians at the national and state levels were unsure of the best course of action [18]. As a result, schools across the country were quickly closed. The design of curriculum and instruction, the reopening of schools, and the creation of content for at-home learning all underwent centralized decision-making during the early months of the epidemic. However, the spread of the COVID-19 virus showed clear regional differences during the duration of the pandemic. As a result, educational facilities found in areas with low transmission rates were forced to remain closed.

18.6.2 Education Scenario in India

Decentralization often results in a major improvement in the effectiveness, accountability, transparency, and receptiveness of service providing when associated to centralized systems. Delivery of education that is decentralized has the potential to be more efficient, better represent local goals, inspire contribution, and, ultimately, recover reporting and excellence. Decentralization could increase efficiency, which appeals to governments with limited resources. Beneficial cost recovery programs like community funding have developed as a mechanism for central governments to unload some of the financial burdens of delivering education services. The process of decentralization can typically greatly improve the efficacy, transparency, answerability, and receptiveness of service supply when compared to centralized systems. Decentralized education delivery may increase efficiency, and better represent local goals and support.

18.6.3 Education Tools

Virtual reality (VR) has proven to be a highly successful technology in supporting and enhancing the processes of teaching and learning. Multiple studies and scholarly articles have provided evidence suggesting that a significant proportion of students retain information from their virtual reality experiences. Consequently, these findings have led researchers to infer that virtual reality environments are more memorable compared to traditional laboratory-based presentations [19]. Ultimately, the utilization of laboratory-based methods, which are deemed to be less efficacious in terms of learning, can result in graduates experiencing deficiencies in fundamental knowledge and practical skills. Consequently, this may provide challenges for them in effectively addressing future workplace issues. In order to tackle the aforementioned concerns, a novel technique is proposed, which utilizes virtual reality (VR) as a means of facilitating teaching and learning. One of the major challenges in providing high-quality educational experiences is the cost associated with accessing relevant content. Educators often face the challenge of limited access to contemporary technological resources in their professional endeavors, including but not limited to expensive robotics apparatus, electrical components, chemical reagents, medical supplies, and similar items. Consequently, developing nations and communities worldwide may utilize their copies, which manifest as three-dimensional representations incorporating identical physical attributes implemented using virtual reality technology. The study of appraising digital libraries is still in its early stages. The following topics are still being researched: who should assess, when should one evaluate, what should be evaluated, how should one evaluate, and why should one evaluate. It describes appraising digital libraries as a difficult procedure that poses both conceptual and practical problems [20]. The other factors that contribute to the challenge include technical complexity; a wide range of materials, uses, and users; and a scarcity of evaluation methodologies. Any evaluation begins with the assessors' conceptual framework, which must be aware of the system's aims as well as the demands and behaviors of its users. In psychology, evaluation is a type of sense formation.

Open education at a distance makes extensive use of modern information and multimedia technologies, which transforms the traditional teacher-centered instructional strategy. Through satellite television broadcasts and computer networks, students have the opportunity to learn. These platforms facilitate the accessibility and distribution of instructional materials for individual study. China's long-distance open education has grown substantially and produced notable results over a lengthy period of study [21]. However, the current network teaching resources and management platform for remote education exhibit a broad range of differences and are incompatible. Therefore, government agencies and institutions of distance education are committed to conducting research on the incorporation of "cloud computing" technologies. In addition to assimilating and incorporating cutting-edge network technology from both domestic and international sources, they aim to enhance the mix of teaching resources for distance open education while incorporating cutting-edge network technology from both domestic and international sources [22].

18.7 Conclusion

This chapter presents the use of decentralization and distributed systems in the education sector and discusses how decentralization policies in India lead to increased educational accountability by transferring decision-making power to local education committees. Education committees consist of parents, teachers, and community members, and they are responsible for overseeing the day-to-day operations in the education field. India has undoubtedly witnessed a wide range of reforms leading to the decentralization of educational governance during the last 10–15 years due to the recent pandemic, the popularity of mobile devices, and the 5G network. According to the report, committees have been successful in reducing absenteeism among students by serving as decentralized representatives of parents and the community. Teachers and members of the education committee both valued financial transparency, which has helped to ensure that institution funds are used deficiently and effectively. This emphasizes the necessity of implementing a decentralized structure in the system of education management. Further research has revealed that it was challenging to solicit donations from neighbors and parents because they were from low-income families and were unable to contribute, and many parents did not understand the value of education for their children's capacity to make an economic contribution. Even while community donations are essential to achieving the growth goals of education, the government must consider the parent's financial capacity. The decentralized strategy may cause inequality within the same neighborhood to increase as current impoverished schools continue to decline. The government must subsidize the areas where parents are unable to contribute to guarantee that all children have equitable access to high-quality education.

References

1. Dhar, T.N. "Decentralized Management of Elementary Education: The Indian Experience." In Govinda, R. *Decentralization of Educational Management: Experiences from South Asia*. UNESCO International Institute for Educational Planning, Paris, 1997.
2. Misra, R.P. and Nataraj, V.K. "India: Blending Central and Grass-roots Planning" in W.B.Stohr and D.R.Fraser Taylor *Development from Above or Below?* John Wiley, 1981.
3. Mathur, K. Small Farmers Development Agency in India: An Experiment in Controlled Decentralization. In *Rural Development in Asia*, Sterling Publishers, New Delhi, 1985.
4. Govinda, R. and Varghese, N.V. *District Level Plan Implementation Strategy with Particular Reference to School Improvement*, International Institute for Educational Planning, Paris, 1994.
5. Report of the CABE Committee on Decentralization in Education, Ministry of Human Resource Development, Government of India, New Delhi, 1995.
6. Maclure, R. School reform in Burkina Faso: The limited prospects of decentralization and local participation, *Can. Int. Educ.*, Vol. 22 No. 2, 1993
7. R.Govinda, Dynamics of Decentralized Management and Community Empowerment in Primary Education: A comparative Analysis of Policy and Practice in Rajasthan and Madhya Pradesh. In R.Govinda and Rashmi Diwan, *Community Participation and Empowerment in Primary Education: Indian Experience*, Sage Publishers, New Delhi (Forthcoming).
8. Weiler, H.N. Legalization, Expertise, and Participation: Strategies of Compensatory Legitimation in Educational Policy, *Comp. Educ. Rev.*, Vol. 27, No. 2, 1983, pp. 259-277.
9. Vinay K. Kantha, V. and Daisy Nariain Dynamics of Community Mobilization in a Fragmented and Turbulent State. In R.Govinda and Rashmi Diwan, *Community Participation and Empowerment in Primary Education: Indian Experience*, Sage Publishers, New Delhi (Forthcoming).
10. Richard Crook and James Manor, *Democratic Decentralization*, Operations Evaluation Department, World Bank, Washington D.C. 2000.

11. R. Govinda, “Capacity Building for Educational Governance at Local Levels” Paper presented at the *International Consultation on Educational Governance at Local Levels*, held at UNESCO, Paris, 27-28 February 2001.
12. Cohen, S.S., J.W. Dyckman, E. Schoenberger, and C.R. Downs. *Decentralization: a framework for policy analysis*. Berkeley, 1981. p. 46-48.
13. Slater, D. 1989. Territorial power and the peripheral state: the issue of decentralization. *Dev. Change*, (20): 501-531.
14. Lauglo, J. and McClean, M. *The control of education: International perspectives on the centralizationdecentralization debate*, Heinamann, London, 1985.
15. Jacques Delors “Education: the necessary Utopia” (pp. 13-35) In *Learning: The Treasure Within Report of the International Commission on Education for the Twenty-first Century*, UNESCO, Paris, 1996.
16. Weng, F.Y. (2003) Centralization and decentralization in educational governance in Taiwan, in Mok, K.H. (ed.) *Centralization and Decentralization: Educational Reforms and Changing Governance in Chinese Societies*. Hong Kong Comparative Education Research Centre, University of Hong Kong and Kluwer Academic Publishers.
17. Welch, A.R. (2005) Korean higher education in international perspective: internationalized or globalized? in K. H. Mok & R. James (Eds.), *Globalization and Higher Education in East Asia*. Singapore: Marshall Cavendish Academic.
18. Alam, T., & Benaida, M. (2020). Blockchain and internet of things in higher education. Tanweer Alam, Mohamed Benaida.” Blockchain and Internet of Things in Higher Education.” *Univers. J. Educ. Res.*, 8, 2164-2174.
19. Chen, X., Zou, D., Cheng, G., & Xie, H. (2020). Detecting latent topics and trends in educational technologies over four decades using structural topic modeling: A retrospective of all volumes of Computers & Education. *Comput. Educ.*, 151, 103855.
20. J. Y. L. Thong *et al.* Understanding user acceptance of digital libraries: What are the roles of interface characteristics, organizational context, and individual differences? *Int. J. Hum.-Comput. Stud.*, (2002).
21. Liu, Y., Fan, X., Zhou, X., Liu, M., Wang, J., & Liu, T. (2019, May). Application of virtual reality technology in distance higher education. In *Proceedings of the 2019 4th International Conference on Distance Education and Learning* (pp. 35-39).
22. Zhao, Y., & Watterston, J. (2021). The changes we need: Education post COVID-19. *J. Educ. Change*, 22(1), 3-12.

Index

Also of Interest

By the same editors

ELECTRIC VEHICLE DESIGN: Design, Simulation and Applications, Edited by Krishan Arora, Suman Lata Tripathi, and Himashu Sharma, ISBN: 9781394204373. This book will serve as a definitive guide to conceptual and practical knowledge about the design of hybrid electrical vehicles (HEV), battery electrical vehicles (BEV), fuel cell electrical vehicles (FCEV), plug-in hybrid electrical vehicles (PHEV), and efficient EV charging techniques with advanced tools and methodologies for students, engineers, and academics alike.

INDUSTRIAL CONTROL SYSTEMS, Edited by Vipin Chandra Pal, Suman Lata Tripathi, and Souvik Ganguli, ISBN: 9781119829256. This volume serves as a comprehensive guide in the journey of industrial control systems with a multidisciplinary approach to the key engineering problems in the 21st century.

MODERN AUTOMOTIVE ELECTRICAL SYSTEMS: Theory and Applications, Edited by Pedram Asef, Sanjeevikumar Padmanaban, and Andrew Lapthorn, ISBN: 1119801047. Presenting the concepts and advances of modern automotive electrical systems, this volume, written and edited by a global team of experts, also goes into the practical applications for the engineer, student, and other industry professionals.

NANODEVICES FOR INTEGRATED CIRCUIT DESIGN, Edited by Suman Lata Tripathi, Abhishek Kumar, K. Srinivasa Rao, and Prasantha R. Mudimela, ISBN: 9781394185788. Written and edited by a team of experts in the field, this important new volume broadly covers the design of nano-devices and their integrated applications in digital and analog integrated circuits (IC) design.

EXPLAINABLE MACHINE LEARNING MODELS AND ARCHITECTURES: Real-Time System Implementation, Edited by Suman Lata Tripathi and Mufti Mahmud, ISBN: 9781394185849. This cutting-edge new volume covers the hardware architecture implementation, the software implementation approach, and the efficient hardware of machine learning applications.

MACHINE LEARNING TECHNIQUES FOR VLSI CHIP DESIGN, Edited by Abhishek Kumar, Suman Lata Tripathi, and K. Srinivasa Rao, ISBN: 9781119910398. This cutting-edge new volume covers the hardware architecture implementation, the software implementation approach, and the efficient hardware of machine learning applications with FPGA or CMOS circuits, and many other aspects and applications of machine learning techniques for VLSI chip design.

INTELLIGENT GREEN TECHNOLOGIES FOR SMART CITIES, Edited by Suman Lata Tripathi, Souvik Ganguli, Abhishek Kumar, and Tengiz Magradze, ISBN: 9781119816065. Presenting the concepts and fundamentals of smart cities and developing "green" technologies, this volume, written and edited by a global team of experts, also goes into the practical applications that can be utilized across multiple disciplines and industries, for both the engineer and the student.

DESIGN AND DEVELOPMENT OF EFFICIENT ENERGY SYSTEMS, edited by Suman Lata Tripathi, Dushyant Kumar Singh, Sanjeevikumar Padmanaban, and P. Raja, ISBN 9781119761631. Covering the concepts and fundamentals of efficient energy systems, this volume, written and edited by a global team of experts, also goes into the practical applications that can be utilized across multiple industries, for both the engineer and the student.

Electrical and Electronic Devices, Circuits, and Materials: Technical Challenges and Solutions, edited by Suman Lata Tripathi, Parvej Ahmad Alvi, and Umashankar Subramaniam, ISBN 9781119750369. Covering every aspect of the design and improvement needed for solid-state electronic devices and circuit and their reliability issues, this new volume also includes overall system design for all kinds of analog and digital applications and developments in power systems.

Green Energy: Solar Energy, Photovoltaics, and Smart Cities, edited by Suman Lata Tripathi and Sanjeevikumar Padmanaban, ISBN 9781119760764. Covering the concepts and fundamentals of green energy, this volume, written and edited by a global team of experts, also goes into the practical applications that can be utilized across multiple industries, for both the engineer and the student.

Check out these other related titles from Scrivener Publishing

HYDROGEN ELECTRICAL VEHICLES, Edited by Mehmet Sankir and Nurdan Demirci Sankir, ISBN: 9781394166381. Hydrogen electrical vehicles are an essential component of the "Green New Deal" and this book covers cutting-edge technologies designed for fuel-cell-powered cars.

PAVEMENT ASSET MANAGEMENT, By Ralph Haas, W.R. Hudson, and Lynne Cowe Falls, ISBN: 9781119038702. Whether you are new to managing pavement assets or a seasoned practitioner, this book provides the most up-to-date information in the field on the history, technology, practice and look ahead.